THE WOMAN WHO COULDN'T WAKE UP

THE WOMAN WHO COULDN'T WAKE UP

Hypersomnia and the

Science of Sleepiness

QUINN EASTMAN

Columbia University Press

New York

Columbia University Press
Publishers Since 1893
New York   Chichester, West Sussex
cup.columbia.edu

Library of Congress Cataloging-in-Publication Data
Cataloging-in-Publication Data is available from the Library of Congress.
ISBN 9780231194648 (hardback)
ISBN 9780231550918 (ebook)
LCCN 2023002759

Cover design: Henry Sene Yee
Cover images: Shutterstock

CONTENTS

THE WOMAN
WHO COULDN'T
WAKE UP

INTRODUCTION

The first time I experienced having my eyes close and legs buckle while standing up, I was an undergraduate working in a biochemistry lab. The night before, I had stayed up late to finish assignments for my classes. After I nodded off a few times, even while someone was speaking to me, I was jokingly accused of having narcolepsy. It was the first time I can remember anyone using the word in a conversation; I had never met anyone with narcolepsy.

What I actually had is called "insufficient sleep syndrome," which goes away after enough rest. I can remember a fender bender in graduate school after a late night in the lab and another time falling asleep at a stoplight, releasing the brakes just enough to nudge the car in front of me. My wife recalls my falling asleep during a scientific lecture and then audaciously asking a detailed question when the lights were turned on. None of this makes me special—many people have had similar experiences. In my late thirties, I was diagnosed with obstructive sleep apnea, a common sleep disorder among Americans. When I visit support groups for people with rarer sleep disorders, I tell them that I view what they have through a keyhole; I've experienced "sleepy," but not to the same degree. This book is about people dealing with something more extreme than what I have.

Just before I started working at Emory University in 2007, the nursing professor Kathy Parker gave a lecture describing her patient Anna, a young lawyer whose life was being taken over by sleep. Despite powerful stimulants and multiple alarm clocks, Anna could sleep more than thirty hours at a stretch. Forced to stop working, she was losing weight because she didn't stay awake long enough to eat. The resulting headline in the university newspaper was: " 'Sleeping Beauty' Case Awakens Hope for Disorder."

Baffled, Parker and other researchers searched for an explanation. They had ruled out conditions such as sleep apnea and narcolepsy and diagnosed Anna with idiopathic hypersomnia, a poorly understood neurological sleep disorder. Following a hunch of Parker's, they glimpsed signs of something in Anna's spinal fluid, which they thought might be causing her sleepiness. This substance's chemical effects resembled those of benzodiazepines, the class of antianxiety and anti-insomnia drugs that includes Valium and Xanax. With that clue, the researchers found a treatment for Anna and convinced the pharmaceutical company Roche to help. Soon afterward, Parker left for another university, and it took years for the remaining group at Emory to publish their work.

I eventually met Anna at a talk given by the neurologist David Rye. When the findings of Rye, Parker, and colleagues appeared in a scientific journal in 2012, it was my job to explain them in an announcement aimed at news organizations. For me, what caused idiopathic hypersomnia or how to treat it didn't matter so much at the time. What changed my mind were the emails and phone calls: "Is there a study I can be part of?" or "My daughter has this. Can you help her?"

I was seeing a small fraction of the avalanche of pleas that Rye and his colleagues were receiving. It went on for months. Then I learned that people with conditions resembling Anna's had organized online and were planning a conference in Atlanta. They had formed a community on social media based on what they shared: debilitating sleepiness that wasn't explained by some other medical condition.

The 2014 Living with Hypersomnia conference was my first contact with this irreverent, often frustrated group. Members bonded over feeling misunderstood by friends and coworkers. They traded morbid jokes and recommendations for alarm clocks. Because of their chronic sleepiness, some were reluctant to drive more than a few miles or had given up driving entirely. A substantial number of people with idiopathic hypersomnia (abbreviated as IH) had to abandon their jobs or school or had applied for disability benefits. Once awake, people with IH don't often fall asleep suddenly, but more than one told me: "Waking up is the hardest thing I do all day."

If you don't know someone with IH, you may be thinking: "Sometimes I have trouble getting out of bed," or, "My teenager moans and groans almost every morning." People with IH get dismissive reactions all the time, even from their families. The father of one teenager with IH described being in denial, considering his son "just a lazy kid." What changed his mind was finding out his son fell asleep not only during lectures in school but also while passing papers back to a classmate. The father, a Lutheran minister, took his son to the family doctor. To better understand his son's condition, the doctor urged the father to try to take three naps in one day. "I couldn't do it, as much as I loved sleep," he said.[1]

As a neurochemical state, sleepiness continues to be difficult to pin down and dissect. EEG (electroencephalography) can indicate with high confidence whether someone is asleep or awake, so you might think that concerns about subjective symptoms that emerge when studying conditions like depression don't apply. But even though being asleep can be determined objectively, sleepiness cannot. There is no rapid biological test for sleepiness. In the doctor's office, it is measured by questionnaire: Do you often fall asleep while watching television or at stoplights? In the clinic, the standard test for narcolepsy or IH involves having someone take five naps throughout the day. The variable measured is the speed at which someone enters sleep. A large number of people in our sleep-deprived society will meet the criteria for this test by dozing off quickly enough each time, but people with IH can do so even after a week of apparently sufficient rest.

At the 2014 conference, I could see that Rye and his colleagues had given people with IH energy, hope, and some new options. Their research also contained a potential explanation for why some people with IH were sleepy: the "somnogen" in their spinal fluid. At that time, it had not been identified. It was known informally as "sleepy stuff" or even "sleepy juice." Rye's collaborator Andy Jenkins was in the process of figuring out what the somnogen was.

This puzzle intrigued me. I had spent a large part of my time as a graduate student in a walk-in refrigerator purifying proteins—the same task facing Jenkins. I wasn't working in the laboratory anymore, but maybe I could give the search for the somnogen a nudge. Perhaps assembling the clues and presenting them to the professionals, like a private investigator does for the police in the movies, would help. More realistically, I could be there when they resolved the mystery. My postdoctoral supervisor once told me: "You can follow a disease, or you can follow the molecules. The molecules may lead you to unexpected places."

For a century, the molecules that make us and our animal relatives sleepy have captured the scientific imagination. As one early example, French scientists reported in 1913 that spinal fluid from sleep-deprived dogs induced sleep when injected into another animal; spinal fluid from rested dogs did not. Today, the roles of several molecules that regulate sleep are well worked out.

Reading the literature told me that what Parker, Rye, and Jenkins had stumbled upon was weirder. A tantalizing hint toward the origins of Anna's sleepiness came from flumazenil, the antidote to benzodiazepines that she relied upon every day. Since its discovery almost four decades ago, doctors have tried using flumazenil to wake up other sleepy people, with little consistency. What flumazenil was counteracting was neuroscience's equivalent of sunken pirate gold: once glittering and glamorous, but later obscured by the complexity of the brain. It was striking how researchers from different fields—hepatology, addiction medicine, and neurology—had become champions of this odd, obscure drug.

I learned that IH occupied an awkward place in sleep medicine. Neurologists who study sleep are accustomed to thinking about narcolepsy, a disorder with a longer history. They tend to regard IH as part of the "narcoleptic borderland." Narcolepsy was their homeland, and other disorders were viewed in relation to it.

Rye was arguing that some people diagnosed with narcolepsy should be grouped together with those with IH. He wanted to redraw the map of how those disorders were classified, based on the mechanism he proposed to explain Anna's and others' sleepiness. His ideas about IH were at the edge of mainstream sleep medicine; his peers didn't think there was enough solid evidence behind them.

Doctors have used many drugs without knowing exactly how they worked. But to convince neurologists and sleep specialists that Anna's case was not an anomaly, someone would have to identify the sleepy stuff. They'd have to demonstrate that it was abnormally active in some people with IH. A rigorous study would have to show that enough of them responded to an agent that counteracted it. That might take years. I realized that Rye's ideas could be wrong, and I had to evaluate alternative explanations for IH that did not depend on his theories.

By the time I started collecting material for this book, organizations such as the Hypersomnia Foundation and Hypersomnolence Australia had been established to advocate for people with IH. Drawn by this activity, pharmaceutical companies had begun clinical trials that included people with IH, but the companies' success wouldn't necessarily show that Rye was correct. The medications that helped Anna didn't always work for others. Anna told me: "My story may be inspiring, but it's not everybody's story."

For me, writing this book started out as a scientific detective story, but it turned into something that was more about the coalescence of a community. As a former lab rat, I'm more comfortable discussing molecules rather than personal identity. But I learned that names people give to their conditions are important. At a support group meeting in Atlanta, I asked people with IH what they wanted to be called and what they thought were the important aspects of their symptoms. A few people with IH had embraced the term "sleepyhead"; others disliked it. One said: "It sounds like something you'd say to a toddler." The organizer of the support group argued that "sleepyhead" did not capture the full extent of the condition. In particular, it doesn't include the "brain fog" that can envelop people with IH, making them indecisive and prone to mishaps. Brain fog also seems to respond less to stimulants; combining the two leads to a "fake awake" situation, in which someone's eyes are open but they are not alert. Still, brain fog is not specific to IH. The symptom appears in depression, multiple sclerosis, and other disorders, including long COVID. In comparison, IH seemed to be purer than other neurological or psychiatric disorders because it involved excessive sleepiness and sleep duration, without other prominent symptoms.

Many IHers (an agreed-upon term) experience something objective but difficult to verify quickly: they need considerably more hours asleep every week than the average person. Moment-to-moment sleepiness is more subjective. It's connected to many variables: lack of sleep, time of day, meals, position of the body, light, and temperature, as well as mental activity and effort. When nonspecialists talk about how they feel after a long day, sleepiness can blend together with tiredness and fatigue. Still, sleepiness is a different experience from the physical fatigue someone might feel after playing sports or climbing a mountain.

Other neurological disorders characterized by subjective symptoms have had difficulty being recognized as something "real." Earlier in the twentieth century, people with multiple sclerosis were misdiagnosed with hysteria because their symptoms were regarded as psychosomatic.[2] According to the sociologist Joanna Kempner, migraine headache once had a "legitimacy deficit" because of the subjective aspect of headache pain and because the stereotypical migraine patient was female.[3] Perhaps IH will undergo a similar evolution. A molecular marker for migraine-related pain called CGRP (calcitonin gene-related peptide) could present a possible parallel with the "sleepy stuff" somnogen.[4]

Recently, the COVID-19 pandemic and the phenomenon of long COVID have led to renewed attention to postviral chronic illnesses and other conditions with uncertain status, such as myalgic encephalomyelitis / chronic fatigue syndrome (ME/CFS). IH has followed a different course because of its historical relationship with narcolepsy—it was acknowledged within mainstream medicine yet received little attention.

Does discussion of IH take something that everyone experiences sometimes—drowsiness or feeling groggy after waking up—and turn it into a disease, requiring treatment? On the margins, maybe. Debate continues among sleep medicine specialists on the proper borders between everyday sleepiness and IH or narcolepsy. Consider how much persistence it often takes for someone to receive an IH diagnosis. Those years of asking doctors for help should not be seen as simply a ploy to obtain stimulants. While many physicians are comfortable prescribing conventional stimulants, the need for alternatives is clear.

Looking ahead, the pharmaceutical industry has been developing several newer wake-promoting medications—an emerging arms race of weapons against sleep.

In a sense, we are all struggling against the environment we created. Chronic sleep deprivation leads to long-term health consequences: higher rates of cardiovascular disease, metabolic problems, and cancer. People in industrialized societies have collectively lost sleep since the appearance of artificial lighting and electricity, and that loss has accelerated in recent decades. We can learn from IHers, who need more of a scarce commodity.

This book follows the emergence of IH as a diagnosis, a sleep disorder, and an FDA indication. At the same time, the value of focusing on the "sleepy stuff" (or stuffs)—the physical currency of extreme sleepiness—is that it might cut through concern about disease mongering and uncertainty about how IH should be defined and is thus potentially relevant to a wider population: all of us.

CHAPTER 1

ANNA SLEEPS A LOT, AND WE DON'T KNOW WHY

Some syndromes have languished for an appreciable time for lack of a satisfactory name. An unfortunate consequence of naming can be the mistaken impression that we understand the condition.

—Victor A. McKusick, 1969

Anna Sumner knew she was taking a risk. She was in the epilepsy monitoring unit of a university hospital, ready to try a medicine suspected of provoking seizures. Her doctors were prepared, just in case. She had EEG electrodes attached to her head, and her nervous parents were waiting outside.

The medicine was flumazenil, an antidote against an overdose of sleeping pills. But Anna had not taken any sleeping pills. Instead, her doctors thought her body was making its own. This might explain why she needed multiple alarm clocks in the mornings and why she could doze through a phone ringing or someone shaking her. How did she get here?

In 2004, after finishing law school, Anna started working at a respected law firm in Atlanta. There she had difficulty managing her body's extraordinary requirements for sleep. She could routinely sleep for twelve or fourteen hours and still wake up without feeling refreshed. "It came to the point where if I had a choice between sleeping or eating, I would rather sleep," she said later. "At the same time, when I did, it was not restorative or satisfying." Anna recalled her predicament in an interview recorded at Emory, the medical center in Atlanta where she sought treatment. "I was always thinking, how can I get more sleep?" she said in her faintly Southern-accented voice. "If I don't wash my hair this morning, can I get thirty minutes more sleep? If I don't eat lunch today at work, can I come home early and go to sleep?"[1]

Her doctors diagnosed her with something called idiopathic hypersomnia. Anna interpreted this mouthful as "she sleeps a lot and we don't know why." It seemed like she was the only one with this condition. For narcolepsy, a disorder in which people experience overwhelming sleepiness, support groups existed. There were even arty movies depicting people with narcolepsy. But who had heard of idiopathic hypersomnia?

When Anna came to Emory in the summer of 2005, she first underwent an overnight sleep study. Petite, slim, and not yet thirty, she was unlikely to have obstructive sleep apnea, usually assumed to be more common among older over-weight men. Her overnight study confirmed that she did not.

The results of a test the next day, when she was asked to take several naps, were more illuminating. This was Anna's first Multiple Sleep Latency Test (MSLT), but she would grow used to being examined in this way. Beforehand, a technician glued electrodes to her scalp and placed more probes on her eyelids, face, and legs.

Coincidentally, a jackhammer was rattling outside that day as part of a construction project. She wondered whether that was part of the test but had no trouble dozing off. For each of five naps, she fell asleep quickly.[2] The average delay was less than three minutes: the "twilight zone" of sleepiness. A healthy person will take at least ten minutes to fall asleep on average, if they manage to do so all five times.

During her naps, Anna did not enter REM sleep—the rapid-eye-movement dreaming phase. This meant that she probably did not have narcolepsy, which is rarer than obstructive sleep apnea. People with narcolepsy tend to enter REM sleep soon after a nap begins. In addition, people who have the more distinctive form of narcolepsy, narcolepsy with cataplexy, can experience muscle weakness when they have strong emotions. Anna has never experienced this symptom.

"I was told that I couldn't have narcolepsy, because I didn't fall out and collapse," she said. Her doctors looked for factors that would explain her sleepiness, such as thyroid or liver problems or other metabolic oddities. Although her iron levels were a bit low, nothing stuck out. The lack of any other explanation meant that her doctors arrived at a diagnosis of idiopathic hypersomnia, abbreviated as IH.

As an initial effort at treatment, her doctors prescribed the same stimulants that someone with narcolepsy might receive. No rigorous clinical studies had been performed to test whether anything was effective against IH, and there were no FDA-approved drugs for the indication.[3] Anna was first given the "smart drug" modafinil, which is supposed to be gentler than what came next: amphetamines. For a while, these medications helped her make it through the day. But to sustain the effects, her doctors had to increase the doses. They ended up giving her enough to make her feel twitchy and uncomfortable. She lost weight and developed high blood pressure and a "somewhat erratic sleeping pattern," so she was advised about keeping a regular sleep schedule.

Something within her body was resisting the stimulants. Although Anna took a slow-release form of amphetamines at bedtime, her roommate had trouble waking her and said that she sometimes appeared confused. She began experiencing crashes, in which she would sleep for more than thirty hours at a stretch. It happened every few months, but by the beginning of 2007, it increased in frequency to every week. The longest crash lasted for more than fifty hours. "That was the scary part," she said. "If I went to bed, I didn't know if I'd fall off the map and wake up two days later." In April, she almost slept through an important hearing. It prompted her to take leave from her job, troubling for an ambitious young lawyer. She and her caregivers had six months to figure out what was going on, after which a formal disability determination would be necessary.

Anna kept a photo from this time period, one in which she's lying in a hospital bed smiling and wearing a T-shirt for the fictitious band Spinal Tap. In this uncomfortable procedure, known formally as lumbar puncture, her doctors obtained a sample of her cerebrospinal fluid, which surrounds the brain and spinal cord. The procedure represented a way to probe what was going on in her brain without slicing into it. Sampling her spinal fluid allowed her doctors to definitively rule out narcolepsy and to discover something else.

The anesthesiology researcher Andrew Jenkins tested Anna's spinal fluid and found that it contained a substance that behaved chemically like benzodiazepines, the class of drugs that includes Valium and Xanax. Doctors use some benzodiazepines to keep people relaxed during medical procedures. According to Jenkins, Anna had a level of sedative-like stuff in her body comparable to someone undergoing a colonoscopy or having her wisdom teeth removed—all the time.

The presence of a benzodiazepine-like substance in her body was why her doctors, led by the neurologist David Rye, wanted to try the antidote: flumazenil. Several years before, Rye had treated a couple patients with similar symptoms with flumazenil. It woke them up, but the effect wore off quickly. Flumazenil was something that anesthesiologists or emergency room physicians usually had, but in tiny amounts. Its manufacturer had stopped producing it, and it was not approved by the FDA for any purpose besides its role as a benzodiazepine countermeasure. On top of that, in a fraction of the cases in which flumazenil was used, there were reports of panic attacks, seizures, even cardiac arrest.[4] Still, it was worth a try. "At that time, I had no dependents; it was just me, so I felt I could afford to take a risk," she said. "And this disorder was threatening to take away my career and even my brain." Anna said that when the researchers found the unusual chemical activity in her spinal fluid, she felt a sense of relief; she had been worried that she was either "lazy or crazy." Before the flumazenil experiment, she was examined by psychiatrists twice, to check for the possibility that she was depressed or inventing her symptoms. Her parents had approached a neurologist outside Emory, who cautioned that she was probably taking drugs recreationally. With her family, Anna managed to keep a sense of humor about her situation. "My brothers turned it into a joke," she said. "We lobbied for naming it the 'Sumner stupor,' because stupor is what it felt like. The only thing that could wake me was a dog's tongue up my nose."

SLOW UNDERCURRENT

Anna grew up as the daughter of a lawyer and a judge outside the small town of Winona, Mississippi. Articles in the local newspaper describe her playing tennis, riding horses, and getting good grades. Her elevated need for sleep began to emerge in her last year of high school, when she started taking naps after morning chapel. At Princeton, she told herself that she was working hard and her body needed extra rest. She was hiding how much sleep was encroaching upon her life. She would choose naps over eating lunch, exercising, or time with friends. Every evening, she returned to her dorm room after dinner to go to sleep for a couple hours. If her parents called, she had her roommate tell them she had gone to the library. Even so, they noticed that when she came home on holidays, she spent most of her time in bed.

At some American universities, students joke that they are forced to pick two of three things: good grades, a social life, or enough sleep. To observers, it might have looked like Anna did not have to sacrifice. She joined a sorority and social clubs and still ranked at the top of her class. Over the years, Anna kept a list of events she slept through: a date, a concert, a friend's wedding. She hid them well; when her sleep disorder was eventually diagnosed, friends told her they thought she had avoided some gatherings because of social anxiety.

After graduation, Anna moved to Bangkok to teach English. In a hot climate, taking siestas didn't seem too strange. Then she spent a winter in London, working on a novel. The weather was different, but her sleepiness stayed with her. Before she applied to law school, she underwent tests to rule out causes of fatigue such as mononucleosis or anemia. At Duke, Anna coped well enough with naps between classes; she then clerked for a federal judge in New Orleans.

"In law school, as long as you show up for your commitments, nobody will bother you," she said. "I never liked coffee, but I did drink lots of Diet Coke."

When she started as a junior associate in Atlanta, the flexibility in her schedule she had previously enjoyed was gone. Napping did not seem like an openly acceptable option. This was the land of billable hours; there were partners and clients to satisfy. She started putting in seventy-hour weeks. Even if she did take a nap, she found it made her feel worse. "That's when it finally hit me," she said.[5] "This is not how you're supposed to feel."

Fast forward a decade. While preparing to write this book, I visited Anna at her law firm's offices in Atlanta. We bought sandwiches for lunch and sat on benches outside. The book mostly won't be about you, I told her cautiously. Although others had called her a "sleeping beauty," Anna said the vision of her as a passive damsel in distress didn't apply: "I think I'm actually an example of a patient advocating for herself, collaborating with her doctors, and not saying 'OK' when she's told to just suck it up."

When the nursing professor Kathy Parker started discussing her case in public, Anna was uncomfortable. She didn't want to be a poster child. She didn't want her full name to be used or the name of her firm. But she became frustrated with the obscurity of idiopathic hypersomnia and the impasse in research, so a few years later, she agreed to participate in promoting the Emory team's findings. As a result, she has received many phone calls from people with similar problems, asking for advice.

After meeting other people with experiences like Anna's, I wanted to learn more about IH's back story and why it had been in the shadows for so long. IH was not a sudden epidemic but rather a slow undercurrent. Sleep medicine has flourished in the last few decades, but until recently, the boom has left out people with IH. Many in the field of sleep medicine considered the category of IH a hodgepodge: people made sleepy by a variety of factors, ranging from genetics to infections or head injuries. The problem was inherent in the name: *idiopathic* means "arising spontaneously or having no cause."

OVERDIAGNOSED OR OVERSHADOWED?

The first person to identify IH as something coherent and distinct was Bedřich Roth, a neurologist at Charles University in Prague. In 1956, Roth published a paper on several patients with the symptom of "sleep drunkenness," persisting long after they woke up.[6] Here's how Roth described an eighteen-year old training to be a locksmith, whose sister, mother, and aunt had similar conditions:

He falls asleep on the tram, in the cinema, at a concert, in the doctor's waiting room. The sleep sometimes takes only 10 or 15 minutes, but usually takes about 5–6 or even 16 hours. In the evening he falls into a deep sleep immediately. His family says it is almost impossible to wake him up in the morning; he often falls asleep again. Waking up takes about 15 minutes—the patient staggers as if he was drunk. Sometimes he even falls down. He is very rude and vulgar, unlike his normal behavior, and doesn't perceive anything during this time.

Acknowledged by his peers as a pioneer, Roth established one of the world's first sleep labs in the 1950s.[7] Resources were limited in Czechoslovakia, but he was able to collaborate with leading sleep researchers in Western Europe and North America. In the 1970s, Roth gave IH its name, proposing sleep drunkenness as one of its principal features.[8] He died in 1989, just as Czechoslovakia's communist government was fading away and the study of sleep disorders was transitioning from descriptive to molecular.

Roth thought IH could be as prevalent as narcolepsy—usually estimated as around 1 in 2,000 people—or multiple sclerosis, now thought to appear in more than 1 in 1,000. Other sleep specialists did not adopt his perspective. IH was overshadowed as other disorders, such as narcolepsy and sleep apnea, became better defined. Current estimates of IH's prevalence are several times lower (currently, around 1 in 10,000 in the United States), although the numbers are rising.

After Roth, the French neurologist Michel Billiard is cited by Roth's Prague-based colleagues as having done the most to refine the concept of IH as a separate sleep disorder.[9] Billiard complained that the IH category was seen as a basket for every sleepy patient for whom an explanation was not available: "Idiopathic hypersomnia is frequently overdiagnosed due to a persistent tendency to label as such hypersomnias that do not fit the criteria of either sleep-disordered breathing or narcolepsy."[10] For IH, this tension between a coherent disorder and a leftover category has existed for years.

RELATIONSHIP TO NARCOLEPSY

Let's lay out some basics. The details of where IH starts and other sleep disorders end may sound arcane to nonspecialists. Still, these distinctions have had tangible effects on patients' lives, affecting how doctors, friends, and relatives view them.

"Hypersomnia" broadly means "too much sleep": chronically feeling sleepy during the day or needing excessive amounts of sleep. "Hypersomnolence" refers to the symptom rather than a disease. According to surveys, substantial fractions of the population of the United States experience excessive daytime sleepiness (about 5 percent) or the need for long sleep periods, enough to interfere with daily life (more than 1 percent).[11]

Some people feel sleepy during the day because they experience insomnia at night. We are focusing on the opposite of insomnia: when sleep occurs readily but never seems like enough. A more difficult distinction to make is whether hypersomnolence comes from a psychiatric condition such as depression. IHers describe their excessive sleepiness as an external force that interferes with their ability to perform desired activities, rather than a companion of their mood.

As currently implemented in sleep clinics, a crude definition of idiopathic hypersomnia is excessive daytime sleepiness, not sleep apnea, not narcolepsy, not lack of sleep, and not anything else. For someone to receive an IH diagnosis, common causes of daytime sleepiness such as obstructive sleep apnea or metabolic diseases should be eliminated. Hypersomnolent patients have shown up at sleep clinics and discovered to have vitamin D or B12 deficiencies.[12]

Clinicians start with questionnaires such as the Epworth Sleepiness Scale, which asks how likely someone is to fall asleep during activities such as reading, watching television, or driving. Then they are supposed to have something "objective" to gauge a patient's sleepiness. No biological test for IH exists, defining who's in the club and who is not. Instead, extreme sleepiness is measured operationally: how quickly someone can doze off, given several opportunities during the day.

Remember the five naps Anna was asked to take? They were part of the Multiple Sleep Latency Test, standard for diagnosing both narcolepsy and IH. Someone who goes to sleep quickly and enters REM sleep enough times (twice) has narcolepsy. Someone who goes to sleep quickly enough but does not enter REM sleep more than once has IH. Those who feel sleepy during the day but during the test don't doze off fast enough are in limbo; their status depends on how much sleep they say they need and their physician's judgment.

In sleep medicine, it's difficult to discuss IH without referring to narcolepsy, which has received more attention and has well-established patient-support organizations. This book highlights how people with IH felt their needs were not being met and created new organizations in response. That said, there is no need to create a competition between the two conditions. At the local level, groups that support and advocate for people with narcolepsy have included people with IH in their efforts for years. A woman who organized a support group for narcolepsy in Washington, DC, told me that people with IH started showing up because narcolepsy was the closest to their own diagnosis that they could find.[13] "There is so much overlap in symptoms that people with narcolepsy have always, in my experience, welcomed people with IH with open arms," she said.

The push for more recognition and awareness for narcolepsy has its pitfalls. Although television shows and movies have included fictional characters with narcolepsy, they are often portrayed for comic effect. One example: perennial bumbler Homer Simpson, diagnosed in a 2015 episode. A few celebrities, such as the late-night television host Jimmy Kimmel, have disclosed their narcolepsy diagnoses.[14] Even so, surveys have found that fewer adults in the United States recognized the term "narcolepsy," compared with other chronic diseases, and didn't understand narcolepsy's main features.[15]

Most of the time when sleep scientists talk about narcolepsy, they mean narcolepsy with cataplexy, or narcolepsy type 1. Cataplexy is when someone experiences muscle weakness in response to strong emotions. Full-body cataplexy can make someone collapse, but it is often more subtle, involving only the face or part of the body. One of the most reliable triggers for cataplexy is laughter, complicating social life.

In defining a disease, we can focus on the symptoms or the molecules. For narcolepsy type 1, the symptoms and the molecules implicated are well understood. For IH, as well as narcolepsy type 2, the symptoms are defined, but the molecules are not. Take a look at the current landscape in table 1.1. Several symptoms for narcolepsy type 1 tend to come together: excessive daytime sleepiness, disrupted nighttime sleep, cataplexy, sleep paralysis (waking up but then being unable to move), and vivid dreams that bleed into waking time, seeming like hallucinations. The last three have been thought to involve partial intrusions of REM sleep into the awake state.

Narcolepsy type 2 is an in-between category, lacking cataplexy but having more of the other features of narcolepsy type 1. While having excessive daytime sleepiness, people with IH have fewer of the REM-related symptoms of narcolepsy. A distinguishing feature between IH and narcolepsy tends to be the effect of naps. People with narcolepsy type 1 often say a fifteen-minute nap makes them feel much better, but with IH, naps are generally long and not refreshing.

Sleep drunkenness is a distinctive symptom of IH, although it is not present in all cases—Anna said she did not experience it. Sleep drunkenness, which can be independent of IH, has been defined as a period of confused clumsiness and slurred speech after waking up. It is sometimes thought of as a stronger, more extended version of sleep inertia, which most healthy people have experienced: a temporary groggy feeling upon waking. In sleep drunkenness, the transition out

TABLE 1.1

SYMPTOMS	NARCOLEPSY TYPE 1	NARCOLEPSY TYPE 2	IDIOPATHIC HYPERSOMNIA
Excessive daytime sleepiness	Yes	Yes	Yes
Sleep paralysis and hallucinations	Yes	Sometimes	Occasionally
Cataplexy	Yes	No	No
Disrupted nighttime sleep	Yes	Sometimes	Occasionally
Naps are restorative	Yes	Sometimes	Occasionally
Severe sleep inertia or sleep drunkenness	Occasionally	Sometimes	Most
Long sleep periods, more than eleven hours per day	Occasionally	Occasionally	Sometimes

Source: Adapted from the International Classification of Sleep Disorders (ICSD-3), with additions by the author.

Note: Psychiatrists have their own version of IH, hypersomnolence disorder, in the *Diagnostic and Statistical Manual of Mental Disorders*, fifth ed. (*DSM-V*). Their version doesn't separate out type 1 and type 2 narcolepsy, and hypersomnolence disorder can coexist with depression, unlike IH.

of sleep is incomplete, and the brain is stuck in a half-awake state. A related phenomenon is automatic behavior, or performing routine tasks like a zombie, which seems to occur as a result of extended time awake and pressure for sleep.

Brain fog comes up a lot in discussion with IHers, although it appears in other chronic illnesses. Brain fog can mean problems concentrating, mixing up words, and even sensory alterations such as blurred vision. One physician diagnosed with IH described brain fog as approaching a state of delirium, interfering with her capacity to do anything substantive: "I constantly say my brain is broken, because that's exactly what IH has done to me."[16]

The recommended amount of sleep for healthy people is seven to eight hours per night, but some people diagnosed with IH spend large amounts of time sleeping per day: more than eleven hours. Combined with constant sleepiness, their increased need for sleep can squeeze out the time or ability to keep a job or participate in family life. Currently, the category includes those who display long sleep times and those who don't, although in practice, many with IH attempt to use stimulants or other medications to limit their excessive sleep.

IH's symptoms don't usually include muscle or joint pain or a severe response to exercise. This separates the disorder from another whose sufferers have felt maligned and neglected: myalgic encephalomyelitis, also known as chronic fatigue syndrome (ME/CFS). There is some overlap, in that some people with IH report symptoms of autonomic nervous system dysfunction, such as numbness in the extremities in response to cold or dizziness upon standing.[17] Still, *fatigue* and *sleepiness* are different entities. What IH and ME/CFS have in common is that people with the disorders say they feel misunderstood or invisible.

The sense of invisibility comes because to friends, family, and coworkers, IHers may not look like they have a chronic illness, so they have to fend off skepticism or disbelief. To be sure, people with IH diagnoses don't obtain them for fun. Many IHers have gone for years diagnosed with something else. Others find the *idiopathic* label insulting: a shrug of dismissal by a doctor. One sleep specialist said he had learned in medical school that idiopathic meant "idiotic for the doctor and pathetic for the patient."[18]

LONG NEGLECTED

At a conference held in Prague in 2017, I observed signs of an upswing in attention to IH. Bedřich Roth's associates were the local hosts of the World Sleep Congress, held at the Communist-era Palace of Culture overlooking the city. There was debate about what IH is and how many people might have it. Although more treatment options were available compared with the past, sleep researchers disagreed on the relevance of diagnostic tests and whether IH is actually one, two, or several entities.

At a symposium titled "Idiopathic Hypersomnia: A Neglected Disorder," the French sleep researcher Isabelle Arnulf sought to quantify the neglect by counting the number of published papers in PubMed, the U.S. National Institutes of Health database. Over the last fifty years, narcolepsy was mentioned in roughly fifteen times more papers than IH (figure 1.1). Was it simply because IH was less prevalent? Arnulf noted that the rarer episodic sleep disorder of Kleine-Levin syndrome also had more papers than IH. Years ago, she said, the term "hypersomnia" was used loosely—to describe the effects of sleep apnea, for example. Over time, usage became more precise.

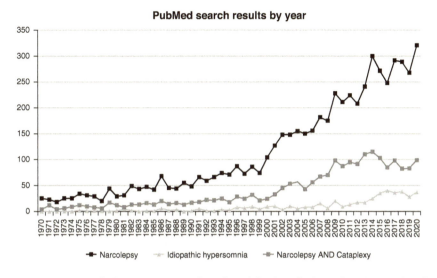

FIGURE 1.1. Graph showing papers in the PubMed database for "narcolepsy," "narcolepsy" *and* "cataplexy," and "idiopathic hypersomnia" from 1970 to 2020.

Arnulf, the head of a center for sleep disorders in Paris, said the lack of attention to IH in the past was because of its status as a category of inexplicable leftovers. IH lacks the dramatic symptom of cataplexy, seen in narcolepsy type 1. Influential experts sometimes have described IH as milder, compared with narcolepsy type 1, and the possibility of remission was left open.[19] IH also sounds like an extension of something most people have experienced. Many people feel drowsy during the day, even without a sleep disorder, because of their demanding schedules or the enticements of screens and artificial light. It can be hard for someone unfamiliar with IH to take it seriously.

Years can separate the onset of symptoms from an IH diagnosis. Many IHers report they began to feel sleepy after an infection or other illness in adolescence or early adulthood, a timeframe resembling that of narcolepsy, while a few say their symptoms were present in early childhood. Doctors and family members can confuse idiopathic hypersomnia with attention deficit disorder or depression or suspect malingering or drug abuse. To doctors, people being treated for IH may generate concern because they can require prodigious amounts of stimulants to stay awake during the day. Some IHers find conventional stimulants or modafinil helpful, but a fraction can't tolerate them or, like Anna, found them unsustainable.

Until recently, pharmaceutical companies have not pursued regulatory approval for the indication of IH. As one industry consultant told me, IH was perceived as a "garbage can diagnosis," too rare or poorly defined to be worth the investment of a clinical trial. Narcolepsy was more legitimate and accepted—it had a head start of around one hundred years.

The distinctive symptom of cataplexy explains why neurologists identified narcolepsy much earlier. In an 1880 publication, the French physician Jean Baptiste Edouard Gelineau, who coined the term *narcolepsy*, described a patient who collapsed when experiencing a variety of positive emotions: "If he is closing a good business deal, if he sees a friend, if he speaks with a stranger for the first time, or if he receives a good hand while playing cards."[20] Although it might look like sleep, cataplexy is not sleep; people experiencing it maintain awareness and remember what's being said around them.

Narcolepsy type 1's biology is relatively well understood today. Recent research has established that narcolepsy type 1 occurs through an autoimmune mechanism, where the immune system attacks parts of the body.[21] It appears to be analogous to type 1 diabetes, in which the immune system destroys cells in the pancreas that secrete insulin. Similarly, in narcolepsy type 1, immune cells eliminate a group of neurons in a specific region of the brain. Those neurons produce another important small protein called hypocretin or orexin, which helps keep people awake and alert. The definitive diagnostic marker for narcolepsy type 1 is a low or undetectable level of hypocretin/orexin in someone's cerebrospinal fluid.

Compare that wealth of knowledge with the vagueness of IH. Roth and others who studied IH have noticed that the condition appears to be inherited in a substantial fraction of cases.[22] This observation agrees with the reports of some IHers that their sleepiness was present in childhood. But despite advances in genomics, no markers for IH have emerged for clinical use.

In clinical practice, the MSLT, which assesses how fast someone can fall asleep in a series of four or five naps during the day, is supposed to guide diagnosis. The MSLT was developed in the 1970s as a procedure for measuring daytime sleepiness. As a research tool, it worked well for understanding the effects of sleep deprivation on healthy people, and it also became a standard test for narcolepsy type 1.[23] However, for many people with IH, the results of their MSLTs have limited their options and not helped them get treatment.

The distinction between IH and narcolepsy type 2 may seem arbitrary. However, the category of type 2 narcolepsy carries information for patients who could be on the way to type 1; sometimes cataplexy begins to appear years after someone starts feeling sleepy. In the United States, testing for hypocretin/orexin in spinal fluid, which could resolve uncertain cases, has not been widely available.

Patients diagnosed with narcolepsy and IH were often treated with the same medications, such as modafinil, methylphenidate, or amphetamines. But the distinction between the disorders meant that one group of sleepy people could get recognition and insurance coverage, while the others had more difficulty. In the United States, several medications had been approved by the FDA for narcolepsy, but none were approved for IH (spoiler alert: until 2021). The *idiopathic* label left IHers out in the cold, sending discouraging signals to their insurance companies, which balked at paying for expensive drugs. Depending on their doctors or the out-of-pocket costs, they might retake the MSLT in the hopes of getting a revised diagnosis.

At the Prague symposium on IH, several researchers criticized the MSLT because it often wouldn't give the same result if performed twice.[24] Some argued that the categories of narcolepsy type 2 and IH could be merged, since current diagnostic tests couldn't reliably separate them. Another proposal was to split IH in two, putting people who slept for long amounts of time in a separate category and joining the others with narcolepsy type 2.[25] Researchers in Europe, such as Arnulf, had taken the view that overall excessive need for sleep was more important than how quickly someone fell into it. They had established alternatives to the MSLT, based on observing how long patients would stay asleep in one continuous overnight stretch or in longer periods of enforced rest.[26] Still, sleep labs in the United States were accustomed to the MSLT, and time-consuming tests would cost more money.

In 2009, Arnulf and a colleague published a study of the second-largest number of IH patients ever compiled: seventy-five.[27] More than half regularly slept for more than eleven hours every night. But most of that group did not fall asleep faster than the eight-minute threshold during an MSLT—highlighting the limitations of the test. Ultimately, the diagnoses were based on the patients' reports of their symptoms rather than any definitive biological or physiological marker.

The historical authorities on IH, Roth and Billiard, have suggested that people with IH represent the extreme end of the category of long sleepers: those who feel best if they sleep nine or more hours per night. Although it is difficult in a demanding world, if long sleepers get enough of what they need, they don't feel drowsy during the day and can function normally. In contrast, people with IH do not seem to ever catch up with their hunger for sleep—naps tumble into a bottomless pit. While IHers describe what they experience when awake as feeling like perpetual sleep deprivation, IH's effects on the brain may not be the same. As Billiard wrote in 2015, without more specific criteria or diagnostic tests, "the question as to whether idiopathic hypersomnia exists *sui generis* as an extreme normal variant or with its own unique biological underpinning remains open."[28]

A POLITE STANDOFF

Part of the reason I was in Prague was curiosity about possible fireworks between David Rye and his critics. Rye has been central to IH's emergence in the last decade. His team's research, beginning with work on Anna's case, added fuel to a debate about IH's origins and status in relation to other sleep disorders. Until that point, I had been listening to just one side of the debate, and I wanted to hear what others had to say.

Rye was a hefty, imposing man; his colleagues said he looked like a football linebacker. (He was in fact the coach of his son's hockey team.) He seemed to enjoy provoking his peers with wisecracks, and he was a keen advocate for "actually listening to the patient." In slides at conferences, he liked to show himself in a bleak Icelandic wilderness, having visited the island many times to collaborate with geneticists there.

After hearing his patients describe their relationships with sleep, Rye felt the need for a simplified, alternative classification scheme for narcolepsy and hypersomnia. The scheme ended up on T-shirts, sold at patient-organized meetings. The design (figure 1.2) showed apples and oranges, with apples representing hypersomnia and oranges representing narcolepsy. Under narcolepsy it said, "Seized by sleep," and under hypersomnia it said, "Consumed by sleep." The T-shirt design emphasized that people with narcolepsy are known for falling asleep suddenly but often don't spend more time each day asleep compared to a healthy person. In narcolepsy, the boundary between sleep and wake is less stable, while people with hypersomnia find that sleep is taking over their lives, hour by hour.

Rye and his supporters had more than T-shirts; they had an idea. Based on his collaboration with Jenkins, Rye proposed that the prolonged sleep and difficulty waking seen in Anna and others were linked to a "somnogen" present in their spinal fluid. They believed this sleep-inducing substance was something produced within the body and that it acted on the brain in a way that resembled benzodiazepine drugs.

Rye had been arguing that clinicians' views of IH had been shaped by the tools they had to treat it—the first-line medicines were generally spillovers from narcolepsy. It is easy to think of stimulants as pushing the brain to go faster, compensating for some lack of drive. Rye described giving high doses of stimulants to Anna as "trying to drive a car with the parking brake on. We needed to release the brake, rather than push the gas harder."

For some people with IH and related disorders, Rye proposed that it was necessary to wipe away a sleep-promoting factor with flumazenil or something

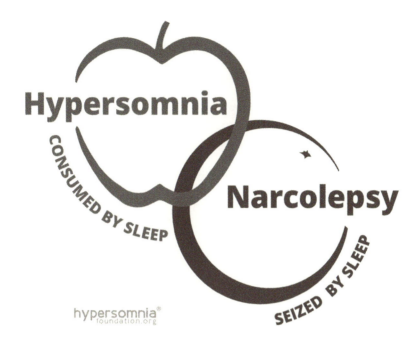

FIGURE 1.2. A simplified view of the relationship between narcolepsy and hypersomnia.

Source: Design by Katie Ratcliffe. Courtesy of Hypersomnia Foundation.

similar. If his theories were correct, a mechanism, if not the ultimate cause, would be defined for some apparently *idiopathic* cases. In the next few chapters, we will explore the basis for his ideas.

However, at the time of the Prague meeting, Rye had not convinced key figures in the sleep medicine field. "Not enough evidence," they said—and they had a point. The biochemical identity of the somngen was uncertain. A test the field would want to see, a randomized controlled trial of flumazenil, had been performed, but the results were underwhelming and unpublished. Rye's peers were inclined to be cautious. Over the past forty years, others had published similar papers, analyzing samples of cerebrospinal fluid from IH patients and finding alterations in various brain chemicals. Those past findings were not confirmed and had fizzled out in confusion.[29]

The year before the conference, it was Rye's turn to face doubts. A group from France, led by the neurologist Yves Dauvilliers, reported that they couldn't replicate Rye and Jenkins's results.[30] The challenge was on Rye's mind when he gave his talk. In a packed, low-ceilinged room, he described how his team's findings had been duplicated by a lab in Australia.[31] He said: "There has been some question about whether our results were reproducible!"

The situation looked like a polite standoff, but elsewhere, the ground was beginning to shift. Other sleep specialists in the United States had started to prescribe flumazenil or related drugs "off label" for IH. Following Anna's example, hundreds had tried them. Even if the neurochemistry was vague, their experiences were real.

CHAPTER 2

THE DOCTORS AND GABA

Anna's extreme condition forced me to depart from conventional wisdom. What if her extraordinary need for sleep were caused not by an absence of wakefulness, but by the presence of sleepiness induced by some other biochemical agent—the Sandman's mythical dust?

—David Rye, "Why Is This Young Woman Tired All the Time?"

Let's return to Anna Sumner in the spring of 2007. She had been taken off the stimulants that made her uncomfortable and provoked jitters and heart palpitations. She was on leave from her job. At this point, Anna was sleeping around eighty-five hours per week, and it was difficult for her to stay awake while watching television or doing something quiet. "On top of the sleepiness, my issue was feeling fuzzy and mentally slow. I would have to re-read things. I was worried that this was just part of who I was," Anna said. "You could physically sit me up, and I would respond to questions, but I would have no recollection of answering them."

Anna recalled taking a MARTA train home after one doctors' appointment, still with EEG wires attached to her scalp, feeling disoriented—"like a space alien." Since April, her mother Ward had been staying with her, making sure that she ate enough. Because of concern for safety, her mother drove Anna to her appointments at Emory's sleep center, then located at a faded former nursing home. The person they saw the most often was a nurse practitioner, Kathy Parker (figure 2.1), who was closely involved in both devising a treatment for Anna and

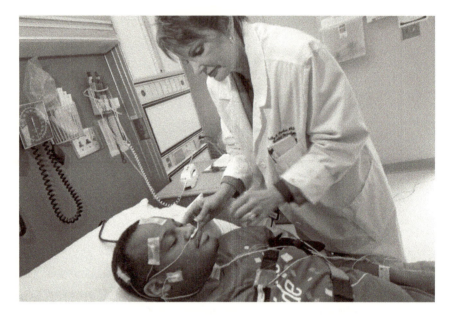

FIGURE 2.1. Kathy Parker preparing a patient for a sleep study.

Source: Age Fotostock America.

in convincing a pharmaceutical company to help. "Kathy had two daughters around Anna's age, and I think there was an emotional connection," Ward Sumner said. "She was a force. She pushed and pushed."

PHYSICIAN EXTENDER

Describing Parker as a "physician extender," a term sometimes used for nurse practitioners, understates her accomplishments. Her father, a chemical engineer, drilled her and her sisters on the periodic table when they were little. When she was hospitalized for a kidney condition as a five-year-old, a nurse she admired inspired her to pursue nursing as a career. Parker's specialty was treating patients with kidney disease, and she became interested in sleep while working on a dialysis unit at the Atlanta Veterans Affairs Medical Center, close to the Emory campus. "When you take care of dialysis patients from day to day, there's not much you can do to fix them, so the focus is more on quality of life," she recalled. "They would complain bitterly about their ability to sleep, and it seemed to be worse on dialysis days."

Parker's observations, along with the experience of being sleep deprived after having her first child, nudged her to go back to school. For her PhD at Georgia State, which she finished in 1990, she analyzed her kidney disease patients' sleep and dream patterns.[1] Soon after finishing her degree, Parker walked into Donald Bliwise's office at Emory and asked to join his efforts. Originally a PhD psychologist, Bliwise was part of a cohort of sleep researchers who entered the field when it was young. Captivated by research on dreams as an undergraduate, he had trained with leaders in the field at the University of Chicago and Stanford and then was recruited to Emory to develop the sleep medicine program in the Department of Neurology. His own research focused on sleep disturbances connected with aging and with neurodegenerative diseases such as Alzheimer's and Parkinson's.

In a newspaper profile of Parker, Bliwise said he viewed their relationship as mentor-mentee at the start and later began to see her as a peer.[2] Parker carved out a niche probing the interactions between dialysis and sleep, finding that her patients' sleep could be improved by slightly lowering the temperature of dialysis fluids, whose warmth appeared to fuel their insomnia.[3]

Parker also joined day-to-day operations at Emory's sleep clinic. This involved a step back—at least, temporarily. "I asked Don and Dave whether I could come into the clinic, just to observe," Parker said. "At the start, I would sometimes run errands for them, or take care of blood work, or help transfer patients." Parker was used to more autonomy working at the Veterans Affairs Medical Center. Under Georgia state law, however, nurse practitioners were more dependent on a supervising physician than in other states and could not prescribe medications such as opioids or stimulants on their own.[4]

In 2001, Parker passed an exam given by the American Board of Sleep Medicine. She was one of only a few nurses in the country with this credential, which meant she could evaluate and score overnight sleep tests.[5] Patients with kidney disease were referred to her, and she began seeing others with a variety of conditions.

In her clinical duties, Parker's supervisor was David Rye, who had come to Atlanta at the same time as Bliwise. Asked about her first impressions of Rye, Anna said: "I remember this big tall guy in the doorway. He was very matter of fact, just saying: 'Let's try this.' She [Parker] and Rye were the first medical professionals I met who didn't think I was making it up, or told me I just had to live with it. They were great bookends. The two of them were very different, but he really paid a lot of attention to what I was experiencing. It seemed to consume him—he wanted the answer."

RESTLESS SPIRIT

Although Rye later became known for Anna's case, he was occupied with other research concerns that summer. He recalled later: "Kathy came in and said Anna's back, she's still really sleepy and she's worried she might lose her job." For Rye, insight into Anna's condition required a change in thinking: away from focusing on molecules in the brain that keep us awake and attentive and toward those that clear the path to sleep. At the time, his major research interest was not hypersomnia but restless leg syndrome (RLS). From 2004 to 2007, Rye was chair of the medical advisory board for the Restless Leg Syndrome Foundation.

Since Rye had RLS on his mind so much, it is not surprising that Anna was initially treated for it, before the flumazenil experiment. Anna called herself "an inveterate leg wiggler," to the point of annoying people around her, and her overnight sleep test picked up the signs. She was prescribed iron supplements and pramipexole, an RLS drug. However, calming her legs did not improve her sleep situation.

RLS is more common than IH and different in both its clinical and molecular features. Those differences can help us understand what Rye was used to thinking about, before IH took his career in a different direction. When it did, he would draw upon his RLS connections.

RLS disrupts sleep, but it also occurs when someone is awake. It's relatively straightforward to track one aspect of RLS, involuntary periodic limb movements, by attaching an accelerometer to someone's ankle. RLS also has a sensory component: discomfort in the legs, which intensifies in the evening and is relieved by movement. People with RLS can describe their feelings as like worms burrowing, spiders crawling, or a tingling wave that never stops. Those sensations keep them pacing at night or drive them to kick their bedmates.

Rye has estimated that around 10 percent of the United States population has RLS. For a fraction of that group, it bothers them enough to affect their quality of life.[6] Variations in several genes contribute to the risk of developing RLS, as well as iron deficiency, explaining why menstruating individuals are more likely to have it. RLS was first described by the English anatomist Thomas Willis in the seventeenth century, although the Swedish neurologist Karl-Axel Ekbom is the one who defined it in the 1940s.

Rye became involved with clinical studies of RLS in the late 1990s, when pharmaceutical companies began repurposing drugs that were FDA approved for Parkinson's disease. He had a personal connection to RLS as well. In 2001, he discovered that he experienced periodic leg movements, after he spent time lying in bed with a broken ankle. While his leg was encased, he had the repeated urge to move his foot. He medicated the feeling away with one of the drugs he had been testing. He then fitted himself with an ankle bracelet and recorded his leg kicking thirty times per hour while he slept. He would point out the symptom in fellow travelers when he flew on airplanes with colleagues.[7] He noticed that a few spoonfuls of ice cream tended to trigger acute attacks, for both him and his patients.[8]

RLS was not well known until television advertising brought it to greater awareness, and skepticism lingered. Some news articles from this time ask: "Is RLS Real?" It was the topic of jokes on *Seinfeld*; Rush Limbaugh mocked RLS as fake on his radio show. As a response to trivialization by others, some in the RLS community have proposed renaming it Willis-Ekbom disease, although the term has not seen widespread adoption.[9]

By disrupting sleep, RLS increases the risk of high blood pressure, and severe cases can become debilitating, interfering with daily life. However, critics have cited the disorder as an example of "disease mongering," inflation through

FIGURE 2.2. David Rye in 2008.

Source: Jessica McGowan / New York Times / Redux.

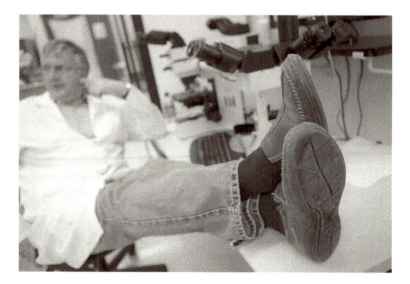

FIGURE 2.3. Rye's restless legs.

Source: Jessica McGowan / New York Times / Redux.

commercial promotion.[10] Not everybody with wandering legs needs to take medication, they said, especially when the drugs can have side effects such as nausea and dizziness.

Rye's status as the RLS guy meant he was often called upon to explain it. He provided cheerful counterpoint in a 2006 *Washington Post* article, which said that RLS was being pushed onto the public through TV advertising.[11] "I don't know of any evidence that it's being over-diagnosed," Rye told the *Post*. "I look at the positive side of it rather than the doomsday view. I think it helps to make a diagnosis."

In the summer of 2007, Rye was finishing work on a *New England Journal of Medicine* paper with geneticists in Iceland.[12] His collaboration with DeCODE, a company that harvests Iceland's well-recorded genealogy for biomedical research, had begun a decade earlier. He had developed the diagnostic tools used in the study and sent two members of his lab to Iceland to test ankle monitors. The *NEJM* paper identified the first genetic risk factor for periodic limb movements in sleep, present in more than 60 percent of the Icelandic and American populations. The paper attracted media coverage, and Rye's comments suggested he viewed it as vindication for his specialty. He said: "We now have concrete evidence that RLS is an authentic disorder with recognizable features and underlying biological basis."

During this period, RLS wasn't only a research interest for Rye; it had become part of his identity. Because RLS was so common, he hesitated to call it a disorder, describing it instead as a trait. In an interview with the *New York Times*, he riffed expansively: "This isn't just restless leg, it's a restless curiosity, it's a restless mind, it's a restless spirit." [13]

FROM DOPAMINE TO GABA

In Anna's case, RLS wasn't a useful explanation. What struck Rye and Parker was how she described craving sleep constantly. It was clear that the medications Anna had been taking were not helping. Stimulants such as amphetamines are sometimes called "sympathomimetic," in that they mimic the effects of adrenaline, making someone irritable and jumpy and increasing blood pressure and heart rate. High doses of amphetamines put stress on Anna's body and led to periodic crashes. The rebound sleep Anna experienced is a well-known feature of amphetamines, although the duration was extreme. Parker wondered whether Anna's diagnosis of idiopathic hypersomnia was appropriate because in the literature, stimulants were reported to be beneficial for the majority of cases.[14]

In addition to stimulants and RLS drugs, Anna had been taking an antidepressant, as well as a beta-blocker to control high blood pressure brought on by amphetamines. This was a lot of different drugs. Some may have been making her sleep situation worse, which is why Parker wanted to wipe the slate clean and start over. "We had never had a patient quite like her before," she said. "I thought to myself that she's going to sleep her life away if we don't do something. I felt like the answer was there, but we weren't asking the right question."

Parker made a chart of the neurotransmitters affected by the drugs she and Rye had been prescribing to Anna. For stimulants such as amphetamines, dopamine and its chemical cousin norepinephrine are critical. Amphetamines cause dopamine and norepinephrine to be released from storage inside brain cells. Also, by inhibiting enzymes that would normally clear the neurotransmitters away, the drugs make dopamine and norepinephrine stick around longer at the junctions between brain cells, stimulating circuits that help keep us awake.

Dopamine's complex relationship with sleep and wake was one of Rye's research specialties, since RLS is often treated with drugs that supplement or mimic dopamine. People with a casual interest in neuroscience have probably heard of dopamine because of its association in popular media with pleasure and addiction; as one writer put it, dopamine is "the molecule behind all our most sinful behaviors and secret cravings."[15] A software consulting firm (Dopamine Labs) was named after it, based on the business-friendly idea that people using mobile phones are searching for something that will trigger a squirt of it.

However, neuroscientists say that dopamine doesn't simply trigger a burst of pleasure or attention; it has several roles, depending on what part of the brain is involved.[16] In one area called the nucleus accumbens, it promotes a feeling of reward. In Parkinson's disease, cells that usually produce dopamine in the middle of the brain deteriorate and die. At that location, dopamine is needed for initiating and controlling movement. Dopamine-related signals can both wake someone up or make them sleepy, depending on the receptors engaged. Asking what dopamine or any other neurotransmitter does is like asking the meaning of a violin or flute in a piece of music. It depends what notes they play and in what context.

MIDNIGHT INSPIRATION

In the spring of 2007, Anna was exploring whatever was available, including changing her diet and alternative-medicine approaches such as acupuncture. For a while, Parker had tried giving Anna donepezil (Aricept), which boosts another neurotransmitter, acetylcholine. Anna reported no noticeable effect of donepezil, sometimes prescribed to patients with Alzheimer's disease, on her mental fog or sleepiness.

While Parker was searching for options for Anna, at some point she woke up in the middle of the night. She wrote the word GABA on a piece of paper and circled it. She said later: "We had tried to tinker with many of the neurotransmitters in her brain, but not GABA." In relation to dopamine and norepinephrine, GABA (gamma-aminobutyric acid) belonged on the other side of Parker's chart. GABA is the main inhibitory neurotransmitter in the adult brain. Its dominant effect is to make brain cells less likely to fire. Many drugs thought of as sleeping pills, such as barbiturates and benzodiazepines, enhance the action of GABA. Several injected and inhaled anesthetics function in a similar way. Alcohol also strengthens GABA signals, although it acts on other neurotransmitters too. All of these drugs make GABA signals stronger. Parker wanted to go in the opposite direction. "We had spent a lot of time trying to push Anna's brain to wake up," she said. "Maybe what we needed to do was make her less sleepy."

We can think of neurotransmitters such as dopamine like pepper. They are loaded into packets called vesicles and stored inside the cell until electrical signals release them. Outside the cell, the vesicles dump out their contents so that a neighboring neuron can take a whiff. The effects depend on a receptor on the neighboring cell's surface, which the neurotransmitter fits into. The receptor's altered shape triggers some action inside the cell: an ionic passageway opens or an enzymatic machine starts churning.

GABA comes in vesicles too, but instead of acting like a spice, it's more like an ice cube or sour cream. It soothes and calms. GABA has this effect by opening a set of gates spanning the cell membrane. The gates, named GABA-A receptors, come in barrel-shaped bundles of five subunits, which can open or close by changing their orientation. When the gates are open, chloride ions flow into the cell. The resulting accumulation of negative charge pushes the neuron away from sending an action potential, the electrical pulse that carries a signal to another neuron. (GABA also has a second set of GABA-B receptors, discussed in chapter 15.)

Signals from GABA may have an overall "dimmer knob" effect on brain cells, as well as inhibiting regions involved in arousal, but that does not mean that flooding the brain with GABA, or otherwise mashing the brain's GABA buttons, will produce healthy sleep. Even though several anesthetics enhance GABA's actions, general anesthesia can't replace sleep; the anesthetized brain does not pass through the complex oscillations thought to make sleep restorative. As with dopamine, no single neurotransmitter controls sleep and wake by itself.

NANOSCALE CRAFT

Given the inadequacy of what she and Rye had already tried, Parker saw counteracting GABA as an option worth investigating further. Parker had been discussing Anna's case with her daughter Kathryn, who was working in the laboratory of the psychiatrist Kerry Ressler, who studied anxiety and post-traumatic stress disorder. The younger Parker relayed her mother's interests, asking: "Who studies GABA at Emory?" Ressler suggested Andrew Jenkins, a young anesthesiology researcher who had been at Emory for just a few years. "I was willing to work with anyone. I was calling people across the country," Parker said. "It was luck to be matched up with Andy."

Trained as a biophysicist, Jenkins's expertise was not in clinical or even animal research, but he knew a lot about GABA receptors. His work, together with others', helped break down an old idea concerning anesthetics. Despite their use in surgery since 1846, how anesthetics dissolve consciousness was unclear until the twenty-first century. The observation that anesthetics' potency can be predicted by their affinity for olive oil, as opposed to water, led to the proposal that anesthetics function by seeping into the lipids in cell membranes. At Imperial College London, Jenkins worked with Nick Franks, whose lab was in the process of dispelling that greasy idea.[17] Franks and his colleagues have shown that several anesthetics have interactions with specific sites on GABA receptors. Not all anesthetics act this way, but commonly used ones such as propofol and sevoflurane do.

In 2006, Jenkins gave a cheeky interview to a graduate student newsletter.[18] He described his time at Imperial, where he completed university and graduate school, as the United Kingdom's equivalent of the Massachusetts Institute of Technology. His experience there was dominated by "beer, rugby, and an insane lecture and lab schedule." Franks ran his lab in an understated way, sparing with praise. At some point, Jenkins was told: "Andy, you're not the smartest guy in the lab, so you'll have to work harder than the other guys to make up for it."

Jenkins encountered several dead ends during his time in graduate school. One of his projects involved experiments on pond snails, another the breeding of alcohol-insensitive guppies. Eventually, he did make headway on anesthetics, but he had to invest a lot of hours. He was the first person in his lab to learn a painstaking technique called "patch clamping." "It requires a tremendous amount of patience," said Adam Hall, a scientist from a neighboring lab who taught Jenkins the technique. "Andy really took to it, with a physicist's precision."

Developing the patch clamp earned two German scientists the 1991 Nobel Prize in Physiology or Medicine, and the technique supports the foundations of modern neuroscience. Patch clamping involves grasping a cell with the tip of a thin glass tube, which holds a *patch* (a small section) of the cell membrane. The patch contains a handful of molecules—GABA receptors or others—that let charged ions through their gates. Inside the tube, an electrode allows the experimenter to monitor the current and voltage.[19]

Patch clamping measures a very small electrical current. How small? A milliampere will produce a tingling sensation as it flows through a fingertip. A microampere is thousand times smaller and represents the current that flows while a finger touches a modern phone's screen. With another thousand-fold squeeze, a nanoampere comes into view. Motion sensors and smoke detectors operate at this level of sensitivity. The currents measured with patch clamps are usually less than one nanoampere.

HUGE POTENTIATION

Prompted by her daughter's conversation, Parker called Jenkins and explained the struggle with Anna's sleepiness. She asked if he would be willing to help. Jenkins had just a few people in his lab and was doing hands-on experiments, exploring science that looked promising.

At the time, Paul Garcia, an MD/PhD interested in neuroscience, was working off and on in Jenkins's lab. Garcia, then beginning his anesthesiology residency, had a connection to Rye as well, having shadowed him as a medical student. Garcia said he had helped revamp Jenkins's lab to handle clinical samples. "The system was all set up," Garcia said.

Parker already had a sample of Anna's CSF when she called Jenkins. In a repeat of her diagnostic workup, Anna had fallen asleep quickly in a multiple nap test, even after sleeping for fourteen hours the night before. Anna's lumbar puncture was performed in the hospital, since she tended to "get the vapors" and pass out, according to Parker. A headache often results after a lumbar puncture, with the intensity related to how much fluid is removed.

Jenkins took charge of the first set of experiments on her CSF. His notebook from May 2007 read: "Huge potentiation." If he squirted a bit of GABA onto a patch of kidney cell membrane, it let in a tiny amount of current, which his patch clamp apparatus was set up to detect. Anna's cerebrospinal fluid contained a negligible amount of GABA, but whatever was in it made the effect of the GABA he provided more than twice as strong. Jenkins would write that her CSF contained a level of GABA-enhancing activity "equivalent to that caused by general anesthetics at loss-of-consciousness concentrations."[20] He was astounded at the neurochemical state Anna had been living in.

There weren't many "off the shelf" options available for diminishing GABA signals. Several relevant compounds, such as picrotoxin or cicutoxin, are known more as poisons than as beneficial drugs. They even have "-toxin" in their names, because taking away the calming influence of GABA can result in a seizure.

The emergency room antidote flumazenil stood out as available and FDA approved. How the idea to try it with Anna solidified was remembered differently by the main players. Parker said she had already thought about flumazenil before contacting Jenkins: "Why go through all of that, if there wasn't something we could do for her?" Jenkins said a specific remedy was not discussed initially but emerged as an option after his experiments suggested it.

When Jenkins added flumazenil to a patch clamp experiment, it reversed the GABA-enhancing effects of Anna's CSF. He used a high concentration of the drug, possibly surpassing what can feasibly be introduced into the nervous system of a living person.[21] Even so, his laboratory findings raised the question: would flumazenil accomplish the same thing in Anna?

ON MY RADAR

Flumazenil, developed by Hoffmann-La Roche, was thought to be helpful in two situations. The first is when someone is sedated for an uncomfortable medical procedure. To make the patient relaxed and the process less memorable, a normal practice is for the doctor to give someone a sedative, such as diazepam (Valium) or midazolam (Versed). If the patient receives too much, the sedative can inhibit parts of the brain that control breathing, which is dangerous. Flumazenil can reverse the sedation.

A second situation calling for flumazenil is when someone arrives at the hospital and appears to have overdosed on benzodiazepines. According to data from the American Association of Poison Control Centers, flumazenil is used in this way about two thousand times per year. It is analogous to naloxone, the fast-acting antidote against opiate overdose.

At the time the Emory group was considering it, flumazenil had already accumulated a weird side story. Flumazenil could revive people with liver failure who had been in a coma minutes earlier. Some studies had explored whether it could be a cognitive enhancer—that is, whether it could make people smarter. A few addiction specialists latched onto it, claiming that flumazenil could help people with withdrawal symptoms.

Flumazenil's effects pointed to an unsolved puzzle. Neuroscientists still think that flumazenil doesn't do much by itself. It elbows benzodiazepines out of their slots on GABA receptors, but its direct effects on those receptors are weak. That explains why endozepines, short for "endogenous benzodiazepines," were proposed to exist. In concept, they resemble endorphins, messengers produced by the body that tickle the same pain-relief receptors that morphine does. If the body produces its own painkillers, it could make its own calming or sleep-inducing molecules too. In some situations, such as hepatic encephalopathy, a term for the effects of liver failure on the brain, the nervous system seems to accumulate benzodiazepine-like chemicals. These substances might be what flumazenil was kicking out of the way.

Jenkins was familiar with previous research on endozepines and hepatic encephalopathy. After Imperial College, he had worked in the United States as a postdoc with Neil Harrison, another important player in defining the mechanisms of action for anesthetics. One of his former lab-mates had been testing whether fragments of heme, breakdown products from decaying red blood cells, might be partly responsible for symptoms of hepatic encephalopathy.[22] This prepared him for his collaboration with Rye and Parker. "That was going on next to my bench for a few years," he said. "It was the reason I was ready for Kathy's call. Those ideas were already on my radar."

But a curious episode in Italy had cast a shadow over the endozepine idea. Neurologists from the University of Bologna had published several papers on a disorder they called "idiopathic recurring stupor." People from Italy and other countries had been falling asleep suddenly, without adequate explanation. Flumazenil woke them up, so their sleepiness was attributed to endozepines. A cluster of cases was revealed to instead come from someone slipping sedatives to their neighbors.[23] "Dave and I talked about those papers," Parker said. "We agreed they were very quirky. Initially, we did wonder if Anna's condition was self-inflicted. We did blood tests and urine tests. She was clean."

There was enough logic behind others' use of flumazenil that Parker and Rye were willing to pursue it. Rye had briefly woken up a couple patients using flumazenil in the 1990s. One had elevated ammonia levels—a sign of liver trouble. Rye recalled these experiments as taking place "in a closet." Also, because of his familiarity with Parkinson's research, Rye knew that a colleague in Houston, Bill Ondo, had completed two small-scale studies on flumazenil's effectiveness against Parkinson's motor symptoms.[24] While Rye was skeptical about endozepines, he was attracted to trying something first and figuring out the hows and whys later. "Kathy is a process person," Rye said. "I'm a results person."

TREATMENT OR TOXIN?

Rye and Parker also were aware this line of thinking was not risk-free, because flumazenil's use in emergency departments had become a topic of debate. Along with naloxone, flumazenil was once proposed as part of a "coma cocktail" for reviving intoxicated people, but medical opinion had turned against it.[25] A 2004 paper summarizes the problem: "Flumazenil—Treatment or Toxin."[26] The risks of flumazenil were often not worth the potential benefits, compared to introducing a tube to aid a patient who was having trouble breathing.

Many papers showed that flumazenil can whisk away a benzodiazepine's sedative effects and restore alertness. However, there were occasional reports of seizures and other adverse events. For most of these, researchers concluded that other drugs, such as tricyclic antidepressants, were responsible or that the antidote had unmasked underlying problems, such as epilepsy or panic disorder.

In a few case reports, it's difficult to say why things had taken a wrong turn. In a small town in Ireland, a doctor was performing an endoscopy on a thirty-year-old woman suspected of having an ulcer.[27] She was given a large dose (15 milligrams) of diazepam. After the procedure, the woman displayed shallow breathing, an indicator that the dose may have been too high. Alarmed, the doctor provided oxygen and half a milligram of intravenous flumazenil. The woman promptly woke up, but a few minutes later she lost consciousness and began to have convulsions, which progressed to status epilepticus: a seizure lasting several minutes, carrying with it an increasing mortality risk. Once the seizure was quenched, she did survive and recover. The woman did not have other medical problems and had not taken other drugs. Conclusion: watch out—a message many doctors absorbed over the years.

Benzodiazepines make brain cells less excitable, but the body and brain adjust to the drugs' presence. It's what makes them so addictive. When they wash out, brain circuits are then more excitable—producing withdrawal symptoms. Flumazenil makes the washout occur even faster than quitting cold turkey. It explains why flumazenil can provoke panic attacks and seizures in people who had been taking benzodiazepines chronically.

In her 1979 book *I'm Dancing as Fast as I Can*, the television producer Barbara Gordon explained what happened when she tried to quit Valium. Gordon had started taking the drug for back pain and became dependent on it to stave off anxiety. When she stopped abruptly, she experienced withdrawal symptoms, which she described vividly: "My scalp started to burn as if I had hot coals under my hair. Then I began to experience funny little twitches, spasms, a jerk of a leg, a flying arm, tiny tremors that soon turned into convulsions. I held onto the bed, trying to relax. It was impossible."[28]

In 2007, Anna had been living in an abnormally sleepy state for years. If there was a benzodiazepine-like substance weighing down her brain, what would happen if it was suddenly stripped away?

CHAPTER 3

THE ANTIDOTE

To justify the use of an off-label treatment, there is one and only one person to bear in mind: the patient. But disposing of the other interests in the delivery of medicine, for example the pharmaceutical company that makes the product, the doctor who prescribes it and the government or insurance company who pays for it, is not an easy task.

—David Cavalla, *Off-Label Prescribing*, 2015

Momentum around trying flumazenil with Anna was building. Prompted by Rye, Jenkins and Garcia looked into the medical literature on the danger it might pose. Garcia read a compilation of clinical data on seizures that had occurred in connection with flumazenil.[1] He concluded the risk was minimal, since Anna wouldn't have other drugs in her system. Jenkins wasn't so sure. "A range of things could have happened," he said later. "Anything from feeling crummy all the way to grand mal seizure."

Anna's previous experiences had prepared her to take a risk, and she was fully aware of potential complications. As an attorney, she had dealt with cases when hospital procedures had gone wrong. One of the cases she had worked on as a junior associate in 2005, before her sleepiness became such an obstacle, involved defending an obstetrician from a lawsuit by a woman who had a gruesome miscarriage.[2] She was still willing to go ahead, telling her doctors: "I am the most informed consent patient you are going to get."

The setting for the experiment—the epilepsy monitoring unit in Emory University Hospital—was on the third floor, away from busy Clifton Road. Each bland-looking room had a camera and an extra console for looking at EEG recordings. If a seizure did occur, Anna was already in the hospital, rather than at the sleep lab, which was more than a mile away. Parker was attentive to Anna, holding her hand at the start, but inside, she was "a nervous wreck." Her maverick proposal was finally going to be carried out. "I think we didn't fully grasp the gravity of what she was doing, until we saw all those specialists standing there," Anna's father, Jim, said.

The flumazenil experiment attracted spectators, some of whom didn't strictly need to be there for medical reasons. Jenkins, for example, wanted to see the human side of his experiment. Anna's parents recalled him as charming. Photos her family took (figure 3.1) show Anna's colorful pajamas, offset by EEG wires attached to her head, along with stark white walls and a box of crackers to snack on. Epilepsy patients and their families often hang around for days, waiting for a seizure so that doctors can determine what part of the brain it comes from.

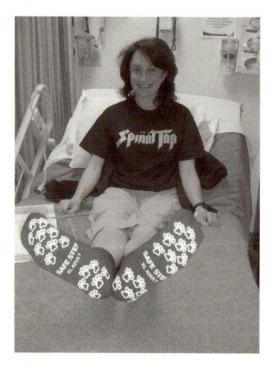

FIGURE 3.1. Anna at Emory, wearing her Spinal Tap shirt.

Source: Courtesy of Anna Sumner Pieschel.

Anna occupied herself with the psychomotor vigilance test, a way of tracking her alertness by timing how fast she could react to racing numbers on a screen. When researchers test the effects of sleep deprivation on reaction time and extrapolate to long-distance truckers or airline pilots, they use the psychomotor vigilance test.[3]

It's designed to be very simple, almost mind-numbing: one hundred challenges, randomly spaced, over ten minutes. It's also difficult (but not impossible) to fake by consistently hitting a button a fraction of a second slower.[4] Anna called the test "the world's most boring video game," but it was important for the team's ability to validate what they were doing. Rye and Parker intended it as an objective measure of Anna's alertness, in contrast to the questionnaires she filled out.

Over two days in June 2007, flumazenil was introduced into Anna's system intravenously. The doctors monitored her vital signs and EEGs. The effects were not immediately dramatic, because she started off with low doses. The anticipation might have contributed to a placebo effect, since Anna knew she was getting a real drug—although many other drugs previously had little effect.

When the dose of flumazenil reached two milligrams, Anna exclaimed to her parents: "I feel alive!" She started talking rapidly and sent an email to friends and family. She said later: "The best way to describe it is that my eyes opened, after being half-closed for so long. It was as if a force grabbed my eyelids and pulled them upwards." Subjectively, her eyes did seem a little brighter. And her reaction time, as measured by the racing numbers, was a fraction of a second faster. According to Parker, Jenkins broke out in tears, saying: "I sit all day behind a bench with rats and finally I get to see my work make a difference for someone."[5]

As exciting as it was, the effect wore off after a few hours. Several sleepy months would go by before Anna would be able to access the same relief.

AN EXTENDED EXPERIMENT

Parker calculated that Anna would need one or two milligrams of flumazenil every hour: several times the amount that reverses midazolam sedation in a healthy person.[6] Flumazenil is usually given intravenously because enzymes in the stomach and liver break it down if swallowed. To deliver flumazenil as it comes from the manufacturer into her body, Anna would have to wear a pump infusing the drug directly into her body. That seemed cumbersome and would carry a risk of infection.

Parker wanted to formulate the drug in a different way—as a nasal spray or delivered under the tongue. For access to a large quantity of flumazenil, she would need the cooperation of Hoffmann-La Roche. Parker began trying to reach executives there in July and August, to her initial frustration. "They would not put me through," she said.

Anna's personal connections opened doors. Through a friend from school, she reached out to George Abercrombie, Roche's chief executive officer, who had been a pharmacist in a small town in North Carolina before embarking on his management career.[7] Abercrombie surprised Parker by calling her and referring her to an employee at Roche's New Jersey headquarters named Bob Baker, whose official title was professional product information director. Baker's job was to explain the fine print to health care professionals who needed information about Roche's products. "I like odd questions, so I decided to take this one on myself," Baker recalled.

After talking with Parker, Baker wasn't quite sure what to do. Using flumazenil to wake up people who weren't taking benzodiazepines seemed a bit fishy. He also wondered how Anna's case compared to other cases of sleep disorders. "What convinced me was talking with Parker and Rye about how thoroughly they had evaluated Anna and all the tests they had done," Baker said. "I went back to my boss and said: 'We should do something to help this young lady.'"

Two main challenges were apparent. The first was physical. Flumazenil was scarce. The drug was not used that much and hadn't been manufactured since around the time the patent expired in 2003.[8] It was available as a generic, but not at the scale Parker was asking for. What she had requested was a substantial fraction of the existing world supply of flumazenil. Baker had to beat the bushes to locate some in Roche's laboratories in Basel, Switzerland.

The second challenge was regulatory. The initial June experiment, when Anna was given flumazenil intravenously, was considered a discretionary "off label" decision by her doctors. Although flumazenil was not approved for sleep disorders, as a physician, Rye had the professional capacity to decide to try it. But if Roche was going to get involved, the company would have to seek guidance from the Food and Drug Administration.

As defined by the FDA, "compassionate use" is a pathway for a patient with a serious or life-threatening condition to try an experimental product outside of the context of a clinical trial, when there are no comparable or satisfactory therapies available. This process, sometimes called "expanded access," was formalized in the 1980s in response to people with HIV infections pushing for access to experimental drugs. Through expanded access programs, Roche had provided HIV/AIDS medications to thousands of people in the 1980s and 1990s.[9]

The request for flumazenil was different: there was no active clinical trial, and the company was opening its vaults just for Anna. The drug was already approved as a benzodiazepine antidote, but for Anna, the dose, the timeframe, and the indication—chronic treatment for a sleep disorder—were all different from flumazenil's approved use. They were distinct enough that the FDA advised that the Emory team would need to make an "expanded access IND" (Investigational New Drug) application. This type of request is typically submitted by the patient's physician and needs to be approved by the drug manufacturer as well as the FDA and an IRB (Institutional Review Board).[10]

According to a 2017 report from the Government Accountability Office, the FDA grants hundreds of requests similar to Anna's every year, and the vast majority are single patients. Most are seeking access to anti-infective, antiviral, hematology, or oncology drugs, with relatively few coming in the areas of neurology, psychiatry, or anesthesia.[11] The FDA has repeatedly taken steps to simplify the "complex and cumbersome" expanded access process, the GAO report says.

Patients' families sometimes mount social media campaigns to convince companies to grant them expanded access requests.[12] With Roche management on board, Anna did not need to do so, but her request still ruffled some feathers inside the company. "There was no process for handling this type of request," Baker wrote in his contemporaneous notes. "Some 'old timers' had never heard of Roche doing this sort of thing."

In August 2007, Anna wrote to Baker, saying that with flumazenil, "I felt awake for the first time in years. It felt incredible. I was animated, could focus, and simply felt *alive*," and "We are desperate to get Romazicon [flumazenil's brand name then] powder." She outlined a scenario in which she would undergo surgery to implant a device to pump flumazenil into her body. She concluded: "In short, without Romazicon in some formulation, I will be permanently disabled." In her letter, she also mentioned that the Emory team had begun to look into three other patients, including her youngest brother James, who might have similar conditions.

"There was no way I could get enough pure flumazenil powder from Switzerland for three people, so I advised them to stay focused on Anna," Baker said. He began to consider the "what ifs." If success with one patient led to something bigger, who would own the rights? If something went wrong, who would be liable? Who would monitor for adverse events and report them to the FDA?

According to Baker's notes, "there was some debate about who was the Principal Investigator and whether she had the credentials to be such." Despite not having a physician's title, Parker ended up completing a master course in FDA and IRB paperwork. In a way, she was devising an experimental study for just one person, with no comparisons possible between Anna and any control group. For several years, Anna was the only person in the world taking flumazenil long-term for a sleep disorder.

SLEEPING BEAUTY AT THE BACK

In October, before the regulatory approvals had come through, Parker gave a lecture on Anna's case at Emory. Anna sat in the back of the room listening but did not participate. It was the first time anyone from the Emory team had talked about Anna in public. Parker identified her as an attorney, using her first name only. "We have a sleeping beauty here," Parker said. "She can't stay awake for more than six hours. She can't go out on a date. She had to take leave from her job."

Parker gave a name to Anna's condition, endozepine-induced recurrent stupor, which embraced the endozepine terminology despite the controversy over cases in Italy. It also matched the language Anna used in her letter to Baker. Parker acknowledged Anna's diagnosis of idiopathic hypersomnia but said it was only used "when we really don't know what the issue is."

This blip of publicity generated concern when an article appeared in the university newspaper, leading to inquiries to Parker from *People* magazine. Nobody on the Emory team was ready for the media spotlight at that time. Rye and Jenkins didn't want more attention before they could gather more information and define what was behind Anna's condition.

The flumazenil finally arrived in February 2008. Parker laughed while recalling how a tin with white powder in a plastic bag arrived at the Atlanta airport from Switzerland. She asked someone at the FDA, with whom she had been conferring for weeks, to send customs officials a message: it was not cocaine. To Parker, it seemed more like gold.

The container made its way to Emory, where Parker had the pharmacy produce lozenges small enough to slip under the tongue. Anna came back to the hospital for another trial. This time, Anna had T-shirts emblazoned with the word "Annazam" (analogous to diaze*pam* and midazo*lam*) made for her and her family. If her doctors believed that her sleepiness was caused by a benzodiazepine-like substance, she would embrace it. But Kathy Parker was nervous. "The night before we started, my greatest fear hit me," Parker said. "What if it doesn't work? It was a sickening thought."[13]

At the hospital, Anna watched television and knitted, again with EEG electrodes on her head. The effect was subtler than the first time around, but she stayed awake for hours. Parker felt a sense of relief.

Over the next couple days, Anna discovered that having the level of flumazenil in her system swing too drastically made her nauseated or groggy. Parker asked the hospital pharmacy to mix up a cream containing the drug, to be applied to Anna's forearms. The cream was originally blue, making Anna remark that she felt like a Smurf.

This combined regimen seemed to do the trick. It evened out the fluctuations and helped her wake up in the morning. During the day, Anna kept a supply of lozenges in a bracelet around her wrist, so they could be taken as needed. Having spent so much time in bed over the past year, she was physically weak and initially had trouble walking across a parking lot without feeling tired.

In April, she went back to work part-time and told others she was gaining strength and able to get by with ten or eleven hours of sleep. She could watch entire movies or join friends for dinner for the first time in years. On top of that, she could drive, something that hadn't felt safe when sleep was such an overpowering presence in her life.

BREAKING THE MOLD

In May 2008, Kathy Parker accompanied Anna to Roche's headquarters in New Jersey to thank employees there. An internal newsletter lists employees who had a hand in overcoming logistical and regulatory obstacles: forty people in ten departments. Talking with them was one of Anna's first chances to tell her own story. "The people at Roche helped save me. You have given me back my friends, my family, my work, my hobbies and my ability to drive, which, to a Southerner, is no small thing," Anna told the group. Parker was effusive, saying: "Watching Anna get her life back has been the most fulfilling thing I have ever experienced. I have had grants, awards, papers published, and more, but this was the peak experience of my entire career."

Despite the heartwarming tone of the occasion, Baker was not eager to spread the news beyond his company. When he was dealing with the Emory researchers, Baker asked his Roche colleagues whether they were interested in repurposing flumazenil for sleep disorders. Despite its life-changing effects with Anna, they declined. "After the luncheon in Nutley, someone asked me if we were going to put out a news release or article in the Newark newspaper," Baker told me. "While it would show a generous side of pharma that few people are aware of, I explained that we could not. First, Roche did this for Anna because it was the right thing to do—not for public relations. Second, if we did, the FDA would view the article as an off-label promotion of the drug. And third, it would drive more people to send letters to the CEO to ask for it."

Back at Emory, David Rye had a set of concerns that overlapped Baker's. He grumbled about what he perceived as Parker's victory lap. She asked him: "Why aren't you more excited about this?" "We need to look at the larger picture," he recalled telling her. "It's very likely to be true for other people besides Anna. There's a lot more to this."

Rye wanted to figure out how Anna's case fit in with established categories—or perhaps broke the mold. He had high hopes for flumazenil and the test for GABA-enhancing activity. He thought those two tools, despite their limitations, could redraw the map of sleep disorders. A new category might include people who conventionally belonged in several baskets: idiopathic hypersomnia, narcolepsy type 2, or others. The patch clamp assay—or some refined version of it—could indicate whether someone belonged in the new category and would be likely to respond to flumazenil.

The Roche newsletter contained clues about what was going on at Emory. One was a shift in language, suggesting a search for an appropriate name for the new category; Parker said she preferred to call Anna's condition "endogenous hypersomnia" rather than "recurring stupor," the term she had used in her lecture. The term "endogenous hypersomnia" was not found in the International Classification of Sleep Disorders and resembles "essential hypersomnia," a term used by Japanese researchers to describe cases of excessive sleepiness distinct from narcolepsy.[14] The newsletter mentioned thirteen probable endogenous hypersomnia patients besides Anna. That number had increased from three since a year before, when Anna wrote to Roche asking for help.

At the Emory sleep center, then based at a faded former retirement home, Rye had been sorting through patient charts and digging through freezers to retrieve samples from patients he had already examined. He was conducting clinical research on a shoestring, without nursing staff or dedicated space. Recruited patients came in on Saturdays in order to avoid his regular clinical schedule. Meanwhile, in the laboratory, Jenkins and sometimes Garcia were testing patient samples via patch clamp. They wanted to know: how many sleepy people had GABA-enhancing substances in their spinal fluid, like Anna did?

FUTURE DIRECTIONS

Kathy Parker's time at Emory was waning. Leaders at the University of Rochester had started recruiting her to become dean of nursing, a position she began in August 2008. Once Anna had recovered, she felt like her role in the hypersomnia story was complete. "I'm not a basic scientist," she said. "I was less sure about what needed to happen next."

Rye had more of an inkling about future directions. Although he gave no public lectures on Anna's case, like Parker did, through documents generated around this time, it is possible to discern his plans. Rye had two main goals: test flumazenil in other people with sleep disorders resembling Anna's and identify the somnogen, the GABA-enhancing substance in patients' cerebrospinal fluid. The first project got off the ground; the other ran into difficulties, which are explored in chapter 6.

In March 2008, after Anna started her lozenge and skin cream regimen of flumazenil, a provisional application for a patent was filed on behalf of Parker, Rye, and Jenkins. The full application came a year later.[15] While it is full of technical language and legalese, the patent also reveals something about the inventors' scientific thinking. It contains the basics of Anna's story, recipes for flavored flumazenil lozenges, and a plan for a clinical trial. Casting a wide net, the patent covers the use of flumazenil and several other GABA-receptor antagonists to treat the broad category of "excessive sleepiness and sleep disorders associated with excessive sleepiness."

The "endogenous hypersomnia" patients Parker mentioned when visiting Roche show up in the patent, too. Two besides Anna had idiopathic hypersomnia, and two had narcolepsy without cataplexy. Five others had "medically refractory sleepiness," a term suggesting either (1) diagnosis with sleep apnea or restless leg syndrome, but having persistent symptoms; or (2) subjective sleepiness unmatched by falling asleep quickly enough in a Multiple Sleep Latency Test. One woman had hypersomnia that fluctuated along with her menstrual cycle, and another man had the rare episodic Kleine-Levin syndrome, with a disoriented mental state lasting days or weeks at a time. Even though other characteristics separated them, Rye reasoned that these patients' conditions all reflected an increased need or drive for sleep. Grouping them together showed Rye's willingness to color outside the boundaries set by the rest of the sleep medicine field.

All of these people had levels of GABA-enhancing activity in their CSF that were higher than controls, according to Jenkins's patch clamp assay. For five of them, the patent and a 2009 National Institute of Neurological Disorders and

Stroke grant application[16] include reaction time measurements, before and after intravenous flumazenil.

Remember Anna's complaints about the "world's most boring video game"? Out of the five, her reaction time actually improved the least with flumazenil. One of her comrades had an average reaction time of almost two seconds—about a quarter of a second is typical. We can interpret this as intermittently nodding off or being almost asleep while taking the test. The higher dose of flumazenil chopped one and a half seconds off this person's average reaction time, even though it didn't change her subjective sleepiness much. Her response on the 1–7 Stanford Sleepiness Scale went from 6 ("Sleepy, woozy, fighting sleep; prefer to lie down") to 4 ("Somewhat foggy, let down"), while Anna's changed from 6 to 1 ("Feeling active, vital, alert, or wide awake").

Besides Anna, who were these sleepy people? For the sake of their privacy, we don't need to know everything about them, but since Rye was taking a conceptual leap in putting these people together in the same group, knowing something about them might help us understand the category he was outlining.

PRACTICE WHAT YOU PREACH

As with restless leg syndrome, Rye felt a personal connection to what he was seeing in Anna and others. Around this time, he was diagnosed with hypothyroidism, a common condition that can drain someone's energy and produce excessive daytime sleepiness. And as an empathetic physician, he was touched by his patients' struggles.

One was a nurse working at Emory, whom he first encountered in 2001. In this book, she is called Valerie.[17] Rye has said that she had helped reorient his thinking. At first, he looked at Valerie's sleepiness through the prism of restless leg syndrome. But after talking with her several times, he thought that some other explanation was necessary. "I was always telling other doctors to listen to their patients," he said. "In this case, I realized, maybe I should practice what I preach."

Sleepiness seemed to run in Valerie's family. "That was my normal," she said. She had often taken long naps as a child, but her sleepiness intensified after she reached adulthood and gave birth to her daughter. When her daughter reached high school, her daughter also often fell asleep in class and needed an Adderall prescription to manage in college.

Valerie initially met Rye through her mother, whom he diagnosed first with RLS and then later with narcolepsy type 2. Her mother had leg pains that led her to walk on a treadmill at night. In addition, she drank coffee throughout the day yet frequently nodded off, exclaiming: "It's hard enough to wake up once a day. I don't want to do that twice."

While living in Florida in the 1990s, Valerie regularly had to drive across Tampa Bay and would often have to stop at a familiar convenience store to take a fifteen-minute nap and load up on caffeine and sugar so that she could continue somewhat safely. Once while driving on Interstate 75, she fell asleep for several seconds and woke up just in time to swerve when traffic stopped in front of her.

Valerie's RLS symptoms were sporadic, becoming more intense every few months. Even after treatment with pramipexole under Rye's supervision, sleepiness continued to be a problem for her. She found she could manage weekend shifts as a staff nurse, when she'd stay moving and on her feet, but a regular five-day work week, along with taking care of her daughter, would leave her exhausted. To cope, she tried modafinil, but it gave her heart palpitations. "Even coffee made me uncomfortable," she said.

When Valerie participated in a test of intravenous flumazenil with Rye, she felt more awake and her reaction time measurements improved as well. "It was like realizing that day-to-day I am really living in somewhat of a fog, and the flumazenil made everything seem sharper and clearer," she said.

FAMILY CONNECTIONS

In addition to Valerie, another connection suggesting a genetic basis for the new sleep disorder category was Anna's brother James, who seemed to have a condition like hers. James has also participated in Emory studies; his and Anna's parents, along with a third sibling, have not reported similar issues.

For both James and Anna, their excessive sleepiness arrived around late adolescence. James noticed his around the time when his sister's difficulties were intensifying. He recalled occasional instances of what he called "sleeping heavy" throughout college: having intense dreams that seemed to last years, then waking up feeling tired. For a short time, he stayed with Anna in Atlanta when she was struggling with her job as an attorney, and he witnessed her crashes. Hearing Anna discuss her own epic dreams and perpetual drowsiness made James wonder if he had something similar.

James, a *Teen Jeopardy!* contestant, has a record of academic achievement similar to his sister's. In 2003, he graduated from Yale, where he studied linguistics and learned to read several languages, including Greek and Turkish. His path after university led in a different direction. He taught himself how to use animation software and began making elaborate, phantasmagorical videos. Collaborating with a friend from school, he created videos to accompany an album by an experimental rock group. He then traveled around the country to present the video at music and art festivals.

After moving to the Northwest, James sprained his ankle. He found that he was "sleeping heavy" more often, which he initially attributed to being less mobile. By 2008, in Portland, his sleepiness had intensified; he began dipping his head into a bucket of cold water several times a day. It got the blood flowing for a short time, at least.

In Atlanta, James had taken (and sometimes lost) various jobs as a pest exterminator, rideshare service driver, or bookstore clerk. He managed with conventional stimulants, which he had begun taking in college for attention deficit disorder. But he found that those medications didn't help him regain the ability to concentrate. The change caught him off guard, because he thought he generally needed less sleep than his peers. He said in an email: "Additionally, I'd always been very self-driven with art, and I found from 2008 on I was incredibly dull-witted and sluggish and could produce very little."

Rye had little to offer people like James, Valerie, or others who were struggling with sleepiness and found available medications unsatisfactory. It gnawed at him. However, Anna's arrangement with Roche and the FDA was considered exceptional, and the supply of flumazenil available outside of that channel was limited. It took several years before Rye began surmounting these obstacles.

CHAPTER 4

RYE VERSUS MSLT

This is a somewhat roundabout way of saying that the sleep-onset REM period contains the essence of all that we have previously called narcolepsy. Conversely, we would suggest that those patients without cataplexy or sleep paralysis who also fail to show sleep-onset REM periods in laboratory tests probably do not have narcolepsy.

—Dement, Rechtschaffen, and Gulevich,
"The Nature of the Narcoleptic Sleep Attack," 1966

When David Rye began his scientific and medical career, he was not set on becoming a champion of under-recognized sleep disorders. His traits of being scholarly yet stubborn and independent of established authority emerged early on.

Rye grew up in Detroit and attended an all-boys Jesuit-run high school. He still avidly follows the Detroit Red Wings hockey team and cites his Jesuit education as important for learning how to be a critical thinker. The son of an industrial engineer who worked in the automotive industry, Rye studied chemistry on a merit scholarship at Wayne State University in Detroit, and he lived at home until his senior year. To make ends meet, he worked sweaty twelve-hour shifts at a Ford assembly plant one summer. During his PhD studies in Chicago, he also worked as a bouncer at a bar in Chicago called the Hangge-Uppe. Given his height, we can imagine him peering down at bar patrons coming in from nearby Division Street.

When Rye started in the University of Chicago's MD-PhD program, his initial attraction was to neuroanatomy, rather than sleep per se. In the early 1980s, the field of neuroscience was expanding rapidly, and the now-elaborate picture of which brain cells use what neurotransmitter was starting to unfold. For his PhD research, Rye chose to train with the pathologist Bruce Wainer, who was developing tools that were making it possible to map circuits within the brain with greater precision than ever before. Wainer was a relatively junior faculty member who was able to give Rye attention. Working with Wainer, Rye's fellow MD/PhD student Allan Levey had generated monoclonal antibodies, then a relatively new technology, allowing the labeling of neurons that synthesized the most well-known neurotransmitter: acetylcholine.[1]

Rye joined the lab soon afterward. The young scientists had fun together, organizing golf outings in Chicago and on some Friday afternoons mixing drinks in the lab using the same blender they used for homogenizing brain tissues. "At first, it was just the two of us," said Levey, who later became chair of the neurology department at Emory. "We spent an enormous amount of time simply looking at microscope slides. It was like a gold mine of information."

In rats, Levey was mapping the connections of the thalamus, a central structure located above the brainstem that acts as a gateway for sensory information. He was inspired by a 1984 paper by the scientific titan Francis Crick, who had turned from DNA to the problem of consciousness. How does a collection of cells conjure up a sense of self, with memory and selective attention? Crick proposed the reticular complex, a net of cells surrounding the thalamus, as an organizing hub, directing an "attentional searchlight."[2]

AWAY FROM THE PACK

Wainer's main interest was in studying Alzheimer's disease. He was adept with antibodies and immunology but not primarily a neurology expert.[3] Levey and Rye soaked up knowledge of the brain from others at the University of Chicago. They both cited the neuroanatomist Rainer Guillery, head of the neuroscience program and an investigator of the thalamus, as a teacher. "It was an environment where we could go where we needed to learn," Levey said.

At the time, a major theory driving research on Alzheimer's was that problems with acetylcholine were central, and much attention was being put on the degeneration of acetylcholine-producing cells in the forebrain. The Alzheimer's field wasn't yet the funding behemoth that it would later become, but to avoid the crowd, Guillery suggested that Rye find a different region of the brain to specialize in. "He's had to carve out his own niche," said Levey, who became an expert on Alzheimer's. "For me, it's been easier."

Characteristically, Rye wanted to make his own way. Here, we can glimpse the origins of his interest in "how the brain keeps the lights on." "I wanted to stay away from the pack," Rye said. "I had access to a useful set of tools. Where was there another area of the brain that predominantly used acetylcholine?"

The answer: the pedunculopontine nucleus, or PPN, a bundle of neurons in the pons, part of the brainstem. The PPN was thought to be a component of the reticular activating system, the network of neural circuits in the midbrain and brainstem that keeps us awake. Rye worked with slices of rat brain, staining the PPN with antibodies and tracing acetylcholine-producing neurons' paths on microscope slides.[4] In his thesis, Rye suggested that the PPN corresponds to neurons that initiate REM sleep, known to be sensitive to acetylcholine.

Even after Rye had left Wainer's lab, he would sometimes return and hold court, smoking cigarettes and giving advice on what current students should be pursuing. Stick to neuroanatomy rather than follow the latest trends, he told them. A fellow Wainer trainee recalled: "I thought he would not mind working in an area where others thought that something else was more important."[5]

Rye said he learned a great deal during graduate school from Clifford Saper, then a young faculty member at Washington University–St. Louis, who moved to Chicago in 1985. He visited Saper's lab to learn neuron-tracing techniques, and they kept in touch. Rye sometimes asked for Saper's help in editing and trimming his papers on neuroanatomy, but on one of those papers, Rye was stubborn. "I took out my red pen and cut the length in half," Saper recalled. "The next version was still too long. I eventually ran out of red pens and told him: 'OK wise guy, send it in.'" The reviewers wrote back: "This manuscript would have been wonderful in the time of Charles Darwin."

During his neurology residency, Rye worked under the supervision of Jean-Paul Spire, director of the University of Chicago's sleep disorders clinic. Spire had wide-ranging interests in neurology and electrophysiology; in newspaper articles, he was depicted probing the brain's responses to sounds to detect brain tumors and monitoring a patient's EEGs and other neurological signs during epilepsy surgery.[6]

Rye's first research paper on human sleep was a case report on a young woman who had a stroke affecting her left pons. The location of her stroke provided a chance to test theories about the regions of the brain controlling REM sleep, which at that point had mostly been evaluated in cats by the French sleep researcher Michel Jouvet.

Over several months in 1989, the paralysis on the right side of the woman's body improved. Rye and Clete Kushida, a medical student, studied the woman's REM sleep, supervised by Spire.[7] They noticed a disruption of REM-sleep EEG patterns coming from the left side of her brain, although she continued to have rapid movements of both eyes.

The experience deepened Rye's interest in sleep neurology, although he did not focus primarily on REM sleep in his later research. Instead, he saw the junction between the study of movement disorders and sleep as attractive. As a resident, Rye worked in several Chicago hospitals, where he encountered a variety of patients with Parkinson's and related movement disorders. Through these experiences, he developed a hunch that "Parkinson's was unlikely to shut off when someone is asleep." Rye's later work on restless leg syndrome and REM sleep behavior disorder, when people act out their dreams, both grew out of that hunch.

PARKINSON'S AND ATLANTA

Rye first met the neurologist Mahlon DeLong, who would recruit him to Emory, on a visit to Baltimore as a medical student, when he was scouting out places to apply for residency. At Johns Hopkins, DeLong was in the middle of transformative research.

In the early 1980s, several young people in California were poisoned by a contaminant in synthetic heroin, reproducing the symptoms of Parkinson's.[8] DeLong and his colleagues were using the same chemical to create a model system for Parkinson's in monkeys. In this model system, symptoms such as tremors and muscular rigidity could be alleviated by surgery. This led to the reemergence of surgery, and later electrical brain stimulation, as tactics for the treatment of Parkinson's.

The symptoms of Parkinson's were viewed, up to that point, as resulting from tissue damage that couldn't be repaired: a loss of function. One of DeLong's insights was that degeneration in one area of the brain, the basal ganglia, resulted in a gain of function, or excessive activity, in other areas: the subthalamic nucleus and globus pallidus. By restraining either with a surgical lesion or with electricity, a balance of signals could be restored.

The late 1980s and early 1990s were a time of rapid expansion at Emory. Even before *Science*'s publication of DeLong's monkey surgery experiments in 1990, leaders at Emory sought him out as a chair of neurology. When DeLong arrived, he brought with him several colleagues from Johns Hopkins, bolstering Emory's neurology department. With dedicated funding from the American Parkinson's Disease Association, Emory became a hot spot for the study of Parkinson's and other movement disorders.

Before the 1990s, Emory had provided a base for sleep researchers such as the psychologist David Foulkes, who studied dreams in children. Gerald Vogel, the first to publish observations on REM sleep in narcolepsy while at the University of Chicago, was based for many years at an Emory-affiliated psychiatric hospital, the Georgia Mental Health Institute. While there, Vogel conducted landmark experiments on the relationship between REM sleep and depression. However, neither was involved in clinical care for sleep disorders, which is what DeLong wanted to build up.

DeLong remembered Rye as "bright and creative" and recruited both Rye and Bliwise to Emory around the same time. Rye also had an offer from Harvard, where his friend Saper moved in 1992. However, Atlanta and its airport looked like a better option for Rye. Convenient airport access could allow his wife, Catherine, to continue managing her family's residential construction business in Chicago.

Rye's early emphasis on Parkinson's made practical sense. The PPN, his favorite region of the brain, was emerging as a vulnerable area in neurodegenerative diseases such as Parkinson's, and it was close to and connected with the basal ganglia. In Rye's first few years at Emory, he was supported by an American Parkinson's Disease Association fellowship, along with NIH grants overseen by DeLong and Bliwise.

Rye had ambitions for himself as thinking and publishing ahead of his peers. As a role model, he admired Percival Bailey, a neurosurgeon and psychiatrist who taught at the University of Chicago in the 1930s. Bailey built the foundations of medical knowledge of brain cancers with his careful studies of anatomy and pathology. According to a colleague, Bailey "was feared because he was so blunt and outspoken."[9]

SNOW WHITE AND THE DWARFS

In his first decade at Emory, Rye recruited several graduate students to work in his laboratory. They took advantage of the same monkey model of Parkinson's that DeLong and his colleagues had used. In the lab, Rye would demonstrate his skills with animal surgeries but was also willing to pitch in with mundane tasks. When students and technicians needed to keep monkeys awake at night for experiments, Rye would join in, periodically flicking lights, playing the radio, or crinkling paper to make sure animals did not doze off.

Among students, his lab had an intimidating reputation because his students, MD/PhDs especially, took a long time to complete their degrees. Yet one of Rye's former graduate students, Gillian Hue, said she did not regret her experience training with him. Originally from Jamaica, Hue had started working as a technician with one of Rye's collaborators and was unsure about graduate school. He encouraged her to apply, squeezing in her application after the deadline. She joined the band of Rye's acolytes. For an opulent party at the Society for Neuroscience meeting one year, she and other members of his lab dressed up as Snow White and the dwarfs—with Rye as Snow White.

Hue encountered a series of frustrations with her research in rats and mice on dopamine and the spinal cord—much of it went unpublished outside her thesis. She was more interested in a teaching career, but that did not lead to conflict with Rye. After finishing her doctorate, Hue went on to teach psychology at a local college and later returned to Emory. "I learned from him how to be a scientist," Hue said. "He asks bigger questions than are possible to address in incremental papers. Sometimes that conflicts with how science is currently practiced."

THE NARCOLEPSY PENUMBRA

After his first decade at Emory, Rye did not stay focused on the PPN. He was drawn to restless leg syndrome, and his personal interest in the condition converged with his knowledge of dopamine neurochemistry. But at the same time, starting from when Rye arrived in the early 1990s, his clinical services were in demand. Emory's status as a referral center in the Southeast meant patients with a variety of sleep disorders were coming through the door.

By the time Rye encountered Anna Sumner, he had been treating patients with narcolepsy for more than a decade. Some of them stuck in his mind, such as an airline pilot who had successfully managed without mishaps for twenty-five years. His condition only surfaced when the pilot wanted to move from shorter to longer flights, and when it did, it threatened to derail his career. When the pilot experienced cataplexy during a test in a flight simulator, his examiners thought he was having a heart attack. An observant cardiologist referred the pilot to Rye.

In 1997, Rye, Bliwise, and others had compiled a survey of forty-one narcolepsy cases, including the airline pilot. About half had been diagnosed after the age of forty, and most with later-life onset did not have cataplexy. In some individuals, sleepiness appeared decades before they began to experience cata-plexy.[10] At the time, the Emory paper went against conventional thinking about narcolepsy, usually observed to have its onset during childhood or adoles-cence. "This has always been a theme in our clinical research," Bliwise said. "We saw that the spectrum of patients with narcolepsy was large. We thought that the penumbra was where the action was."

While exploring this territory, Rye gradually became dissatisfied with the Mul-tiple Sleep Latency Test, the daytime nap procedure used to diagnose narcolepsy and idiopathic hypersomnia, and, more generally, with the categories it generated. He later adopted a rebellious stance with respect to the International Classifica-tion of Sleep Disorders—the consensus guidebook for the sleep medicine field.

To understand where Rye's criticisms came from, we have to back up and away from Atlanta. The concept of IH has evolved at the edges of narcolepsy, and run-ning through narcolepsy's history has been a tug of war over how to define it.

IS CATAPLEXY REQUIRED?

In the first half of the twentieth century, the term narcolepsy was sometimes used loosely to refer to the symptom of overwhelming sleep, or "sleep attacks." Ger-man-, French-, and English-speaking neurologists disagreed on whether narco-lepsy should be considered a disease of its own or simply a sign of some other condition that made someone sleepy. Some mistakenly thought narcolepsy was related to epilepsy or originated in psychological conflicts, with dubious psycho-analytic interpretations.

Among neurologists, encephalitis and traumatic brain injury were recognized as possible causes of narcolepsy. The distinctive symptom of cataplexy, which did not appear in everyone with narcolepsy, did suggest a neurological origin. But an unresolved question was whether cataplexy was necessary for a patient to receive the narcolepsy label. The Australian-born neurologist William Adie wrote one of the clearest arguments for one side of this debate: "To my mind cataplexy in narcolepsy is as characteristic of the disease as the sleep attacks themselves. Given a case with sleep attacks alone the diagnosis is difficult, for sleep attacks indis-tinguishable from those of true narcolepsy occur as a symptom in many dissimi-lar diseases; but if definite cataplectic attacks are present as well the diagnosis is made certain, for the combination is seen in no other condition whatsoever."[11]

Others publishing around the same time thought narcolepsy should be defined more broadly, based on short attacks of sleepiness plus other symptoms. The

British neurologist S. A. Kinnier Wilson dismissed the idea that narcolepsy should be thought of as one disease and referred to "the narcolepsies."

In the mid-1950s, neurologists at the Mayo Clinic solidified the clinical picture, describing the "narcoleptic tetrad": excessive daytime sleepiness, cataplexy, sleep paralysis, and hypnogogic hallucinations. However, all four symptoms were not needed for a narcolepsy diagnosis. In the Mayo compilation of 241 people with narcolepsy, only a few displayed all four, and more than 30 percent didn't have cataplexy.[12]

The discovery of REM sleep transformed the debate. Pioneer sleep researchers such as Vogel, along with William Dement, described as the "father of sleep medicine," recognized that when people with narcolepsy went to bed at night, they tended to enter REM sleep immediately or within a few minutes.[13] This didn't occur every time people with narcolepsy were studied in the laboratory, but rapid REM onset rarely occurred in healthy people, who usually took about ninety minutes to enter REM sleep at night.[14]

Moreover, the symptom of cataplexy seemed to go hand in hand with rapid REM onset.[15] Several symptoms of narcolepsy, such as cataplexy, sleep paralysis, and hypnogogic hallucinations, could be interpreted as some aspect of REM sleep—dreaming or loss of muscle tone—intruding into waking time. The central problem was seen as a lack of properly regulated REM sleep. This view was adopted as the definition of narcolepsy at the First International Symposium on Narcolepsy in France in 1975: "A syndrome of unknown origin that is characterized by abnormal sleep tendencies, including excessive daytime sleepiness and often disturbed nocturnal sleep, and pathological manifestations of REM sleep."

At the time, little information was available about the neurological basis of narcolepsy. While investigators such as Dement were beginning to probe brain chemicals in both people and dogs with narcolepsy, REM sleep was the most tangible phenomenon that they had to grasp. The emphasis on REM onset shifted diagnosis away from relying on what patients recalled and toward examination in the sleep laboratory. Sleep specialists could use early REM onset as a way to confirm patients' anecdotal reports of narcolepsy symptoms, such as sleep paralysis and hypnogogic hallucinations.

THE GOLD STANDARD

In the 1970s, Dement's group at Stanford was having people with narcolepsy take single daytime naps to look for sleep-onset REM.[16] However, the multiple-nap structure of the MSLT came from their work on healthy adolescents.[17] His graduate student Mary Carskadon was experimenting with having volunteers go through a ninety-minute day—sixty minutes for wake and thirty minutes for sleep—in an attempt to outrun REM sleep. Dement and Carskadon found that REM appeared more quickly when people were deprived of it. They also observed that their subjects' underlying sleepiness, driven by their internal circadian rhythms, varied throughout the day.

In a phone interview, Carskadon said that she and Dement always had narcolepsy in mind while developing the MSLT, even if her work focused on healthy children and adolescents. She said they were influenced by the work of Bedřich Roth in Prague, who had also reported clinical experiments with single naps and REM onset. "Our goal was always to have a test for the clinic," she said. "Once you have a tool that can measure physiological sleepiness, people can use it in many ways."

The idea behind the MSLT was straightforward. If someone fell asleep faster, that meant they were closer to the sleep state—and thus sleepier—beforehand. The person being tested was monitored while lying down in the dark, eyes closed, with no external obstacle to sleep put in their way.

The MSLT's developers presented it as *objective*,[18] compared with simply asking someone how drowsy they are or measuring performance on an alertness task, which can be influenced by muscle fatigue, practice, or motivation. Someone can fill out a questionnaire however they like, but without the influence of sedatives or prior sleep deprivation, it's more difficult to *try* to fall asleep faster than usual. Other sleep researchers, such as Tom Roth and his colleagues at Henry Ford Hospital in Michigan, were taking similar approaches to study the effects of antihistamines around the same time.[19]

There were existing alternatives to measuring sleep latency, such as pupillometry—tracking the size of the pupil or its spontaneous oscillations, which reflect underlying signals in the nervous system. The Mayo Clinic had tested pupillometry for the diagnosis of narcolepsy beginning in the 1960s,[20] but the approach required expensive equipment and seemed to be less reliable than the MSLT.[21] Similarly, Canadian sleep researchers tested evoked potentials, a neuroelectrical technique sometimes used for diagnosis of multiple sclerosis. But neither approach looked for REM sleep, whose rapid onset was considered an important marker of narcolepsy.

With sleep-onset REM in mind, Dement's group at Stanford adapted the "experimental MSLT"—used for studying healthy young people—into the "clinical MSLT" for narcolepsy diagnosis. Four or five naps gave several chances for REM sleep to appear, within a cutoff of fifteen minutes. Two sleep-onset REM periods, plus an average interval before falling asleep of five minutes or less, were deemed sufficient. The five-minute average was later extended to eight minutes.[22]

In the 1980s, the growing field of sleep medicine adopted the MSLT as a standard procedure for the diagnosis of narcolepsy. Although narcolepsy and cataplexy historically had been tied together, Dement and colleagues wrote: "There are unresolved differences of opinion as to whether a definitive history of cataplexy is a necessary component of the narcolepsy syndrome."[23]

A criticism of the MSLT emerged, reflecting the test's operational basis: subjective sleepiness and the ability to quickly fall asleep do not necessarily match up. In some studies, various treatments for narcolepsy were effective in alleviating subjective sleepiness but not in extending sleep latency. In 1992, sleep researchers from the Netherlands referred to the MSLT as a "paradoxical test": "This leads us to a situation where the very drugs, which are considered as the most potent in the treatment of sleepiness in narcolepsy, do not in fact result in an improvement of what is considered the most reliable test for sleepiness."[24]

Sleepiness has multiple dimensions, many researchers recognized. To extend the technique, Tom Roth and his group developed a variant of the MSLT called the Maintenance of Wakefulness test.[25] A principal difference between the MWT and the MSLT is effort. In the MWT, participants are asked to sit up in a dark room and to try to stay awake, recruiting the parts of the brain that help someone deliberately resist sleep. The MWT has been used for medication studies and also for airline pilots or commercial drivers who need to show that they can stay awake.

Divergent results for objective versus subjective sleepiness highlight an underlying issue. The MSLT measures one part of sleepiness, but it does not capture other aspects, such as those experienced by people with IH: the number of hours someone may need to sleep or the inability to leave sleep behind.

NARCOLEPSY OUT ON THE STREET

As a diagnostic test, what bothered Rye about the MSLT was that it often placed people into categories that didn't make sense if one had more information about their medical conditions. "At the very least, the MSLT is a nonspecific test," he said. "If you start looking at sleepy people besides those with classic narcolepsy, it doesn't help you distinguish them."

Rye and Bliwise were the first to publish papers on sleepiness in people with Parkinson's disease, showing that about 30 percent met MSLT criteria for narcolepsy.[26] With Parker, they also established that a similar fraction of end-stage kidney disease patients also had pathological levels of sleepiness.[27] Their 1997 survey of narcolepsy patients had used the MSLT, but it documented how people who didn't fit the classic picture of the disorder were showing up in their clinic. It concluded: "Increased recognition of N- [narcolepsy without cataplexy] might reflect heightened public awareness to EDS [excessive daytime sleepiness] and its treatment options along with improved diagnostic capabilities."

In 1998, Rye and Bliwise published a case report on a forty-six-year-old woman with an apparent mixture of depression and narcolepsy.[28] She experienced typical symptoms of depression and was prescribed antidepressants. However, she also had gained thirty pounds in the last several months and had displayed sleep attacks at work and while commuting. The woman didn't report cataplexy, sleep paralysis, or hallucinations, but she did enter REM sleep in three out of five MSLT naps. Treatment with bupropion, an atypical wake-promoting antidepressant, alleviated both sleepiness and depression symptoms. Should the woman be seen as primarily having depression or narcolepsy? The boundaries were not clear.

Rye wasn't the first to point out the MSLT's elasticity. In the early 1990s, other sleep researchers proposed that people whose clinical history didn't fit narcolepsy should be labeled "hypersomnia with sleep-onset-REM periods" instead.[29] Neither the distinction nor the clunky language stuck. In the United States, the FDA's 1998 approval of modafinil for narcolepsy—without cataplexy as a necessary requirement—gave physicians and patients an incentive to move toward the narcolepsy diagnosis.

The initial studies establishing the MSLT did not look at patients with sleep disorders besides narcolepsy. They also left out people who might be sleepy as a result of their work schedules. From their ninety-minute-day experiments, the Stanford researchers were aware that people without narcolepsy could display sleep-onset REM periods if deprived of REM in the laboratory. They may have discounted how often doctors would see the same effect "out on the street" (Rye's phrase). It took years to organize studies that would include enough people to map that landscape, and the results were confounding.

A community-based study in Wisconsin examined a group of more than five hundred people, not only those who came to a doctor's office concerned about sleepiness. More than 5 percent of men and 1 percent of women met MSLT criteria for narcolepsy.[30] This didn't mean that there was an unseen epidemic of narcolepsy in Wisconsin but rather that the test roped in false positives. In particular, those with sleep apnea or who worked night-shift jobs showed up as apparently having narcolepsy. Tom Roth conducted a similar population-based study in Michigan, and he concluded: "If someone has a pathological level of sleepiness, you're going to see sleep-onset REM some of the time."[31]

Researchers at the University of Pittsburgh noticed that people with depression displayed more rapid entry into REM sleep, but not as fast as in narcolepsy.[32] Did that mean that depression had something in common with narcolepsy? Perhaps not. The significance of fast REM onset in people who don't have narcolepsy remains unclear. The biological function of REM sleep is mysterious; people under the influence of certain antidepressants or with rare brain injuries can manage without experiencing much REM sleep at all.[33]

In the 1990s and 2000s, sleep specialist clinicians used the MSLT with "unbridled enthusiasm," according to an editorial in the trade publication *Sleep Review*: "In many sleep laboratories, the MSLT became part of the evaluation of virtually all sleepy patients, including those with obstructive sleep apnea, periodic limb movements, insomnia, and circadian rhythm sleep disorders, as well as patients with suspected narcolepsy and idiopathic hypersomnia."[34]

In 2005, a standards of practice committee of the American Academy of Sleep Medicine declared the MSLT "the *de facto* standard" for objective measurement of sleepiness yet warned that its diagnostic value, beyond confirming suspected narcolepsy, was limited.[35]

THE MSLT EXPERIENCE

The MSLT is something people with IH have in common; it is the gateway through which they had to travel to receive their diagnosis (figure 4.1). But listening to discussions at support group meetings or even lurking on social media, it becomes clear that sleep labs don't always implement the MSLT according to current guidelines. Sometimes the patient does not sleep enough beforehand, because they are awakened early by external noise, incoming staff, or changing shifts. This may be a source of overdiagnosis, although a standard overnight sleep test lasts just seven hours, and someone with IH may normally sleep for much longer than that.[36] On the other side of the coin, it is possible for someone who feels extremely sleepy to have trouble falling asleep on command. Performance anxiety seems to play a role.

Normally, the night before an MSLT, an overnight sleep test checks for sleep apnea and leg movements and also measures how fragmented someone's sleep is. An overnight sleep test collects a vast amount of information on someone's EEG

FIGURE 4.1. Both the PSG (polysomnogram) and MSLT (multiple sleep latency test) procedures can be uncomfortable.

Source: Off the Mark Cartoons, Atlantic Feature Syndicate.

patterns and sleep stages, but that information often goes unexamined for narcolepsy/IH diagnosis.

Chronic insufficient sleep can make it look like someone has narcolepsy or IH, which is why recent guidelines call for actigraphy—wearing a device that monitors movement—for two weeks before an MSLT.[37] Guidelines also call for patients to stop taking stimulants or antidepressant medications two weeks beforehand. Antidepressants can distort MSLT results because they suppress REM sleep, but physicians may be reluctant to insist on having someone discontinue antidepressants because of the possibility of withdrawal or worsening depression. Opioid or cannabis use can also confound MSLT findings. "The MSLT was one of the most stressful tests I could go through," said David Kellogg, a licensed hearing aid specialist from Oregon who was diagnosed with IH in 2012.

Several years before his IH diagnosis, David noticed that he was having trouble staying awake while driving to work. An overnight sleep study detected mild sleep apnea, so he underwent surgery recommended by an ear, nose, and throat specialist. It didn't help. He would sleep ten to twelve hours per day and "still woke up feeling like a truck had driven over him." He tried everything he could think of to feel better: exercise, eating better, vitamin D. He changed jobs, stepping down from demanding managerial roles. Sleepiness still threatened to pull him under while watching TV, reading, or even sitting at a red light.

Unable to work, David went on disability, and he had to complete a second MSLT in 2016 to demonstrate that his IH was still present. Beforehand, he stopped both stimulants and antidepressants and experienced withdrawal side effects. In between naps, he was trying to relax but also texting his wife in alarm, worried that his disability payments were on the line. He still fell asleep five times, averaging less than four minutes.

Another person who was diagnosed with IH in a Canadian hospital, and then later in the United Kingdom with narcolepsy, wrote: "The test was torturous: being woken repeatedly from the required naps left me with a violent migraine and I vomited into the clinic toilet."[38]

THE BUSINESS OF BREATHING AT NIGHT

Another gateway some people with IH pass through is diagnosis with obstructive sleep apnea: fleshy, flabby interruptions in breathing that interfere with rest and put strain on the heart. High rates of obesity have made obstructive sleep apnea common in the United States, and the condition is the main revenue source for most sleep medicine practices.

A standard remedy for obstructive sleep apnea is continuous positive air pressure (CPAP): having air blown up the nose to keep the airway open. Someone entering a sleep clinic is likely to see face masks and hoses on display. The first CPAP machines were driven by vacuum cleaner motors, but current devices have become more sophisticated, with internet connections and pressure modulation. Supplying devices for sleep apnea treatment is a multi-billion-dollar business.

This leads us to one of Rye's grumbles about his field. Many sleep specialists in the United States were trained as pulmonologists, not neurologists. According to Rye, they tend to manage sleepy patients with a "hammering everything that looks like a nail" approach, assuming obstructive sleep apnea is the cause.

While CPAP provides benefits to many people, it can be uncomfortable; it has been estimated that more than half of those who are prescribed CPAP eventually abandon it.[39] Complicating matters, untreated sleep apnea is thought to damage brain circuitry, contributing to excessive daytime sleepiness.[40]

We might discount Rye's complaints as arising from friction with colleagues or impatience with issues such as finding a face mask that delivers CPAP properly. Still, his counterparts concede that "pulmonary medicine specialists are often responsible for the diagnosis and treatment of a number of sleep conditions, including several that are not traditionally considered related to respiratory medicine."[41]

Others besides Rye have made similar criticisms. A sleep medicine program director in Maryland wrote in *Sleep Review* in 2014: "For two decades, clinical sleep medicine was somewhat myopically focused on diagnosing OSA, too often with little regard for the patient's long-term adherence, satisfaction, or outcomes."[42]

David Kellogg's experience in Oregon demonstrates how an assumption that a drowsy patient has sleep apnea can play out. David's hourly rate of breathing interruptions was relatively low, but before his IH diagnosis, he still put in a solid effort at making CPAP work, trying different masks and pressures for eight months. At each appointment, he brought in the CPAP memory card to verify that he was using it.

Even after his IH diagnosis, he faced doubts from some health care providers. At the advice of a council of academic physicians in Oregon, David had a second airway surgery that he was reluctant to undergo. He worried that refusing surgery would trigger a loss of disability payments. Afterward, his frequency of sinus infections went down, but he was still constantly tired. He said he felt like a car whose idle had been set too low and whose engine kept stalling. Only after his second MSLT did his new doctors agree that he had IH. "Nobody knew what IH was," he said.

Sleep specialists' reluctance to go beyond an initial diagnosis of sleep apnea is understandable. They may not want to advance to prescribing stimulants, a possible consequence of an IH diagnosis, if someone hasn't made a sustained effort with CPAP. Some recall patients who claimed to have a sleep disorder but seemed suspiciously interested in obtaining a prescription for stimulants.[43]

Doctors have a saying: "When you hear hoofbeats, think of horses, not zebras." That is, look for the expected cause, not something exotic. On average, many more people coming to a sleep clinic will have sleep apnea than narcolepsy or IH. Current estimates say that more than a quarter of men and 10 percent of women in the United States have some level of detectable disruptions of breathing during sleep.[44]

Because it is so common, obstructive sleep apnea has a "borderland" issue akin to that of narcolepsy. With sensitive equipment that measures nasal air pressure, sleep labs have become very good at detecting disruptions of breathing. A 2015 study of more than two thousand middle-aged people in Switzerland found that three-quarters of the men and half of the women had obstructive sleep apnea, under International Classification of Sleep Disorders criteria.[45] Higher than previous estimates, this was an "unrealistically high prevalence" and suggested a need to revise those criteria, the authors concluded.

In the Swiss study, significantly fewer people experienced strong levels of subjective daytime sleepiness: 12.5 percent of men and 5.9 percent of women. There was no statistical association between frequency of disruptions in breathing and Epworth Sleepiness Scale scores.[46] A retrospective review of sleep clinic patients at the Mayo Clinic obtained a similar result: ESS scores do not correlate with the severity of breathing problems, especially for women.[47] People evidently have a wide range of sensitivities to the nighttime arousals that breathing disturbances trigger.

The bottom line is that many people experience excessive daytime sleepiness, but fewer go through a rigorous diagnostic process, exhaustively checking potential causes. Sleep apnea has the most obvious overlap with IH, but other possibilities include depression, hypothyroidism, even hepatic encephalopathy. As we see it today, IH is partly created by deficiencies in the health care system, but potential scientific value may lie in understanding those who fall through the cracks.

CHAPTER 5

BEHIND THE CURTAIN

One might be tempted to view the pre-REM era of sleep and dream research as a dark age illuminated feebly by Freud and the reticular formation, while the post-REM days provide untarnished enlightenment. Not precisely so.

—Eugene Aserinsky, "Drug and Dreams, a Synthesis," 1969

Where did the concept of idiopathic hypersomnia come from? The nineteenth-century medical literature includes descriptions of people who might have been diagnosed with IH today.[1] Early sleep researchers such as Nathaniel Kleitman defined hypersomnia as "uncontrollable somnolence and pathologically prolonged sleep, from which it is sometimes difficult to arouse the sleeper, or to keep him awake for any length of time after he has been awakened."[2]

That said, the clinical profile of IH comes from Bedřich Roth (figure 5.1). He was the first sleep researcher to distinguish IH from other sleep disorders, conceiving it as the other end of a spectrum including narcolepsy. While studying patients with narcolepsy during his medical training and early career, Roth encountered other patients whose characteristics were different. They made him think that sleep drunkenness, or prolonged clumsiness and confusion after waking up, was both a symptom and a key feature of a separate illness, now called IH.

Roth's ideas began to take form before the transformative studies of the 1960s, when narcolepsy was tied to aberrations of REM sleep. They also predate

FIGURE 5.1. Bedřich Roth in 1959.

Source: Courtesy of Jan Roth.

the rise of clinical sleep medicine in the United States and Europe. His situation in Cold War–era Czechoslovakia, separated from other sleep researchers, may have given him the space to develop his ideas—while depriving him of resources at the same time.

FIRST ENCOUNTER WITH NARCOLEPSY IN PARIS

The text of an autobiographical speech, provided by Roth's son Jan, has a section titled "How I Came to Engage in Sleep Research." Roth describes how he first observed someone with narcolepsy at the Salpêtrière hospital in Paris. He spent just two years in Paris, but the experience shaped the rest of his life.

Some of Roth's important learning experiences were in France. We can view him as being an extension of the French tradition of neurology, embodied by figures such as Charcot, Babinski, and Tourette. In particular, Charcot developed the Salpêtrière, once an insane asylum and prison for prostitutes, into a premier center for neurology in the late nineteenth century. He is also remembered for his discredited work on hypnosis and hysteria.[3] Still, Charcot made detailed symptomatic and anatomical descriptions of several diseases, such as multiple sclerosis, amyotrophic lateral sclerosis, and Parkinson's, paving the way for them to be understood scientifically. Decades later, Roth undertook similar tasks for narcolepsy and IH.

During his medical training, immediately after World War II, Roth spent time at a clinic supervised by the neurologists Pierre Mollaret and Georges Guillain. Every Tuesday at the hospital, professors examined outpatients in front of an auditorium of students. On one of those days, Mollaret brought in someone with narcolepsy. "All that the patient complained of was sleepiness, but, using precisely targeted questions, Prof. Mollaret was able to establish that the patient was also suffering from cataplexy," Roth wrote. "I had great admiration for him then."

Mollaret, Roth's primary teacher, was known for his studies of brain anatomy, as well as of infectious diseases such as malaria. During World War I, Guillain and his colleague Jean Barré had identified the autoimmune paralysis named for them (Guillain-Barré syndrome).[4] Roth's thesis was on polio and was titled "Considerations on the Prolonged Treatment of Respiratory Paralyses, with the Aid of an Iron Lung." Roth was also involved in early efforts to treat Guillain-Barré syndrome with penicillin, according to his son.

In 1949, after Roth had returned to Czechoslovakia, he began working at the Charles University neurology department, where two people with narcolepsy were being treated. One of them displayed "peculiar attacks lasting about 5 minutes." These episodes sound like cataplexy, since the patient was unable to move or talk yet perceived what was going on around him. Roth's colleagues thought the attacks were manifestations of hysteria, but he came to his own conclusion: it was a state of dissociation between body and mind, with the body entering sleep while the mind stayed awake.

After studying the two people with narcolepsy, Roth had them come to a seminar, which was attended by neurologists and other physicians all over the country. Aware of his interest in such disorders, his colleagues began to refer patients with pathological sleepiness to see him. By 1952, he was able to report having seen forty-two patients with narcolepsy, all with cataplexy. Descriptions of hundreds more with narcolepsy and hypersomnia would follow. "He felt that he was part of the international community of sleep researchers and wanted to contribute to it," his daughter Anniki Rothova said. "He didn't have access to the same equipment or resources, but what he did have was a unique group of patients."

Roth established one of the first clinics devoted to sleep disorders, beginning in the early 1950s, before many others in Europe or the United States.[5] Roth was a master of EEG interpretation, capable of diagnosing narcolepsy by glancing at a patient's EEG recordings. However, most of his recordings came from daytime naps, since Roth didn't have technical staff, like in a modern sleep lab, to help him observe someone progressing through the cycles of nighttime sleep.

The Charles University neurology department was housed in a former monastery, which was reconstructed in the eighteenth century and then converted into an asylum for the mentally ill. In 2017, Roth's former student Karel Šonka showed a group of visitors a now-unused section of the building, with small, cramped rooms where Roth had conducted sleep studies. Those that work in the rest of the building today say its appearance hasn't changed much in years.

To Šonka, who was much younger, Roth had "a quiet charisma." Roth's friends and family remember a gentle, gregarious man who almost never raised his voice. According to his daughter, he had a subtle sense of humor, playing with language, inventing new words and prodding his family to look at situations in new ways. He would buy sweets for colleagues from a shop across the street when one of his papers was accepted by an international journal.

ESCAPE FROM RUZOMBEROK

Outwardly mild-mannered, Bedřich Roth displayed determination and improvisation as a young man during the wartime occupation of his country. He was born in 1919 in Ruzomberok, a town in central Slovakia. His family was Jewish, observing religious holidays and dietary restrictions. His father Moritz was a lawyer, but the family was not well-off. When the elder Roth died in 1930, from complications after gastric surgery, his wife, Elsa, had to start working as a dressmaker.

In Ruzomberok, the close-knit Jewish community had a synagogue and cemetery, a kosher butcher and a separate school. Many young people, including Roth, were part of Jewish sports clubs and summer camps. Roth loved to hike and ski in the nearby Tatra Mountains, and this affinity stayed with him his entire life.

In the fall of 1937, Roth began his medical studies at Charles University. A year and a half later, Nazi Germany invaded and absorbed the western part of Czechoslovakia. In the separate German-allied state of Slovakia, Jewish students were barred from attending universities. After student demonstrations against occupation, Charles University was closed until the end of the war.

Back in his home town, Roth found a job in a locksmith's shop. In the evenings, he continued to study medical textbooks borrowed from the local hospital. He appears to have narrowly missed being deported to a concentration camp. He and other young Jewish and Roma men were called up for military service, which for them meant hard labor on tasks such as road repair. The first transports from Slovakia to Poland occurred in March 1942.[6] A second round occurred in May. Most of Bedřich Roth's extended family was eventually killed, according to his son.

Despite Ruzomberok's small size, the town was home to a textile plant and paper mills. Roth deserted from military service to make his way back to Ruzomberok, where shipments of lumber were being sent by train to Switzerland. A friend who worked at a sawmill learned the train schedules and bribed the man loading them to prepare a wagon with enough space to hide inside. In August 1942, Bedrich Roth and his cousin Jozef smuggled themselves by train all the way through Austria, ending up in Switzerland. It was an uncomfortable, cramped journey lasting more than a week. The cousins had to stay silent to avoid military guards along the way.

The cousins were first taken to a Swiss prison, then a refugee camp and a labor camp. After several weeks, they were able to contact a representative of the Allied-backed Czechoslovakian government-in-exile in Switzerland. Roth managed to spend two years in Bern continuing his medical studies. As the Nazis' grip on Europe loosened, he made his way to France, hoping to join the Czechoslovak armed forces.

Roth arrived in Paris at the end of 1944, just a few months after that city's liberation from its German occupiers. He spent several days on the cold streets, sleeping in Metro stations, after he did not pass an examination needed for military service. A sympathetic diplomat at the Czechoslovakian embassy helped arrange for a scholarship, allowing Roth to finish his medical studies in France. Roth's training with Mollaret took place in postoccupation Paris, when many buildings had been demolished or lacked heat, and food was scarce. Part of the Pitié-Salpêtrière hospital complex had been taken over successively by the German and then American militaries.[7]

Roth returned to Czechoslovakia at a tumultuous time, as the Communist Party was preparing to take power. People with his background and wartime experiences faced suspicions. Roth joined the Party in 1948, at a time when many academics and professionals were being recruited.[8] Roth's son said his father's affiliation reflected his genuine sympathies, although he had been to France and Switzerland and saw through Communist propaganda.

At Charles University, Roth had an ally in Kamil Henner, chairman of neurology, who shielded him from political trouble and encouraged his research interests. In his early work, Roth used a Grass EEG machine, then state of the art, obtained with Henner's help from UN relief funds.[9]

SLEEP DRUNKENNESS

A man Roth first met in 1950, a nurse with the initials DV, made an impression on him.[10] As a child, DV had experienced infectious hepatitis and stomach problems and suffered daily headaches as a teenager. At age twenty-five, he was well known to his family and coworkers for difficulty waking up. Positioning alarm clocks close to DV's head or banging on a door didn't do the job. "The only way to wake up the patient is by shaking him fiercely," Roth wrote. "This usually makes him partially awake; he usually says something, but then he sleeps on. He must be shaken ceaselessly and dealt with resolutely, until he finally gets up."

Even when DV did get dressed and eat, he did so like an automaton, and when he arrived at work thirty minutes later his coworkers would notice how drowsy he appeared. During his military service, he was often punished for being late to morning lineup. Three years later, DV was working in a hospital, spending nights on the same corridor as several doctors. They were often awakened by his "special English alarm clock" ringing five times in a row, which DV slept through.

Despite this man's difficulties waking up, he was slow to fall asleep in the evenings and did not doze off against his will during the day. Paradoxically, staying up late seemed to help him. Thus, he had difficulty transitioning *out* of sleep, but his total level of sleep demand, in terms of hours per week, wasn't as strong as in other patients Roth had seen.

DV appears in Roth's 1956 paper on sleep drunkenness, which focused on twenty patients, eleven of whom had an independent form of the condition and nine others who experienced it together with narcolepsy or epilepsy. In this paper, he laid out the concept that sleep drunkenness is at one end of a spectrum of sleep disorders, with narcolepsy with cataplexy at the other end. Sleep drunkenness was both a symptom that could appear in other disorders and the marker of a separate entity. The English summary concluded that sleep drunkenness "may appear either in an independent form or within the framework of the narcoleptic syndrome. . . . In the idiopathic forms it is a case of functional insufficiency with at present no demonstrable organic basis, most probably congenital or acquired at an early stage."

Roth was not the first to use the term "sleep drunkenness"; he cites a 1905 paper from the German psychiatrist Hans Gudden; the German word is *Schlaftrunkenheit*. However, Gudden was writing about a different situation, when someone is awakened from deep sleep. Today, sleep specialists would term this sporadic, acute state "confusional arousal," in contrast with the chronic symptom that Roth observed in some of his patients.

GENETICS AND DEPRESSION

A family also appears in the 1956 paper on sleep drunkenness: a mother, two children, and the mother's sister. The mother, a medical student, first came to him at age thirty-three; the trigger of her hypersomnia's onset was mysterious. She would fall asleep while typing and often sleep more than fourteen hours in a day. After waking up, she was disoriented and said she felt as if she had just emerged from anesthesia.

For her depression, the mother was treated with electroshock therapy in Prague. The depression receded, but her sleep disorder persisted for years afterward. The mother's two children both had similar symptoms; one was the locksmith in training from chapter 1. Roth points out that in the older siblings, the illness appeared in their thirties, accompanied by depression, while in the second generation, it appeared around puberty, but without depression.

Genetics emerged as a theme later in Roth's work with his colleague Soňa Nevšímalová. She began working with Roth in the 1960s, quizzing patients about the content of their dreams and conducting neuropsychological examinations. She went on to become a respected clinician and researcher in her own right, studying hereditary neuropathies and neurodegenerative diseases. "It was largely because of Bedřich Roth's influence that I began studying sleep," Nevšímalová said in an interview in her office, where she still kept charts of patients they had seen together.

In the early 1970s, Nevšímalová completed her PhD thesis in genetics, compiling histories and pedigrees of patients with narcolepsy and hypersomnia. She found that patients with IH frequently had relatives with the same condition—more than a third did. The effect was stronger when sleep drunkenness was present. Nevšímalová described two families in which three successive generations developed what she called "essential hypersomnia of the sleep drunkenness type," later named idiopathic hypersomnia. One forty-year-old man could not hear alarm clocks, was disoriented upon waking for thirty to sixty minutes, and often slept for several hours in the afternoons. His mother and teenaged daughter had similar symptoms.[11]

Because of the strong inheritance pattern, Nevšímalová and Roth believed that hypersomnia might be determined by a mutation in a single gene, inherited in a dominant fashion but with incomplete penetrance, not always manifesting in disease.[12] The pattern contrasted to postencephalitic or post-traumatic cases of hypersomnia. Before recombinant DNA and sequencing technology was available, it was not possible to pin down what genes were responsible or determine whether the same gene was mutated in separate families.

Nevšímalová and Roth observed a high rate (26 percent) of depression in people with IH.[13] A similar number (28 percent) of those with narcolepsy without cataplexy also had depression. At that time in Czechoslovakia, people were less likely to consult a psychiatrist than people in Western Europe or the United

States. Depression was a relatively rare diagnosis; its prevalence was estimated at around 1 percent, and depression was generally defined as requiring hospitalization or involving a suicide attempt. Nevšímalová and Roth remarked: "Some of the depressive symptoms might, of course, be due to the inability to cope because of daytime somnolence rather than being a symptom of the same basic neurochemical disturbance."

In some patients, worsening depression went together with stronger hypersomnia, so they were thought to be both features of the same disturbance. Antidepressants presented a puzzle; they didn't alleviate sleepiness in narcolepsy, but they could diminish abnormal sleep states if someone had hypersomnia connected with depression.

Roth himself suffered from bouts of depression for around thirty years and was treated with the tricyclic antidepressant amitriptyline, according to his children. "My father would be the first person to say that his decisions were influenced by depression," his daughter said. "When he was in such a state, all sorts of emotions would be amplified—guilt, his sense of being attacked or perceived fault. But when he was free of depression, he was full of jokes and smiles."

FROM FUNCTIONAL TO IDIOPATHIC

In his early work in Prague, Roth divided his patients into the categories of narcolepsy and hypersomnia in a more intuitive or subjective way than specialists are used to today. According to his system, in narcolepsy, sleep occurs quickly and forcefully, almost irresistibly, but does not last as long as with hypersomnia. Roth did not insist that cataplexy was necessary for an "essential narcolepsy" diagnosis, although it was present in about two-thirds of cases.[14]

With people with hypersomnia, Roth noted that they usually had the ability to sit or lie down before sleep overwhelmed them, but their naps lasted longer than in narcolepsy, and their nightly sleep time was longer and steadier as well.

From the point of view of the twenty-first century, an important element is missing from Roth's classification scheme: REM sleep. It is unclear when Roth first learned about REM sleep. Early on, Roth did cite one paper by Dement on what was later called "paradoxical sleep" in cats,[15] and he began recording eye movements in his own sleep lab in 1965.

Half a century ago, the separation between neurological and psychiatric disorders was not as strict as it is today. In Roth's 1957 book *Narcolepsy and Hypersomnia from the Aspect of Physiology of Sleep*, he divided hypersomnias into the categories of "organic" and "functional." "Organic" meant that the origin was clear or structurally discernible, coming from encephalitis, traumatic injury, or a tumor. "Functional" held everything outside that category, including what would later be called IH.

In neurology, the word *functional* sometimes has a connotation of "merely psychological." For Roth, functional hypersomnia was a big tent. He gave some functional cases the label "vegetative dystonia," because of symptoms like headaches, dizziness, or digestive problems. Others were neurotic, with "hysterical personality" as one contributing factor. Roth was willing to consider physiological causes, such as genetics, endocrine disturbances, or chronic sleep deprivation, for the neurotic type of functional hypersomnia. Over time, Roth reevaluated apparent neurotic cases and decided that most of them belonged in the IH category, because of heritability in some cases, along with the stable character of the symptoms.

Hypersomnia with sleep drunkenness, or simply sleep drunkenness, began as a third category of functional hypersomnia. In the 1950s version of his book, sleep drunkenness gets a separate chapter. At that point, we can see a contrast between how Roth thought of sleep drunkenness and other types of functional hypersomnia. Generally, he recommended psychotherapy or behavioral therapy for most cases of functional hypersomnia, sometimes in combination with the mild stimulant ephedrine. He advised avoiding amphetamines, because of side effects: "Psychoton [amphetamine] is only prescribed by exception in functional hypersomnia, in cases in which all therapeutic methods have not succeeded, and even then only temporarily. If possible, we do not give psychoton in those cases where a larger neurotic component has been detected."[16]

In contrast to other types of functional hypersomnia, with sleep drunkenness, Roth was not shy about recommending strong stimulants, such as amphetamine— the same treatments he favored for narcolepsy. He was also enthusiastic about methylphenidate (Ritalin) and phenmetrazine (Preludin), both of which first appeared in the 1950s. For sleep drunkenness, he specifically recommended a low dose of amphetamine before bed to make waking up easier in the morning.

CAVEAT AND CONTEXT

Would Roth's patients receive the same diagnoses today? Roth started forming his ideas about hypersomnia before it was recognized how common sleep apnea was. He began studying sleep apnea in the 1960s, and he thought it was responsible for a small number of cases in his cohort. In hindsight, Nevšímalová said she thought some cases Roth diagnosed as idiopathic hypersomnia may have actually been sleep apnea. In some cases of functional hypersomnia, Roth did notice obesity and headaches, which are often associated with sleep apnea. However, he reported that half had low blood pressure and half normal; this is less characteristic of sleep apnea.

Putting sleep apnea aside, to appreciate Roth's insights, we need to ask why somebody hadn't identified IH before. At the time, the dominant view in English-language medical literature was that long-lasting cases of sleepiness, which were sometimes given the label *hypersomnia*, were psychiatric disorders. Hypersomnia was considered a symptom, not a disease of its own. Other diseases that enhanced sleepiness were considered to have an endocrine or psychiatric origin. Epidemic encephalitis (discussed more in chapter 11) attracted attention in the 1920s and 1930s but then faded away. Neurologists had enough difficulty wrestling narcolepsy away from psychiatrists and, in effect, decided that hypersomnia was too vague or borderline to bring with it.

The 1950s marked the height of Freudian influence in North American psychiatry. During this period, hypersomnia mainly showed up in English-language medical literature as "psychogenic," and it was portrayed as an escape mechanism. A 1959 paper from the Mayo Clinic in the *American Journal of Psychiatry* summarized twelve cases with long, inappropriately timed and unrefreshing naps, combined with disturbed nighttime sleep.[17] The authors wrote: "The pattern of unusual drowsiness and undue sleep was consistently a means for passive retreat from conflict."

In contrast, Freud and psychoanalysis were not respected in 1950s Czechoslovakia, where "psychiatric research was primarily biologically oriented, mental health promotion was marginalized, and those who suffered from severe psychiatric disorders were confined in psychiatric hospitals."[18] The ideological environment may have enabled Roth to move ahead with physiologically oriented research on narcolepsy and hypersomnia.

More broadly, Roth's contribution was recognizing that there was a group of people for whom hypersomnia was the root of their problems. Their excessive sleep did not resemble narcolepsy and was not obviously tied to encephalitis or traumatic brain injury. As a first step, seeing sleep drunkenness as a distinct neurological symptom, analogous to cataplexy, may have made this possible.

WORLD TRAVELER

The early 1950s, when Roth was forming some of his concepts, was a bleak period for Czechoslovakia, especially for those from a Jewish background. The communist government was consolidating its power with mass arrests, antisemitic purges, and show trials. Some of Roth's cousins and family friends were imprisoned, his son said. Roth was unable to travel abroad, and communications were limited.

Later in that decade, the political situation in Czechoslovakia relaxed enough for Roth to travel. In 1957, Roth was able to spend six weeks in France with Henri Gastaut, an authority on epilepsy. When Roth came to work with him, Gastaut was exploring the EEG signs of altered states of consciousness, such as syncope or fainting.[19]

In Marseilles, Roth refined his knowledge of EEG. With Gastaut's advice, he was able to analyze sleep recordings more rigorously. From their collaboration, a paper emerged on narcolepsy.[20] Examining both Czechoslovak and French patients, Roth and Gastaut concluded that "narcolepsy and cataplexy are transient, non-critical phenomena that have nothing to do with epilepsy. . . . Essential narcolepsy must therefore depend on a functional disorder of the structure responsible for the vigilant state and sleep regulation."

Roth's book on narcolepsy and hypersomnia, translated into German but not English, received respectful reviews. In 1961, Roth published a paper in the then-premier journal *Electroencephalography and Clinical Neurophysiology* on the EEG signs of "lowered vigilance," a transitional state on the way to sleep.[21] He also presented his work on narcolepsy at a conference in Rome. This exposure seems to have brought Roth to the attention of other sleep researchers.

At Charles University, Roth's career prospered. He became the go-to person for others studying external influences—light stimulation, radiation, decompression sickness, and industrial poisons—on brainwaves.[22] In 1963, Roth published a paper with his wife, Nada, a pediatrician, who was studying epilepsy in children.[23] Anniki Rothova recalled that when her mother had medical duties in the evening, "it was great fun" because instead of cooking dinner, her father would take her to a nearby cafeteria. Around this time, Roth and his family acquired a yellow Skoda, for which they'd had to save for several years. It would have been considered small in other countries, but for them, it was a luxury.

CHICAGO AND SANTA MONICA

In 1967, Roth came to the University of Chicago to work with Allan Rechtschaffen, a leading sleep researcher there. At that point, during a temporary political thaw in the mid-1960s, foreign travel was complicated but possible. This was Roth's chance to study people with hypersomnia in the sleep lab overnight, which he had not been able to do conveniently in Prague.

Some of Roth's peers reportedly thought hypersomnia with sleep drunkenness was something peculiar to Czechoslovakia.[24] To confirm that people with hypersomnia could be found elsewhere, Rechtschaffen had classified ads placed in the *Chicago Tribune*: "WANTED—For research studies, patients with clinical hypersomnia and postdormital hypersomnia and postdormital confusion but without narcolepsy. Call Sleep Lab, University of Chicago."

Roth was in Chicago for just a few weeks, staying in the university's housing for international students. He and Rechtschaffen stayed up several nights examining the people they recruited. In these experiments, Roth was able to see that people with hypersomnia did not enter REM sleep quickly and had normal proportions of REM and non-REM sleep; in fact, the organization of sleep was "completely normal apart from its long duration." Roth, Nevšímalová, and Rechtschaffen later wrote: "It is clear that the disorder is not specifically related to any single sleep stage. Rather, the disorder appears as an extension and 'intensification' of normal sleep."[25]

In 1967, Roth was part of a group of notable figures in sleep research: the Rechtschaffen and Kales committee.[26] Roth was one of three from outside the United States, along with Michel Jouvet from France and Ian Oswald from the United Kingdom. They produced a standard handbook to evaluate polysomnography studies, which was used in the field for four decades. The meetings, held in Santa Monica, were contentious, since each lab had developed its own ways of reading the voluminous paper graphs produced by overnight studies. Rechtschaffen barred the door and told attendees they were not allowed to leave until they came to a consensus.[27]

INVASION AND NORMALIZATION

Roth was able to return to Prague and confirm that only narcolepsy with cataplexy involved quick entry into REM sleep, while "independent narcolepsy" (that is, narcolepsy without cataplexy) and hypersomnia did not. In the 1970s, he and Nevšímalová eventually performed overnight studies in Prague with their colleague Le Van Thanh, with similar results to what he found with Rechtschaffen: the REM/non-REM sleep architecture in hypersomnia was preserved.

Roth's work with Rechtschaffen did not appear in *Archives of General Psychiatry* until 1972. Geopolitics had intervened with the 1968 Soviet invasion of

Czechoslovakia. Roth was depressed and nearly immobile for weeks after the invasion, according to his daughter. He hesitated over whether to accept a job offer in Western Europe but ultimately declined.

Roth refused to sign a declaration of loyalty to the new government, something many in Czechoslovakia had to do to keep their jobs.[28] The incoming chairman of the neurology department, Josef Vymazal, protected Roth from losing his post, but he was demoted and could not be listed as a faculty member who interacted with students.[29]

In 1973, Roth's daughter became sick of communist indoctrination and immigrated to the Netherlands, becoming an ophthalmologist. Her departure was a blow and contributed to another period of depression for Roth. He continued to write his daughter regularly.

Political restrictions eventually eased enough for Roth to travel again, although he needed permission from a series of officials, including the minister of health, to leave the country. He participated in landmark conferences, such as the First International Symposium on Narcolepsy, held in southern France in July 1975. Roth's talk at that meeting, titled "Functional Hypersomnia," contains some of his clearest advocacy on behalf of people with IH: "It is necessary to stress, however, that the symptoms of idiopathic hypersomnia are, without any doubt, pathological manifestations. These patients suffer considerably and have many difficulties in their professional as well as their private lives. This is why they so often seek medical help."

INFLUENTIAL FRIENDS

In the 1970s, Roth became friends with other sleep researchers, such as Michel Billiard in France and Roger Broughton in Canada. Broughton, who had also trained with Gastaut, was head of the sleep disorders clinic at Ottawa General Hospital. His research interests ranged from early work on sleepwalking and night terrors to circadian rhythms.

Broughton and Roth forged a professional partnership. Broughton arranged for Roth to speak at international conferences and facilitated his travel. Roth's patients became part of Broughton's surveys on the social and economic burdens of narcolepsy. Roth presented data on his IH patients at the 1978 European Congress on Sleep Research, writing: "It is evident that patients with idiopathic hypersomnia experience very serious effects on a number of parameters including work, education, personality, memory and so forth, as has also been documented for narcoleptics. . . . These socio-economic consequences, in fact, often exceed those of narcoleptics. This is perhaps due to the greater amounts of daytime sleep and of the intensity and duration of diurnal drowsiness."

Broughton helped Roth revise his book on narcolepsy and hypersomnia, which was finally translated into English.[30] Since the 1950s, the science of sleep had made great advances, so they revamped most of the book, and Broughton cowrote five chapters. When discussing IH in the later edition of his book, Roth included a monosymptomatic form, characterized by excessive daytime sleepiness only, and a polysymptomatic form, which also included long sleep time and sleep drunkenness upon waking. Roth wrote: "I originally believed that those patients who, in addition to diurnal sleepiness also suffered from deep and protracted nocturnal sleep and from sleep drunkenness on arousal, had a separate nosological entity, which I called hypersomnia with sleep drunkenness. However, I later came to the conclusion that the two clinical pictures were variants of the same disease."[31]

It is possible to see Roth's imprint on the 1979 DCSAD (Diagnostic Classification of Sleep and Arousal Disorders), whose hypersomnia sections were edited by Broughton. There IH is called "idiopathic CNS hypersomnolence," although the descriptions do not exactly match Roth's.[32] We can also see Roth's influence on publications from Dement's and Guilleminault's group at Stanford. Their 1975 paper on "hypersomnia with automatic behavior," submitted just before the meeting in France, makes no reference to Roth.[33] Jumping ahead to 1986, their comparison of narcolepsy and IH says in the introduction that Roth was "the first to clearly isolate the idiopathic hypersomnia syndrome."[34] At the same time, Guilleminault's identification of "upper airway resistance syndrome" (UARS) as a borderline condition for sleep apnea chipped away at IH. Some patients who might have been diagnosed with IH were diagnosed with sleep apnea or UARS instead, because of clinicians' improved ability to detect disturbances of breathing during sleep.

In the 1980s, Roth and Nevšímalová adapted to new trends in sleep medicine. They and their patients contributed to early genetic studies of narcolepsy.[35] They also criticized the MSLT, the emerging diagnostic standard. Their alternative, tested in a small number of people, took into account both sleep latency and the duration of sleep stages entered during daytime naps.[36]

Roth died in November 1989 of Hodgkin's lymphoma, just a few days before the Velvet Revolution. His friend Broughton has argued that he was "the true father of sleep medicine."[37]

CHAPTER 6

THE ESSENCE OF SLEEPINESS

*In the case of narcolepsy, are there not cases where a metabolic prob-
lem leads to the more rapid and intense production of the hypnotoxin? Research into a toxic activity of this type in the CSF of such
patients would be quite desirable.*

—Henri Pieron, 1927

Let's return to Emory, where as early as mid-2007, investigators
thought that Anna's cerebrospinal fluid contained something that
was making her extremely sleepy. The identity of this substance was
unclear. Because flumazenil worked well for Anna, Parker had invoked the term
"endozepine," which has a history of its own. In the late 1970s, researchers at the
National Institute of Mental Health had been hunting for what they called "the
brain's Valium." Their idea was that benzodiazepine drugs had natural counterparts: *endogenous benzodiazepines*, or *endozepines*, which might reveal something
fundamental about how anxiety was regulated.[1] What they found turned out to
be more complicated, and by the twenty-first century, the concept of endozepines
had fallen out of favor.

Despite language that Rye would later use to explain what they had found, he
and Jenkins thought the GABA-enhancing activity was something distinct from
a benzodiazepine. A plan detailing how they proposed to identify it can be found
in a 2009 grant application.[2] This was one of the first places where Rye unfurled
his somnogen flag: "We were compelled in the course of this work to consider an

alternative hypothesis—namely, that hypersomnia might be caused by excess of a naturally occurring somnogenic substance. Indeed, we have discovered that an endogenous, positive allosteric modulator of GABA-A receptors causes hypersomnia and sleepiness in humans."

Rye and a scientist in his lab, Glenda Keating, had been investigating a molecule called Cocaine and Amphetamine-Regulated Transcript, or CART. They had thought CART might be like hypocretin: important for keeping people awake and missing in narcolepsy. However, measurements of CART in CSF samples from people with sleep disorders revealed that CART levels correlated more with the use of stimulants than with sleepiness. CART was not the next hypocretin.

When Anna's response to flumazenil appeared to be sustained, Rye decided he needed to pivot away from CART. An opportunity came through the Obama administration's 2009 economic stimulus package. The American Recovery and Reinvestment Act included a $10 billion boost for the National Institutes of Health.[3] A buzzword in Washington, DC, at the time was "shovel-ready," meaning that a construction project would generate jobs quickly if funded. In the research realm, the concept was applied to projects expected to yield fast results. If scientists had already gathered a large number of samples and wanted extra funding to analyze them, those proposals were given priority. Rye's successful grant application fit into that framework. The funding climate grew colder after the stimulus package, and he would not receive federal support for hypersomnia research for another six years.

A SHADOWY SUBSTANCE

Jenkins and two colleagues at Emory, the pathologist Jim Ritchie and the pharmacologist Mike Owens, had already been probing what the somnogenic substance could be. When they centrifuged CSF from hypersomnia patients through membranes, what came through was still active in the patch clamp assay, indicating that it was smaller than the holes in the membranes. In addition, when CSF was subjected to the enzyme trypsin, which chews up proteins, its patch clamp activity was mostly wiped away.

Together, these clues suggested that the mysterious substance was a peptide, or small protein. The prospect of a somnogenic peptide resonated with a long line of research in the field, stretching back decades. As one example, celebrated experiments with sleep-deprived goats had identified a sleep-promoting peptide called "Factor S" in their CSF, although it didn't look like the hypersomnia substance was the same thing. In contrast, before Rye and Jenkins, several groups had gone searching for abnormalities in CSF among people with idiopathic hypersomnia. Some looked for changes in dopamine metabolites; others looked for alterations in the neurotransmitter histamine. This line of research had yielded few returns.

Over the next few years, Jenkins was guarded about concluding whether CSF from people with hypersomnia had *more* of an otherwise ubiquitous GABA-enhancing peptide. An alternative was that one of those peptides might be altered and more potent than usual in people with hypersomnia. He and others have described CSF as "the sewer of the nervous system," since they were analyzing waste products, without knowing their precise origin. CSF components can change when someone has an infection or other brain injury and can also vary depending on the site where the fluid is removed.

One experiment hinted that Rye and Jenkins's research might be applicable beyond people with rare sleep disorders. The 2009 patent includes data from rhesus monkeys that Rye and Keating had drawn CSF from at various times of the day: early morning, afternoon, and late evening. (Rye's work on Parkinson's had used monkeys; at the time, he was running out of money to maintain them.) After four monkeys had been kept awake and would normally be asleep, the CSF GABA-enhancing activity in two of them increased above the threshold for hypersomnia in human patients. Although it was never published, this preliminary data reveals how Rye was thinking. Part of the patent speculates: "It appears that this substance waxes and wanes in animals as it does in humans during the normal day-night cycle. These promising results indicate that under sleep-deprived conditions, humans may also benefit from flumazenil or other GABAergic therapy to relieve the symptoms of fatigue they experience from the accumulation of this somnogenic compound."

Identifying the somnogen in hypersomnia patients' CSF was a classic biochemistry problem: purifying a protein out of a complex mixture. When Jenkins started to separate human CSF into fractions biochemically, he glimpsed peptides that were present only in Anna's CSF. The peptides' sequences were not in biomedical databases, suggesting that what they were hunting had not been seen before.

At the outset, Jenkins had to establish whether the "sleepy stuff" was stable over time and would not decay. Could he let a sample of CSF sit around at room temperature for a few hours or even overnight? What if he froze it and rethawed it, or dried it in a vacuum? Both were OK. Were measurements the same between experimenters? With patch clamp assays, it was only possible to measure a small number of CSF samples per day.

Early on, Jenkins compared results with Garcia, and later with his postdoc Amanda Freeman, a former graduate student of Rye's who was asked to help a couple days per week. Accustomed to painstaking electrophysiology work, Freeman found she could get through measurements of three or four patients per day, if everything was set up correctly. "Working on the rig all day—for me, that was a good time," she said.

In contrast to measurements of GABA-enhancing activity, the hunt for the peptides responsible ran into difficulties. After Jenkins retooled his rigs to reduce the flow rate, less CSF was needed per measurement, but CSF from patients was still scarce and valuable. The 2009 grant application included a proposal to measure GABA-receptor activity with fluorescent dyes instead of with patch clamps, which could allow experimenters to process samples more quickly. In the years to come, Jenkins had a tough time convincing graduate students to work on the hypersomnia project; they gravitated toward anesthesia-related projects with more predictable returns.

Colleagues peppered Jenkins with suggestions, like obtaining large amounts of CSF from animals or trying particular candidate peptides whose attributes seemed relevant. At some point, the decision was made to pool CSF samples from hypersomnia patients. That gave experimenters more volume to work with, but the decision implied the active "sleepy stuff" ingredients in each person were the same. Both Rye and Jenkins would later regard this choice as a mistake.

Imagine police detectives on television. Someone witnesses a crime but doesn't know the perpetrator's identity. So the police ask the witness to describe the person she saw to a sketch artist. *How tall was he? What was he wearing?* Like the sketch artist, Jenkins and his colleagues had a few clues: the GABA-enhancing substance's approximate size, what it was made of, and its sensitivity to flumazenil. But they didn't know what it was. The perpetrator, or perpetrators, had escaped into the crowd.

CUSTOMIZED EQUIPMENT

A few years later, when Rye gave a seminar at Emory, the puzzle captured my imagination. From my own biochemistry experience, I guessed the process could be a hard slog. Patch clamping has a reputation for difficulty. Much lore and special tricks go along with the technique. For many years, it resisted being automated, unlike other processes in molecular biology, such as DNA sequencing. A graduate student in California engaged in patch clamp wrote in 2013: "Patch-clamp bugs require a personality with the highest tolerance for frustration I've ever encountered."[4]

Practitioners of the technique have retained some elements of craft (figures 6.1, 6.2). The glass tubes needed for patch clamping are melted and stretched in the

FIGURE 6.1. Patch clamp rig in Jenkins's laboratory at Emory, 2018.

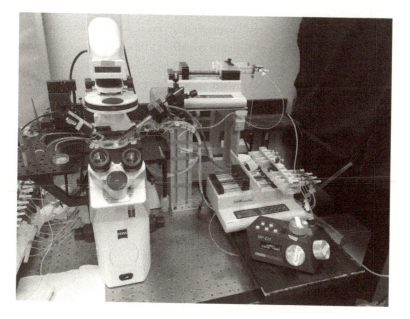

FIGURE 6.2. Patch clamp rig in Jenkins's laboratory at Emory, 2018.

lab, rather than being manufactured and delivered, because of fragility and the possibility of introducing dust. A manual on patch clamping advises scientists to try different locations in a building, seeking the place where the floor vibrates the least.[5] Sometimes, the experimenter surrounds the entire setup with a metal Faraday cage to screen out electrical influences from other pieces of equipment. Some patch clampers fire-polish their pipettes or apply bleach to the silver wire that measures current, aiming to leave just the right coating of chloride on the silver.[6]

I asked Jenkins to visit his lab to see what he and others had been dealing with. In the corner of a large research building, his lab had two active rigs, with components held in place with aquarium glue and glass pipettes resting on modeling clay in covered dishes. Several Post-it notes next to the rigs compiled a list of procedural reminders. A rig included a microscope, a micromanipulator for bringing the glass pipette in contact with the target cell, electrical signal amplifiers, and a thicket of plastic syringes and flexible tubing. Adding up all the components, each rig cost about $100,000.[7] "A rig is a customized piece of equipment, which is constantly being torn down and rebuilt to do different experiments," Jenkins told me.

I was surprised to learn that patch clampers use their mouths to control suction through the glass pipette. In my graduate school years, fledgling scientists were admonished to avoid "mouth pipetting," because of the risk of ingesting some harmful chemical. Patch clampers use their mouths because their hands are busy positioning the glass tip and managing electrical equipment. Jenkins chuckled when asked about this, saying that the cheeks have more sensory nerve endings and provide better control. "It's also partly personal preference," he said. "I never had success with a syringe in my hand. Using my mouth just gives a better feel."

Olivia Moody, who would earn her doctorate working with Jenkins, showed me how she established contact between the glass pipette and the cell membrane, applied suction, then waited for the electrical resistance to climb to gauge whether the seal is good enough. When I asked about the difficulty of patch clamping, she shrugged, smiled, and said: "Mice don't always do what you want either."

SLEEPY PUPPIES

A century before, scientists in France and Japan conducted experiments with sleep-deprived animals, believing that a sleep-inducing substance, or "hypnotoxin," accumulated in their bodies. They put their subjects into extreme, cruel situations, outside the rhythms of day and night. In a way, they were following Koch's postulates: viewing sleepiness like a disease-causing microbe that could be isolated and transferred to another animal. In several countries, authorities overseeing animal welfare would discourage such experiments today.[8]

Starting in 1907, the Japanese physiologist Kuniomi Ishimori and a host of medical students in Nagoya kept puppies awake for up to 113 hours, "by whipping the animals via ropes or by other coercive means."[9] Ishimori's group proposed that the puppies' brains contained substances that induced sleep when injected into fresh animals. Their experiments had limitations, since their whole-brain extracts also provoked shivering, salivation, and tear production in recipients. Still, only the sleep-deprived puppies' brains, and not those of animals allowed to rest, contained the sleep-inducing factor, they reported.

A few years later in France, the physiologists Henri Pieron and Rene Legendre had dogs constrained by leashes so they were unable to lie down. Pieron and Legendre isolated various body fluids from the dogs and discovered that CSF from sleep-deprived dogs contained a sleep-promoting activity. The effect could be seen when the CSF was injected directly into the brains of rested dogs. Pieron and Legendre concluded: "A prolonged waking state produces in the organism, probably as a waste product of cerebral metabolism, the formation of a toxin that causes local cellular lesions and a pressing need for sleep, phenomena which are obviously related."

Nathaniel Kleitman, a pioneer of sleep research in the United States, cited Pieron's work as an inspiration. In another groundbreaking study, Pieron tracked the body temperature rhythms of nurses working the night shift at an asylum. His 1913 book *Le probleme physiologique du sommeil* was considered one of the twentieth century's most influential scientific studies of sleep, because its focus was distinct from philosophical introspection or questions of the soul.[10]

Nobody today knows what the hypnotoxin was. Pieron speculated that it might be responsible for narcolepsy, which is unlikely, given what we know now. Techniques for biochemically characterizing sleep-inducing substances were then primitive, and Ishimori's and Pieron's were probably different. But the concept was influential. Previous experimenters had kept animals awake with exercise, making it difficult to separate fatigue from sleepiness. While Ishimori was little noticed in Europe or the United States, Pieron's experiments with dogs spurred a series of efforts to repeat and extend their results.

If we look at the history of scientists' search for molecules connected with sleep, quests like Pieron's have led to unexpected insights—or dead ends. Some of the molecules now considered the most important for regulating sleep were hiding in plain sight or were discovered by scientists who were not primarily investigating sleep.

Like their predecessors, Rye and Jenkins were making a bet that hypersomnia patients had something inside them embodying their sleepiness—and they could extract it. However, what the Emory investigators were doing was somewhat different conceptually. Rye and his team weren't deliberately putting people with narcolepsy and IH under the stress of sleep deprivation. They thought that what they were looking for was more fixed and long-lasting. Don Bliwise put the distinction this way: "It depends on whether you see sleepiness as a temporary state or a permanent trait. If you see overall need for sleep as a lasting trait, it makes a lot of sense to probe people with hypersomnia and figure out what distinguishes them."

PAPER CLIPS AND RUBBER BANDS

In 1939, researchers at Northwestern University in Chicago reported they had confirmed Pieron's observations regarding sleep deprivation in dogs. However, they argued that "the normal picture of sleep is not produced," because manipulations of CSF produced changes in both body temperature and intracranial pressure. The challenge emerged of untangling the processes that induce healthy sleep from those producing sleepiness driven by fever or illness.

Physiologists at Harvard led by John Pappenheimer followed Pieron in the 1960s, using goats instead of dogs.[11] The shape and thickness of bone at the rear of goats' skulls allowed the Harvard scientists to implant nylon tubes and repeatedly withdraw CSF without anesthesia. The goats were conditioned, with electric shocks initially, to avoid lying down in the laboratory and thereby kept awake for forty-eight to seventy-two hours.[12]

Pappenheimer wrote in *Scientific American* that his lab's initial results, obtained by injecting goats' CSF into cats, "were so striking and of such potential physiological significance that we put other plans aside in order to devote full time to the systematic exploration of the Pieron phenomenon."[13] When extracted CSF was injected into animals' brains, it produced something close to natural sleep, with EEG signatures resembling those observed after the animals were sleep deprived.

Pappenheimer and his colleagues began trying to concentrate and purify what they dubbed "Factor S." An old-fashioned scientist, Pappenheimer preferred bicycles to cars and didn't like making long-distance phone calls. He was "parsimonious in quite an English way in that he wanted to do things with paper clips and rubber bands."[14] "Each assay required 6 hours of recording on two or more rabbits fitted with implanted guide tubes and EEG electrodes," Pappenheimer said in a 1982 lecture. "This part of the research took several years, and there was very little to show for it until there was enough pure product to analyze chemically."

Like Rye and Jenkins, the Harvard group could determine the substance's size and that it was a peptide. Over three years, they collected about six liters of CSF from a colony of goats, but this wasn't enough. They moved on to extracting whole brains of goats, sheep, and rabbits but estimated they would need thousands of sleep-deprived animals. Eventually they turned to human urine, which was collectable at a sufficient scale.[15]

Once they had enough and could confirm that urinary Factor S behaved chemically the same as CSF Factor S, they teased Factor S apart and found it contained surprising components: muramyl peptides, which come from bacterial cell walls and not from mammalian cells. Pappenheimer suggested that the bacterial components "may be regarded as akin to any of the essential amino acids or vitamins which cannot be synthesized by mammalian cells." "There was a lot of skepticism," said James Krueger, who began working with Pappenheimer at Harvard in the 1970s. "I'd go to meetings, and people would be shocked at the idea that something from gut bacteria could be acting in the brain. They thought it was contamination. Grant applications came back rejected for the same reason."[16]

How the microbes that live in our intestines, on our skin, and in other parts of our bodies contribute to our metabolism and brain function was underappreciated at the time. Since they are bits of bacteria, muramyl peptides are highly stimulatory for the immune system and have been used as vaccine-enhancing adjuvants. Krueger proposed that the bits were produced through processing by macrophages, white blood cells that engulf and digest bacteria. In animal experiments, muramyl peptides brought on fever, which could be alleviated with acetaminophen, without blocking their sleep-promoting effects.

Muramyl peptides were potent, but they weren't as fast as cytokines—inflammatory messengers that arouse the immune system—at inducing sleep when injected into the brain. So the bacterial bits were thought to act indirectly. Throughout the 1980s and 1990s, more players entered into Factor S's proposed sleep-triggering mechanism. "When I started with Pappenheimer, we were working with a simple idea," Krueger said. "He thought there was a kind of sleep hormone, a single sleep factor. Now it has become clear that there are many molecules involved."

IMPLICATIONS FOR IH

The most straightforward interpretation of Factor S was that it represented sleepiness associated with infection or inflammation. This is consistent with everyday experience; a mild infection or fever generally enhances sleepiness, pushing down REM sleep in favor of non-REM, but a more intense fever can disrupt someone's sleep. The lethal effects of sleep deprivation in animal experiments may have been caused by bacterial invasion of internal organs.[17] Yet Krueger and his colleagues have accumulated evidence that cytokines induced by Factor S also have an everyday physiological role, outside of fever or extreme sleep deprivation. Their levels vary rhythmically, rising at night, and tweaking their levels in mice can modulate how much the animals sleep.[18]

Clinical studies are underway that may address whether these observations apply to IH. People with IH aren't sleep-deprived goats, but it is possible that some of their sleepiness comes from substances produced by intestinal bacteria or from heightened levels of inflammatory cytokines. People with sleep apnea—who may superficially resemble those with IH—have elevated markers of inflammation, although this is difficult to disentangle from obesity, which often contributes to sleep apnea.

ROMANCES GONE ASTRAY

By the time Factor S was being characterized, several rival sleep molecules had appeared. Two examples demonstrate how sleep's biochemical complexity has defied simple dissection.

In the 1960s, a wave of enthusiasm washed over the sleep research field for the neurotransmitter serotonin. The French neuroscientist Michel Jouvet, who discovered the areas of the brain that control REM sleep in cats, described his and others' research on serotonin as like a love story: a romantic encounter, leading to a honeymoon, followed by divorce.[19]

Jouvet and others found that depleting serotonin by inhibiting its synthesis caused profound insomnia in cats. Anticipating a central role for serotonin in sleep, the Swiss neuroscientist Werner Koella suggested renaming it "somnotonin." But electrical probing of neurons that produce serotonin showed that their activity decreased during sleep and increased during time awake—the opposite of what Jouvet and Koella had predicted. Decades later, the insomnia of serotonin depletion in the cats was found to come from hypothermia.[20] It illustrates how sleep is bound together with metabolism and regulation of body temperature. Researchers still think that serotonin is central to the regulation of sleep, but one molecule is unlikely to be the golden thread coordinating a complex set of brain functions.

Another sleep-associated molecule that did not withstand close scrutiny deserves highlighting: DSIP (delta sleep-inducing peptide). The rationale for pursuing DSIP was the opposite of that for Factor S. Instead of depriving animals of rest, Swiss researchers thought they could synchronize animals' brains and see if they produce something that can cause sleep in another animal. They exploited the discovery that electrical stimulation of the thalamus produced deep slow-wave, or *delta*, sleep—hence DSIP's name.

The Swiss researchers set up a system in which rabbits were electrically submerged into sleep while relevant substances were continuously extracted from their blood via hemodialysis. In the late 1970s, they had accumulated enough of the "sleep potion," as the *New York Times* called it, to decipher biochemically.[21] It also turned out to be a peptide, nine amino acids long.

In the 1980s, DSIP was cited as the most extensively tested sleep molecule. DSIP was different from Factor S, but the two molecules were perceived as competitors. Although a variety of neurochemical effects were attributed to DSIP, doubts began to emerge early on.[22] A peptide would be unlikely to last long enough in the blood or cross the blood-brain barrier if administered peripherally.

In their long trek toward purifying DSIP, the Swiss team was supported by several pharmaceutical companies. DSIP appeared promising enough that Hoffmann-La Roche filed for a patent on it for treatment of addiction withdrawal.[23] As clinical studies proceeded, DSIP's previously reported properties became difficult to explain, and Roche abandoned commercial development of DSIP in the late 1980s. Some of the observed benefits probably came from a placebo effect. DSIP enjoyed an afterlife at a few Swiss clinics, where people recovering from addiction were willing to pay for it,[24] and in Russia, where derivatives of DSIP were tested in clinical trials.[25]

Nobody ever isolated a gene encoding DSIP or a receptor for it. In 2006, Vladimir Kovalzon, a Russian sleep researcher who had studied DSIP at length, suggested that its original discoverers had gotten the peptide sequence wrong. He called DSIP a "still unresolved riddle."[26]

A IS FOR ADENOSINE

For one sleep regulatory factor, there was no need to go hunting for it. The first letter of the genetic alphabet had been in front of us for decades. Adenosine is what's left when phosphates are removed from ATP (adenosine triphosphate), the carrier of chemical energy in cells. In the brain, extracellular adenosine is like discarded wrapping paper and boxes outside a house: a sign that gifts were opened inside. When brain cells send messages to one another, they pack ATP into vesicles, together with neurotransmitters, such as dopamine, glutamate, or GABA. Upon delivery, ATP is left over and gets quickly converted into adenosine. Outside the brain, adenosine has other functions, regulating blood pressure and inflammation. It is found on emergency room crash carts because it can calm a dangerously racing heart.

Adenosine's sleep-inducing effects were observed in the 1950s by scientists in London, who were injecting drugs directly into the brains of live cats.[27] They were testing many substances—acetylcholine, serotonin, and others, too. At that point, they were really just stabbing in the dark; it was difficult to know the physiologically relevant amounts.

In the early 1970s, the neurochemist Henry McIlwain observed that cells released adenosine when they were stimulated electrically. Others showed that when enough adenosine is around, it will inhibit brain cells' firing, sort of like GABA. However, adenosine has its own set of receptors—four, with two types in the brain—and acts through different biochemical mechanisms.

For adenosine to modulate alertness and sleepiness makes intuitive sense, because caffeine, something many are familiar with on a daily basis, interferes with adenosine's access to its receptors. Adenosine seemed to fulfill Pieron's hypothesis that a product of cellular metabolism accumulates in the brain.

Some popular books, such as Matthew Walker's *Why We Sleep*, say that adenosine is responsible for sleep pressure, the force that drives us to feel drowsy after a long time awake, and that adenosine builds up with time awake.[28] This is a nice explanation, but adenosine's actual accumulation in the brain has been difficult to demonstrate, and adenosine is not the only element of sleep pressure.

In the 1990s, scientists at Harvard Medical School proposed this role for adenosine, showing that its concentration progressively increased in the brains of cats kept awake for several hours.[29] These experiments weren't as cruel as before; the cats were kept awake by petting them or by having them play with plastic toy lizards.[30]

The Harvard group originally thought that prolonged wakefulness would elevate adenosine levels throughout the brain.[31] However, in animals, accumulation of adenosine has only been confirmed in one region, the basal forebrain, which is part of the network that modulates alertness. At other sites, adenosine levels appear to stay stable or fall off over time awake. Adenosine may be broken down or recycled too quickly to accumulate noticeably in extracellular spaces.

In humans, the question was tested—just once. Neuroscientists rarely have a chance to peer inside the human brain directly, so they take advantage of the opportunities they have, which come when epilepsy patients are in the hospital for seizure diagnosis. While monitoring the brain for seizure activity, surgeons can leave in place electrodes that are hollowed out, with space to collect a small amount of fluid for chemical analysis. When investigators at UCLA did this, they did not see increases in adenosine concentration. The levels actually fluctuated or decreased over time, even in patients who were kept awake overnight to amplify the likelihood of a seizure.[32]

"We couldn't say much about the basal forebrain, because we were restricted to probing areas that were clinically relevant for the patients' epilepsy," said Jamie Zeitzer, lead author of a 2006 report on the UCLA experiments. "However, we could say that an increase in extracellular adenosine seems not to be a global phenomenon." "Technically speaking, adenosine doesn't accumulate," he added in an email. "Increased extracellular concentrations are due to a mismatch between release and reuptake, where the latter cannot keep up with the former. Adenosine, though, is rapidly removed from the extracellular space."

Despite inconclusive experiments in humans, plenty of evidence supports adenosine's role as a messenger of sleepiness. Even if it's difficult to detect adenosine's accumulation, having more available to the nervous system has a measurable effect. For example, people who have a less active form of an enzyme that breaks down adenosine displayed deeper sleep, with fewer awakenings at night.[33] Those with the less active form didn't sleep longer overall, but they were more susceptible to sleep deprivation.[34] Experiments with mice showed that signals from adenosine appear to be important for the deeper slumber that occurs after insufficient sleep.[35]

Still, the lack of observable accumulation leads to a more fundamental question: where does sleepiness live? In a substance outside cells or within the cells themselves? "It's simpler to conceive of it as one thing, but I don't think it is," Zeitzer said.

Besides adenosine, other molecules have been seen to accumulate in response to sleep deprivation in humans, such as beta-amyloid, the neurotoxic protein fragment connected with Alzheimer's disease plaques.[36] This doesn't mean beta-amyloid is the embodiment of sleepiness, but it does indicate that sleep deprivation impairs the machinery that flushes beta-amyloid out of the brain.

Recent experiments suggest an alternative way of viewing how the brain is responding to the lack of sleep. German researchers have observed that when volunteers went without sleep overnight, one type of adenosine receptor increased in density throughout the brain. That is, the stress of sleep deprivation pushed brain cells to produce more receptor proteins and to keep more of them on their surfaces. When study participants were able to go to sleep, adenosine receptor density went back down again.[37] Incidentally, caffeine mainly works through a different adenosine receptor—so caffeine and sleep deprivation are on parallel tracks. With extended wakefulness, the brain is becoming more sensitive to adenosine. But the molecules determining that sensitivity—the physical embodiment of sleepiness—are embedded in our cell membranes, rather than flowing between them.

SLEEPY FLIES AND SLEEPY MICE

Pieron's and Pappenheimer's ideas continue to inspire modern sleep researchers. But by the twenty-first century, it became possible to use genetic tools, or to comprehensively scan proteins and other molecules in the brains of sleep-deprived animals, rather than having to rely on ingenious schemes to distill the essence of sleepiness. There is also more of an emphasis on the distinction between acute sleep loss (easier to study) and chronic sleep deprivation (pervasive in society).

Using *Drosophila* fruit flies, sleep researchers at the University of Pennsylvania have isolated a sleep-inducing factor called *nemuri*.[38] Hirofumi Toda, a postdoc in Amita Sehgal's laboratory, randomly inserted thousands of individual genes into flies, in such a way that each gene would be overproduced in flies' brains in response to a drug. After screening thousands of genes in this way, they found only one that increased total sleep time. "I had been fascinated with the idea of a somnogen ever since I started working on sleep," said Sehgal, who had previously studied circadian rhythm genes in flies (along with many others). "We had considered trying to get at this biochemically, but then decided to focus on the genetics approach that *Drosophila* is so well-suited for."

The *nemuri* gene encodes a secreted peptide that is also involved in defending flies against bacterial infection. The protein is difficult to detect in unperturbed flies, but its production is turned on in response to infection, stress, or sleep deprivation, either as a result of caffeine or physical agitation. In a way, the sleepy flies recapitulate Pappenheimer's experiments with sleep-deprived goats' CSF, even though the molecule in the goats came from bacteria, rather than being produced by the animals' cells. *Nemuri* may correspond to the inflammation-related cytokines Krueger identified as inducing sleepiness during infections in mammals.

Sehgal has argued that *nemuri* meets criteria for a homeostatic factor, one that helps maintain balance and whose production increases when an animal needs more sleep. Supporting this idea, when the *nemuri* gene is disrupted in flies, they took longer to recover from sleep deprivation; the mutant flies were also easier to arouse and took longer to return to sleep afterward.

Looking more closely at *nemuri* might give us clues about the elusive sleep-promoting peptide in hypersomnia patients. Flies that overproduced *nemuri* slept over three hours per day more than control flies. The flies were also more difficult to arouse by jostling them, a situation possibly analogous to sleep inertia. Unfortunately, there is not a gene in humans whose sequence looks similar enough to make it a counterpart for *nemuri*.

Several alternative peptides identified in vertebrate brains could be somnogen candidates. For example, scientists working with zebrafish—a vertebrate, if not a mammal—have identified a secreted sleep-promoting factor called neuropeptide VF.[39] The relationship between any of these peptides and what Jenkins and his labmates glimpsed in hypersomnia patients' cerebrospinal fluid is still unclear. Several candidates identified by other scientists could match the criteria they laid out. If or when the identity of the "sleepy stuff" comes to light, the same questions will come up as for adenosine and Factor S: how and where does it function in healthy people? Does it become more abundant in response to sleep deprivation, inflammation, or other stresses?

Another recent tour de force of molecular biology comes from Japan, and it raises just as many questions about the neurochemical basis of hypersomnia and sleepiness. The leader of this effort, Masashi Yanagisawa, has a flair for epic projects, and he seems to see himself as the scientific heir of Pieron and Pappenheimer. Intent on "deciphering the neural substrate for sleepiness," Yanagisawa set out on producing mutant mice on an industrial scale and established a facility at the University of Tsukuba that would allow his team of geneticists to do so. They sprinkled genetic flaws across mouse sperm and then systematically screened more than eight thousand of the progeny for altered sleep behavior.[40] They found a line of mutant mice that sleep several hours longer than normal laboratory mice. The mice are named, unsurprisingly, *sleepy*.

The sleepiness in Yanagisawa's mutant mice doesn't appear to come from something secreted outside cells like adenosine or a peptide. Rather, it derives from the altered activity of an enzyme called SIK3, which is located *inside* brain cells. SIK3 is a kinase, a type of enzyme that adds a negatively charged phosphate group to another protein, usually altering its activity. Among the arrays of signaling proteins inside cells, a kinase enzyme selectively turns up a few dials or adjusts a few control levers. Intricate chains of kinases modifying other proteins manage many basic processes in cells, such as cell division.

In the *sleepy* mice, the SIK3 enzyme isn't gone, but it is missing a stretch that another kinase operates on. Sleepiness can't be distilled down to one particular molecule and is not represented by SIK3 itself. Instead, sleepiness is a pattern of all the phosphate alterations SIK3 makes on many different proteins, which are nestled in the synapses of the brain. In fact, mice that are sleep deprived and the *sleepy* mice have patterns of phosphate alterations that overlap and look similar.[41]

Yanagisawa and colleagues have proposed that their *sleepy* mice may be a potential animal model for idiopathic hypersomnia in humans.[42] The mice didn't seem to have alterations in wakefulness or arousal—they responded normally to caffeine or the mental stimulation of being placed in a new cage. They did have an increased need for sleep, along with an exaggerated response to sleep deprivation. But do the brains of people with IH look biochemically like people who are sleep deprived, like the *sleepy* mouse brains did? Just like the situation with adenosine, this question may be difficult to answer directly.

Our picture of the molecules that make up facets of sleepiness will continue to evolve, since adenosine, various peptides, and patterns of phosphate modifications are probably not the only embodiments of sleepiness. Chronic sleep disruption leaves imprints all over the brain.[43] One example: recent research in zebrafish and mice indicates that transient DNA damage, marked by the accumulation of DNA repair proteins, is a component of sleep pressure.[44] The challenge will be to show how these findings apply to people with IH and other sleep disorders.

CHAPTER 7

MY FAVORITE MISTAKE

"Serendipity" is a category used to describe discoveries in science that occur at the intersection of chance and wisdom.

—Samantha Copeland, 2019

In the summer of 2008, while Anna Sumner was still getting back on her feet, something unexpected happened. She developed bronchitis and lost her voice, which is a problem for someone whose professional role occasionally requires her to argue in court. Most of Anna's legal work for corporate clients was conducted behind the scenes, but around this time, she was defending the owner of a home for rescued dogs against noise complaints in Gwinnett County, near Atlanta.[1]

To treat her bronchitis, Anna's internist prescribed the antibiotic clarithromycin, known commercially as Biaxin. After taking it, she couldn't sleep. After three nights of insomnia, she became frightened. "This had never happened to me before," Anna said. "I was concerned that it was some bizarre individual reaction to the medication."

She frantically called the neurologist Lynn Marie Trotti, who had become her main sleep specialist. Rye was in the United Kingdom attending a sleep research conference, but Trotti was not filling in as a one-off substitute. She had first come to Emory after finishing medical school in 2003. During her fellowship year after residency, she had worked with Rye on restless leg syndrome, saw several of his patients, and took over care for some of them. We will come back to Trotti, since she became more central to hypersomnia research at Emory over the next several years. But first: clarithromycin.

If someone who was otherwise resistant to high doses of stimulants developed insomnia, that seemed significant. In medical literature, clarithromycin had been connected with occasional reports of delirium in hospitalized elderly patients and "agitation, euphoria, and insomnia" in otherwise healthy young people with acute infections.[2] How an antibiotic produced these effects was unknown. In some cases, it might have been interfering with the metabolism of other drugs.

Rye and Trotti took what was considered a nuisance or harmful side effect, and turned it into the main effect. Clarithromycin was available outside of the practical and regulatory constraints surrounding flumazenil. It only took a short time for Rye to try clarithromycin with some of his hypersomnia patients. He prescribed it to Valerie, who texted him that she felt surprisingly energetic and that her daughter had to coax her out of the garden in the evening.

Trotti followed in Rye's footsteps a few months later. Initially, she was concerned that the antibiotic would trigger seizures or some other adverse event, such as diarrhea caused by *Clostridium difficile*, a stubborn type of bacterial infection that can colonize the intestines following antibiotic treatment.

In the laboratory, Jenkins and Garcia were inspired by Anna's experience to investigate clarithromycin's effects on GABA signaling via patch clamp. Garcia tried clarithromycin himself after a residency shift; it had some kick, he said. The pair presented their results at an October 2009 anesthesiology conference.[3] They showed that clarithromycin acts on GABA-A receptors, but in a different mode than flumazenil. Clarithromycin blocked those receptors' ionic gates, even if hypersomnia patients' CSF is absent. In Anna's case, what tipped the balance toward insomnia was probably the combination of clarithromycin and flumazenil together.

Trotti and Rye have continued to hold open the possibility that clarithromycin works through some other mechanism besides GABA receptors. At first Rye was uncertain whether clarithromycin even entered the brain, so he had CSF samples sent to an outside chemist to check. In terms of molecular size, clarithromycin was larger and clunkier than flumazenil. Was it perhaps indirectly reducing sleepiness, by acting as an anti-inflammatory agent or altering intestinal bacteria?

Trotti and Rye prescribed clarithromycin to more than fifty of their patients with refractory hypersomnia over the next four years.[4] Almost two-thirds experienced subjective improvements in sleepiness, they reported in *Journal of Psychopharmacology*. This paper is a snapshot of the Emory Sleep Center hypersomnia population before the influx triggered by media attention in 2012. Most patients had tried and rejected two or three other wake-promoting medications, such as modafinil, before being prescribed clarithromycin. More than 40 percent had a history of depression. In terms of formal diagnoses, the patients were a mix of narcolepsy without cataplexy and "primary hypersomnia"—not *idiopathic*, since they were all positive for GABA-enhancing CSF. The patch clamp assay didn't predict who was likely to respond well to clarithromycin. The only strong predictor was age; nonresponders were older.

Most of those who reported a benefit from a two-week trial of clarithromycin continued to take it. Some people did not tolerate the antibiotic because of side effects, such as abdominal cramping, diarrhea, and nausea. Commonly reported was a lingering metallic taste, sometimes described as like "a rusty sink" or "like sucking on quarters all day." Valerie, for example, said that she used clarithromycin intermittently, only when she really needed it, partly because of the side effects and also because she was concerned about growing tolerant and wanted to keep it in reserve.

Although it was readily available from pharmacies, Rye and Trotti did not consider clarithromycin an ideal solution for people with hypersomnia. Prolonged use might promote antibiotic resistance or have other harmful effects on intestinal bacteria. A Danish study of people with heart disease detected an increased risk of mortality connected with clarithromycin, prompting anxiety from Emory patients.[5]

For these reasons, Rye pushed ahead with a clinical trial of flumazenil in its sublingual form. To find commercial partners, he worked with Emory's Office of Technology Transfer, without success. The language around what Anna and others had and who the target market might be remained confusing. An OTT PowerPoint presentation from that time promoting the project was titled "Sublingual Flumazenil Treatment for Recurring Stupor."[6]

Rye and Jenkins drew up plans for a company called Somnolytics, meaning "dissolving sleep." A January 2010 application to the Georgia Research Alliance (a state government–private partnership promoting research) for $50,000 in seed money emphasized the possible advantages of flumazenil. It would be presumably nonaddictive and not subject to restrictions by the Drug Enforcement Administration, in contrast to traditional stimulants. Cephalon's modafinil was described as a model because it was first approved for narcolepsy, then more prevalent conditions such as shift work and sleep apnea.

The scale of the clinical trial was small: ten people received either flumazenil or placebo for a day. Rye intended this study as a first step, which he could use to convince investors or a larger pharmaceutical company to do business. The primary outcome, what researchers tested as a gauge of success, was reaction time measured through the psychomotor vigilance test. The results were disappointing:

no obvious differences between flumazenil and placebo for reaction time or subjective sleepiness.[7]

In a presentation he gave afterward, Rye listed several lessons learned. Possibly because there was no adjustment for Anna's small size, the doses of flumazenil in the trial may have been too low. Most participants did not reach the same level of flumazenil in their blood that Anna had. Also, participants were still using other medications, and a single day of treatment spent lying in bed was probably insufficient to assess changes in symptoms. Overall, the scale of the study, limited by financial constraints, was too small.

CAPTAIN OF THE SHIP

For the flumazenil clinical trial, Trotti was listed as the principal investigator. A condition for the Georgia Research Alliance funding was that Rye not have that role, to avoid conflict of interest with his status as a potential beneficiary of flumazenil's commercialization. For studies of clarithromycin, Trotti took the lead in seeking support from the American Academy of Sleep Medicine (AASM) Foundation and then in being principal investigator for the study. "I was captain of that ship," she said.

Trotti started out in a trainee relationship with Rye but has developed her own expertise and preferences. In comparison with him, she was more cautious and less eager for media attention. She was less likely to tell stories about her unnamed patients. Also, she was less openly critical in relations with their peers and more willing to work within organizations such as the AASM to change guidelines and policy.

Growing up in Canada and then upstate New York, Trotti had wanted to be a psychologist or family physician, having little personal experience with neurology or sleep medicine. She gained some neurology research experience over the summers while attending Rice University in Houston. In addition, she was attracted to the field because of instructors she admired at Baylor, where she attended medical school, and because neurology practice would call for her to maintain ongoing relationships with patients. "I could have gone somewhere else after residency, but I knew that it would be foolish to leave," she said. "What was happening here was too important, too unique."

Trotti has occasionally mentioned being influenced by her father, Lorne Becker, a proponent of down-to-earth family medicine. When she was in high school, she and her family moved from Toronto to Syracuse so her father

FIGURE 7.1. Lynn Marie Trotti (seated) and Prabhjyot Saini (standing).

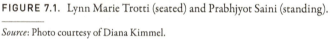

Source: Photo courtesy of Diana Kimmel.

could become chair of the family medicine department at the state university there. He has been active in the Cochrane Collaboration, an organization that organizes systematic reviews, an effort to squeeze out wishful thinking and industry spin. Becker and Trotti have collaborated on Cochrane reviews related to restless leg syndrome and, more recently, wake-promoting medications for IH.[8]

After he returned to Toronto, Becker told another author that his experience in the United States made him appreciate Canadian health care: "The overwhelming good thing about this system is never having your hands tied by the patient's financial status. It's a relief not having to fight with patients to get mammograms because they cost too much."[9] In her own practice in Atlanta, Trotti has struggled against insurance companies' restrictions on her patients' behalf. She told an audience of other doctors that her time in the sleep clinic made her worry about how many people are driving while drowsy: "There are days when I think I'll never get on the highway again."[10]

WHY YOU AND NOT ME?

After the clarithromycin episode, Anna said that she was starting to feel a bit sleepier: "She complained of mild difficulties with sustaining vigilance." As a result, modafinil was added to her daily regimen. Although modafinil had been insufficient for Anna back in 2005, the combination of flumazenil and modafinil worked well for her, and she returned to work full time. Although Anna was thereafter less of a source of drama, she continued to be a reference point for the Emory researchers.

Anna did not come into contact with others with similar sleep disorders until a couple years later. In October 2010, she attended the Narcolepsy Network national conference in Virginia, along with her parents. For people with narcolepsy in the United States, the conference was a chance to meet others who understand their struggles. In the absence of organizations focused on IH, narcolepsy organizations and support groups often seemed like the best fit for IHers. Some had been previously diagnosed with narcolepsy or told others they had "something related to narcolepsy." At the conference, Anna told her story of getting her life back. Some people who were listening to her talk had tears in their eyes.[11] "I was amazed at the community," Anna said. "Some people said, 'Finally, someone is studying this.' But others had some uncomfortable questions, like, 'What makes you so special that you got this treatment, and others won't? Why you and not me?'"

The 2010 Narcolepsy Network conference was one of the first occasions when Rye told his sleep medicine peers about what was going on at Emory. In Narcolepsy Network's newsletter, he shared an update on his research. He mentioned the flumazenil clinical study and vented about how his National Institutes of Health grant applications had been reviewed unfavorably. Rye was quoted as saying: "We have determined that greater than two thirds of patients presenting to us with one of these diagnoses are essentially making their own endogenous anesthetic, i.e. they're making their own Valium, so to speak."

ANECDOTAL ANNA

The manuscript on Anna's case encountered difficulties when it was sent to the *New England Journal of Medicine*, because reviewers associated it with the "idiopathic recurring stupor" research at the University of Bologna from more than a decade before. "One reviewer used a kind of 'straw man' argument," Rye recalled. "They said: 'You haven't explained this other thing, which happened in Italy. So I don't understand what you're telling me here.'" Another referee called the report "merely anecdotal." Rye told Anna, and she was offended at the thought of being anecdotal. To show that his patients weren't sleepy because of known drugs, Rye and the pathologist Jim Ritchie paid thousands of dollars to have urine and blood samples tested by outside forensic laboratories. They also looked for newer drugs, ones not commercially available in the 1990s. All samples were negative.

Nature Medicine rejected the manuscript without outside review. When preparing to try again with *Science Translational Medicine*, the Emory team wanted to give their submission a little more heft. They asked Anna, then back at work, to abstain from flumazenil and other medications for several days and come back to the hospital. At that point, her doctors felt like family, and she was willing to make an extra effort to help. She and Rye had been discussing a documentary about the chess player Bobby Fischer, so they watched it together, ordering takeout over a weekend in April 2012. Coming back to the sleep laboratory was not especially fun. Having to stop flumazenil made her feel grumpy and sleepy all over again. "It reminded me how debilitating this condition actually was," she said. "I felt like a pin cushion because they drew blood every hour."

Anna's alertness responses had already been tested twice, but a third time could remove lingering doubts. The aim was to test whether she had undergone a remission or if the findings were spurious, Rye told the editors. When they gave Anna flumazenil again, they measured her reaction time and EEG responses before and after. Just like four years prior, her reaction time went from slower than average to faster. This extra effort may have put the manuscript over the top.

When he was getting ready to publicize his upcoming paper, Rye did not want to confuse matters by comparing the GABA-enhancing sleepy stuff seen in Anna and others to what had been published before. He said emphatically: "Don't call it an endozepine."

MY BED WAS MY HOME

When the paper highlighting Anna's case was finally published in *Science Translational Medicine* in November 2012, it aroused interest from news organizations and also in people with sleep disorders around the world. Anna's recovery was highlighted on NBC's *Today Show*. "My bed was my home. It was everything," she told NBC. "It was like an addiction. I would hit a point in the day where I thought: 'If I do not go to sleep right now, I will literally not survive.'" A voiceover intoned, "She's able to do tasks she could never finish before." Anna was shown in the Emory sleep lab, at her Midtown Atlanta law office, raking leaves outside her house, and unloading her dishwasher.[12]

In an interview for the program, Rye said: "We've discovered that the body seems to be producing a substance that acts very much like a sedative." The unidentified sedative-like substance had been detected in thirty-one others, and the *Science Translational Medicine* paper included descriptions of several people who had responded well to flumazenil in the hospital, displaying improvements in reaction time and subjective sleepiness. Just two of this group fit then-current diagnostic criteria for idiopathic hypersomnia; others were long sleepers or had narcolepsy without cataplexy or Kleine-Levin syndrome. This group appears to include those described in the 2009 patent and NINDS grant application.

NBC pointed out that Anna was currently the only person with a sleep disorder taking flumazenil long term but that her supply could run out in a year. The alternative of clarithromycin wasn't mentioned, even though more people were taking it at the time. "There is a huge unmet need of people who identify with Anna," Rye told the *Wall Street Journal*, in an article that appeared a couple weeks after the paper's publication.[13] He added that Emory's sleep center had received a wave of calls from interested patients and other sleep clinics—ninety-five in one week. One concerned mother of a college student with IH contacted Emory after reading the *Journal* article and was not able to secure an appointment with Rye until a year later.[14] (The mom was Betsy Ashcraft, who later became a board member of the Hypersomnia Foundation.)

Again, there was confusion about what to call Anna's condition. The *Journal* called it "primary hypersomnia," while the *Today Show* said it didn't even have a name. A Fox News headline writer misled readers with " 'Sleeping Beauty' Gene Proves Beastly for Sufferers," since no specific gene was in sight.[15]

Science asked Rye's friend Cliff Saper about the Emory research, and he referred back to the Bolognese fumble with idiopathic recurring stupor. But Saper said the new paper made a strong case that alterations in GABA signaling had a role in hypersomnia. Rye had his work cut out for him: "Identify the mystery compound, figure out a faster way to detect it, and conduct a larger clinical trial to test the benefits of flumazenil."[16]

PASSING THE TORCH

It's time to introduce someone else who played a pivotal role in forming the hypersomnia community. Like Anna, she has appeared on television and in news articles as a spokesperson. For many people, her experiences may be more relatable than Anna's. Instead of elite law firms, her stories involve feeling extremely sleepy while driving her kids to soccer practice. They also include a long period when doctors thought her sleepiness was something else.

In the 1990s, Diana Kimmel was living in Rockland County, north of New York City. She worked in medical billing and had a side gig in photography. Diana began to perceive that she was "more tired than most people." She would take naps, but she felt embarrassed about it, jumping up when someone came to visit.

Doctors checked her thyroid and iron levels, sometimes her vitamin B12. They would attribute her fatigue to her hectic schedule or her two children or advise her to lose weight. A few offered her conventional stimulants, which made her feel shaky or jittery. Modafinil made her feel like she was having a heart attack. "I stopped complaining to doctors for a while, because I wasn't getting anywhere," she said.

Around 2006, Diana was preparing to move to Georgia to be closer to her parents. While she still had good insurance coverage through work, she decided to have a few tests performed to check out her persistent sleepiness. She recalled that at the time, she felt foggy and would nap during the day, while being unable to sleep at night. An MRI scan discovered a pituitary tumor, which was removed surgically in 2007. "After that was taken out, I thought everything was going to be fine," she said—but she felt worse.

During the day, she would experience recurrent sleep paralysis: she would wake up from a nap and find she couldn't move. While driving, she would lose chunks of time. She would emerge from a mental fog and not realize how she got to where she was. An endocrinologist thought her potassium was low. She bounced from doctor to doctor. Her need for sleep increased. "I would be up to fifteen, sixteen [hours]," she later told a television interviewer.[17] "I would sleep all night long. I would wake up and by nine, ten o'clock, I was back in bed."

In retrospect, Diana having a sleep disorder made sense, given the presence of a pituitary tumor and possible neurological damage associated with it. The pituitary's location next to the hypothalamus may explain sleep disturbances among patients treated for pituitary tumors.[18] It also suggests that her hypersomnia might not be strictly *idiopathic*.

On a day when Diana had to drive her son to soccer practice and it hurt simply to stay awake, she passed out. Her son had to grab the steering wheel to prevent a collision. Even then her car veered across two lanes of traffic and into a pole. She was hospitalized for five days. "That was the beginning of 'What the hell is wrong with me,'" she said.

She began to see one of Rye's colleagues at Emory. His main theory was that Diana's lapses in consciousness were seizures, a possible explanation for times when she would stutter or stare off into space. He prescribed the antiepileptic drug

Keppra (levetiracetam). "I don't think he was a bad doctor, but epilepsy is what he was used to thinking about," she said.

Another time, she was visiting a supermarket, talking to someone behind the delicatessen counter. She felt dizzy and went to the bathroom to splash water on her face. She lost consciousness and woke up in a pool of blood. After that, doctors redoubled their efforts, having her undergo ambulatory EEG to look for seizures. They tried a tilt table test to check whether her lapses of consciousness might be coming from problems with blood pressure control.

This was around when Diana had her first overnight sleep study, to look for sleep apnea. It had been a few years since her pituitary tumor had been removed. At one point, the doctor she had been seeing seemed to grow frustrated, telling her: "You are presenting like someone who is abusing drugs and alcohol."

A call came about another overnight sleep study and an MSLT. The physicians were confused about who referred her, but someone else in Emory's sleep lab—a nurse or technician—suspected that Diana might be one of Rye's kind of people. She knew that a specialty of his was restless leg syndrome, which she thought didn't fit, so she was guarded at first. In between the naps of her MSLT, she heard Rye say in the hallway of the sleep lab: "No wonder she's driving into poles." After the next nap, he asked permission to perform a lumbar puncture to get a sample of her CSF.

Rye diagnosed Diana with idiopathic hypersomnia sometime in 2011. In October of that year, she began taking clarithromycin. It gave her some digestive issues, but she stuck with it. She found out that she had GABA-enhancing activity in her cerebrospinal fluid when Trotti called her about the results. "The good news was that I had a diagnosis, a name for what this was," she said. "The bad news was—and Rye said this straight out—there is no cure, there are no approved treatments, and other doctors are generally unaware of it."

CHAPTER 8

THE ATLANTA SLEEPERS CLUB

The sleepyheads were all keenly interested in flumazenil, of course, but that's not why most of them had made the trip. They were there to meet other people with the same confusing condition. They wanted to get advice, commiserate, laugh at inside jokes—to be heard and understood.

—Virginia Hughes, "Wake No More," 2015

Diana Kimmel eventually stepped into a role that Anna had felt uncomfortable with: public advocate for IH. She said that her drive to do so emerged from conversations with Rye, in which he had explained that little information or resources were available for people with IH. In October 2013, she attended the Narcolepsy Network conference in Atlanta. She liked what she saw and wanted to create something similar for IH. What she had in mind was support groups, national conferences, and a patient advocacy organization. In short, a community. A website for what would become the Hypersomnia Foundation existed, but the rest didn't. Diana said later: "I had attended other conferences before and noticed that there was a huge feeling of unity after attending a conference together. You get to talk to somebody who actually understands how you feel, what you go through on a daily basis, and what your needs and fears are."[1]

Several people made connections at the Narcolepsy Network meeting, such as David Rye's wife, Catherine Page-Rye. During the couple's time living in

Atlanta, she had been both raising their children and regularly flying to Chicago to manage her family's residential construction businesses. After selling her stake in 2008, the idea of helping her husband launch something new was appealing. Members of her family had talked about a "Page sleepy gene" for years, she said.

At the meeting, Page-Rye asked for advice from Michael Twery, director of the National Institutes of Health's National Center on Sleep Disorders Research. He suggested starting a registry: a comprehensive database of patient information that would serve as a resource for others. "We didn't know much about how to do that at the time," Page-Rye said. "It took a while to find the right people."

Page-Rye was able to meet Diana and her supporter, the school administrator Jennifer Beard. A seed for the hypersomnia community can be traced to a discussion they had with the Ryes. Beard, who became a founding board member for the Hypersomnia Foundation, told them: "We need to treat this like it's a real disease. We need to do something like what people with other diseases do."

CONSUMED BY SLEEP

At the Narcolepsy Network meeting, the group could see there was a constituency ready to mobilize. There was some overlap with narcolepsy but differences as well. Recommendations for short scheduled naps, for example, would be inappropriate for those with hypersomnia, for whom naps extended for hours unintentionally. Advice about managing cataplexy—much discussed at narcolepsy conferences—was not relevant for people with hypersomnia.

Diana and Jennifer Beard recorded a talk of Rye's titled "What's in a Name? Understanding the Origins of the Terminologies for the Family of Hypersomnias." He laid out a critique of the Multiple Sleep Latency Test—a moneymaker for sleep labs, he suggested. A few months before, he and Trotti had published a paper on the MSLT's unreliability when patients do not have a narcolepsy type 1 diagnosis.[2] According to their data, someone with IH or narcolepsy type 2 taking the MSLT more than once was equally likely to fall into a different category the second time. Distinctions between the two diagnoses based on MSLT performance mattered little for prognosis or choosing treatment, he said.

In his talk, Rye made sly pokes at the sleep medicine establishment, evoking laughter several times from the audience.[3] For him, it was both a squirt of acid aimed at his peers and a call to action for patients, saying: "You deserve better." Rye argued that some patients had a relationship with sleep that was qualitatively different than people with narcolepsy: "They have a hard time getting up in the

FIGURE 8.1. Diana Kimmel and David Rye at the 2022 Hypersomnia Foundation conference in Charlotte, NC.

Source: Photo courtesy of Diane Powell.

morning. They need five alarm clocks. They need ice cubes in their underwear. This is a different disorder."

On the video, we can hear him working out the diagnostic dichotomy, which became a T-shirt slogan: "These people aren't seized by sleep, these people are frickin' *consumed* by sleep, OK?"

CLOSE TO THE SUN

Only a few months after the 2013 Narcolepsy Network meeting, another conference meant for people with hypersomnia took place at an Atlanta airport hotel. Before this event, it's difficult to point to a distinct hypersomnia community in the United States. It emerged from interactions between Diana and others like her, brought together by online activity over the course of the previous eighteen months. One measure of her impact was her estimate that she'd sent a thousand people to their doctors with information about clarithromycin, but it wasn't all about medications—it was about connecting people.

An extraordinary accelerant to Diana's efforts came in the form of a young man named Lloyd Johnson. Much of the work and planning for the meeting happened in conversations around the Ryes' kitchen table, with Beard and Kimmel as lead organizers. But for the conference, Lloyd became the charismatic personality in front. "We used the fact that people wanted to see Lloyd and made it an opportunity for the conference," Diana said.

Across the world from Atlanta, Lloyd had been going through his own struggles with hypersomnia in Western Australia. He had built a website called Living with Hypersomnia and had attracted a following online in several countries. After Lloyd learned about the Emory research, he embraced it and tried clarithromycin and flumazenil, with the help of doctors in Australia. However, the benefits he experienced from flumazenil, in particular, didn't last. And Lloyd's starring role in the community he helped coalesce was also temporary.

I only met Lloyd Johnson once, at the 2014 conference. Afterward I was able to make contact with him through his mother, who said that he remained incapacitated by his IH. I began to see him as an Icarus figure: someone who flew too close to the sun.

CONFIDENT FUTURE

A feature article about Lloyd by the journalist Virginia Hughes dates his difficulties with sleep to when he was twelve or thirteen, growing up in Perth.[4] He had unsuccessful surgery for what doctors thought was a hip problem, which gave him intense pains in his right leg for months. Together, Lloyd's sleep and pain issues led to him dropping out of high school.

Lloyd's leg pains later receded, and he was able to finish high school and university, studying computer science. He became a fitness enthusiast, completing a triathlon and half marathon. He rode a red motorcycle. Hughes describes a photo of Lloyd and his girlfriend "mid-stride, beaming, the epitome of carefree, healthy youth."

In university, Lloyd delved into neurolinguistic programming (NLP), a framework for psychotherapy and personal growth developed in the 1970s.[5] He started his own business as a life coach and therapy instructor, with a website called Confident Future.[6] A 2012 YouTube video shows him selling a book on neurolinguistic programming; he taught courses on NLP, hypnosis, and Time Line Therapy (a set of techniques for letting go of negative emotions from the past).[7] He maintained a busy schedule, traveling around Australia and Southeast Asia.

By his own account on Facebook, Lloyd's sluggishness returned at the beginning of 2012 with a bout of tonsillitis. His sore throat disappeared, but he felt a gradual increase in fatigue. In April, his doctor diagnosed him with Epstein-Barr virus infection, also called mononucleosis or glandular fever, and prescribed rest. Epstein-Barr virus has attracted interest as a possible trigger for hypersomnia, going back to Stanford investigators in the 1980s.[8]

Lloyd reported being diagnosed with IH sometime in 2012. According to a case report on his treatment, he first underwent an overnight sleep study that detected mild sleep apnea, but a Multiple Sleep Latency Test was not included.[9] Brain scans appeared normal. Narcolepsy and sleep apnea were ruled out because "the patient did not have sleep paralysis, snoring was not a major feature, and he did not have any symptoms suggestive of cataplexy." He tried CPAP (continuous positive air pressure), standard treatment for sleep apnea, but experienced no relief.

Lloyd told his doctors he had the full range of IH symptoms: long and undisturbed sleep at night usually lasting more than ten hours, unrefreshing naps, sleep drunkenness, cognitive problems, cold hands and feet, and heart palpitations. He couldn't drive safely or work; he was shutting down his NLP/hypnotherapy business.

Lloyd would try several standard wake-promoting medications or stimulants—including modafinil, methylphenidate, and amphetamine—resulting in temporary improvements, along with negative side effects such as high blood pressure, nausea, and loss of appetite. He was also prescribed antidepressants, which did not improve his mood. He claimed that in addition to mainstream medications, he had also tried a variety of diets and alternative medicine approaches. "Many help in the short term but the only reliable result has been a lowering of my bank balance," he wrote in October 2012, introducing himself to the Idiopathic Hypersomnia private group on Facebook.

SETTLING THE ONLINE DESERT AND CRASHING A PARTY

Before Lloyd set up his site, little oriented for nonmedical professionals was available online in English about IH. As one example, in 2001 the National Institutes of Health had set up an encyclopedia of rare diseases (rarediseases.info.nih

.gov). At the start of 2013, there were entries for sixteen diseases whose name started with the word "idiopathic," but not IH.[10] Another trusted source of medical information, the Mayo Clinic's website, did not have an entry for IH until 2014. Even then it consisted of one sentence: "Idiopathic hypersomnia is a sleep disorder in which you're excessively tired during the day, either with or without a long sleep time."[11]

The internet wasn't a complete desert for IHers. They could slink into online forums for people with narcolepsy, such as Narcolepsy Network's. The industry-supported site Talk About Sleep had discussion forums for people with idiopathic hypersomnia, but the information on the site made no distinction between IH and the symptom of excessive daytime sleepiness, affecting around 5 percent of the population.[12]

The Idiopathic Hypersomnia Facebook group appears to have been started by someone from the United Kingdom in 2007. The group's founder wrote about having seen other groups for people with narcolepsy but nothing for hypersomnia: "The meds and problems people are discussing are the same as we experience, only difference is there is more research going on into N and I've certainly found its easier for a Narcoleptic to get treatment/help/support." The group surprised its founder by increasing in size to twenty-five members by the end of 2007. By the time Lloyd showed up, there were about four hundred members. He helped start another group called Major Somnolence Disorder.

Facebook groups like Idiopathic Hypersomnia and Major Somnolence Disorder were a portal through which Rye could observe confusion over narcolepsy and hypersomnia diagnoses from the patient's point of view. During his "What's in a Name?" talk, he pulled out his phone and dipped into a discussion in the Major Somnolence Disorder group. Rye recited a woman's complaint about the ambiguous diagnoses she had received based on her MSLT, adding: "Read that Facebook group and you will see people incredibly frustrated with this."

Social media activists also drove people with IH to a 2013 Food and Drug Administration program meant for narcolepsy. They were crashing a party they were not explicitly invited to. Narcolepsy, supported by established patient advocacy groups, was the first rare disease to get attention from the FDA's Patient-Focused Drug Development initiative. The agency held a September 2013 public hearing on narcolepsy and called for additional comments.

At least thirty people who identified themselves as having IH or hypersomnia, out of 175, sent comments to the narcolepsy docket.[13] It is not possible to verify most commenters' diagnoses, and hypersomnia-specific input was not included in the FDA's final "Voice of the Patient" report. Still, these comments indicate accumulating discontent. Dean Jordheim, an American friend of Lloyd's who contributed to the Living with Hypersomnia site, wrote: "I am a long-time sufferer of severe Idiopathic Hypersomnia and because of it I lost my career and my family so I am stuck at home alone, on disability, in misery. Thus, I spend hours every day sharing information with members, talking about their experiences, and doing everything I can do [to] help them maintain a positive outlook on life in a world that does nothing but push them aside or beat them down."

LIFE BETWEEN NAPS

Around the time that Lloyd Johnson discovered the IH Facebook group, he made contact with Michelle Chadwick, a woman with IH from Queensland. Michelle was diagnosed with IH at age thirty-six, after a childhood of sleeping for long periods and never feeling like it was enough.[14] Once, she left a pot on the stove and then dozed off, nearly setting her kitchen on fire. As an adult, office jobs were difficult for Michelle because of the possibility of falling asleep at her desk. "I'm not embarrassed to say that working in an ice cream factory—where it was super cold, very brightly lit, and I was active all day—was far easier than any of the desk jobs I had," she told me.

Michelle was both a sounding board and partner for Lloyd in setting up the Living with Hypersomnia site, but she parted ways with him later. In 2013 she formed Hypersomnolence Australia, the first nonprofit organization devoted to IH, and created Idiopathic Hypersomnia Awareness Week, which continues to take place worldwide. When she contacted Australian sleep medicine organizations, she was told that IH was not an independent sleep disorder or that it was a subcategory of narcolepsy. Although French and Italian narcolepsy advocacy organizations nominally included IH, it seemed to be an afterthought. "I wouldn't have set up Hypersomnolence Australia if I didn't have to," she said.

Besides Michelle Chadwick, Lloyd also received advice from others in the United States, including a physician diagnosed with IH who was editing a Wikipedia page for the sleep disorder. Lloyd's enthusiasm and friendliness was welcomed among people who felt IH had been pushed aside or overlooked.

The Living with Hypersomnia site's first post was in October 2012. The tagline was "for those making the most of life between naps!" The topics that attracted the most comments on discussion forums were "What treatment works best for you?" "Can you still drive?" and "How do you explain IH to someone?" The site had contributors from around the world and was active for about three years, eventually including patient handouts and a peer-to-peer "Dear Doctor" letter, and a directory of more than thirty IH-sympathetic doctors with testimonials.

FANDOM AND SELF-EXPERIMENTATION

The earliest posts on the Living with Hypersomnia site do not mention the Emory research; they were mostly personal stories and humor. But soon posts included explicit instructions ("What you say to your doctor if you want to trial clarithromycin") and petitions to make flumazenil more widely available.

A couple years before, members of the IH Facebook group had noticed listings for the flumazenil and clarithromycin studies on clinicaltrials.gov, but they didn't get much attention. This time, Lloyd was head cheerleader. He urged his followers to nominate Rye for a Lifetime Achievement Award from the National Sleep Foundation.

Throughout 2013, Emory patients shared information about the Living with Hypersomnia site and the Facebook groups with their doctors. Lloyd contributed to the surge in hypersomnia-related online activity both as a social media cheerleader and by experimenting on himself. He made a series of Facebook posts and YouTube videos about his experiences.

Although he displayed bursts of energy while setting up his site, Lloyd was getting bogged down. According to Hughes's story, at one point he stopped eating and drinking for three days, leaving his bed only to stumble to the bathroom. He applied for disability, but hypersomnia was not recognized as an illness by the Australian disability office. He had trouble explaining his illness to family and friends. At the beginning of 2013, according to the case report from his doctors, he was taking lithium because he had reported suicidal thoughts.

He first tried clarithromycin, available with a prescription in Australia for its standard use as an antibiotic. He did get a temporary benefit for several days but experienced gastrointestinal side effects and a nasty taste in his mouth. After persisting more than a month, he gave up on clarithromycin.

Lloyd then discovered that a doctor based in Perth, George O'Neil, had been treating benzodiazepine withdrawal with flumazenil implants.[15] He was surprised that something that seemed so distant was only minutes away. In April 2013, he was able to walk out of O'Neil's clinic with a syringe infusing flumazenil into his abdomen.

In Australia, O'Neil was known for a history of public controversy over his enthusiasm for unlicensed naltrexone implants to treat opiate addiction.[16] Naltrexone is analogous to flumazenil in that it is the pharmacological opposite of a class of drugs that many people become dependent on; it is an opioid receptor antagonist, one lasting longer than flumazenil.

When Anna Sumner was first trying flumazenil, her doctors were concerned the drug might precipitate seizures. Paradoxically, addiction specialists in Australia and Italy have found that a low continuous dose of flumazenil can stave off anxiety and withdrawal symptoms in people who have developed benzodiazepine dependency. The mechanism is not well understood but is consistent with flumazenil acting as a very weak benzodiazepine. The risk of seizures is low, investigators in Italy have reported.[17]

Erin Kelty, a researcher at University of Western Australia working with O'Neil, speculated that in this context, flumazenil might work like naltrexone.[18] Low doses of flumazenil might reset the sensitivity of GABA receptors in brain circuits related to anxiety or withdrawal symptoms, she told me.

With an initial infusion, Lloyd felt great, and he was able to reduce his sleep from more than thirteen hours per day (including naps) to less than eight hours. Four days later, he returned for a longer-lasting implant, costing about $3,000, which was similarly effective—for a while. Kelty, O'Neil, and colleagues calculated that most of the flumazenil in an implant would run out after two weeks.

Lloyd tried additional implants and infusions, but even when he refilled the syringe every few days, the effects wore off. He was becoming tolerant to flumazenil. He found several weeks of abstinence were necessary to have its effect return. In 2014, Lloyd made Facebook posts about shortening the time needed to regain sensitivity to flumazenil by taking a sedative. In early 2015, he wrote: "When I take the flumazenil I take it by itself with no stimulants. It clears the brain fog, allows me to focus properly and my mood is back to happy/normal Lloyd."

Lloyd's online activity tailed off after this point. Because of his poor health, Lloyd stopped maintaining the Living with Hypersomnia site, according to his mother. The last several posts on Living with Hypersomnia came from a friend in the United States. In the hypersomnia community, patient advocacy has—overall—shifted away from personal websites and blogs.

BODY ON THE LINE

Several years later, opinions still differed about Lloyd Johnson's influence in the IH community in the United States. He was accused on social media of being manipulative, of hijacking the nascent IH community for his own goals, and of selling drugs such as modafinil online.[19] Some distaste for the term "sleepyhead" may linger from his enthusiasm for it.

For a time, Lloyd was close to the Ryes, staying at their house when he visited Atlanta and appearing on a "Doctors' Roundtable" podcast with them. Yet they also parted ways with him after the Living with Hypersomnia conference. In an interview, Page-Rye declined to comment on Lloyd's past activities.

Michelle Chadwick has contended that he was more motivated by personal advancement than by building the IH community. Yet Lloyd also retained some supporters. "I am one of the most pro-Lloyd people out there," said Oregon IHer David Kellogg, who credited Lloyd with saving his life with encouragement and emotional support. David recognized that Lloyd was willing to undergo a risky experiment with flumazenil implants. "He was putting his body on the line," he said.

A FREE COUNTRY

Even though flumazenil's mode of delivery was awkward and the benefits were not sustained, Lloyd's personal example was critical in pushing Rye to make the drug more widely available in the United States. Rye did not want his patients to be left behind. He said later: "I was tired of seeing people who had run out of options, whose lives seemed stuck." Several members of the hypersomnia community have said they considered traveling to Australia to try flumazenil implants themselves. Rye told Hughes: "I'm like, 'This is bullshit.' If this guy can do this in Australia and get access to this drug, there's got to be something we can do. We live in a free frickin' country."

In the United States, the first person after Anna to receive flumazenil for a chronic sleep disorder was Danielle Hulshizer, a former schoolteacher and administrator in Georgia who had been diagnosed with IH several years before. Hulshizer described herself as sleepy since childhood and recalled falling asleep in class in high school, but her family and teachers had attributed her sleepiness to her demanding schedule as a competitive figure skater.[20]

Before being referred to Rye by one of his colleagues, Hulshizer was taking large amounts of Adderall—more than 100 mg per day—to compensate for her sleepiness. That level of dosing gave her debilitating headaches and tremors. "I couldn't hold a pen steady," she said. "Putting a spoon to my mouth was difficult. I had heart palpitations."

When she went on medical leave and abstained from stimulants, Hulshizer could only stay awake for a few hours per day. She had trouble driving to the grocery store. Her husband Scot repeatedly pushed Rye for alternatives. With flumazenil, Hulshizer found she could think clearly—an effect stimulants did not have. "Everyone focuses on the sleepiness, but that's not the worst part of IH for me," she said. "It's the cognitive impairment while awake. Not being able to find the right words. I have always described it as being behind a semi-transparent wall." Before she met Rye, she had felt alone. She didn't know other people like her existed. "It was a huge revelation," she said. "When I left his office, I felt like a weight had been lifted from my shoulders."

Rye's records show he first started prescribing flumazenil for "hypersomnolence refractory to psychostimulants" around March/April 2013. He found a compounding pharmacy in Atlanta willing to process flumazenil into under-the-tongue lozenges and a skin cream. When he began contacting potential suppliers of flumazenil, one threatened to report him to the FDA. He consulted a friend at the FDA, who was reassuring but could give no official ruling.

More cautious than Rye, Trotti disagreed with his haste, and they argued about it at the sleep clinic early in 2013. Rye and Trotti had been haggling with the FDA over potential sources of flumazenil for a more extensive clinical study. She then contacted the FDA to ask for permission to prescribe flumazenil for sleep disorders in modes other than intravenous administration. She did not receive a response until November 2013, just after the Narcolepsy Network meeting. The agency indicated it would not interfere with their ability to prescribe drugs but did not provide explicit clearance.

By the end of 2013, Rye and Trotti had prescribed flumazenil to more than fifty patients with hypersomnolence and various diagnoses. Most of those who responded well had refilled the prescription, costing about $250 per month. One person had successfully negotiated with their insurance company to cover the cost. Many more were waiting for an appointment.

Patients' sources of access for flumazenil, compound pharmacies, represent a gray area within the health care system. They are regulated by states, not by the FDA. Their traditional role is to offer drugs in customized formulations that would not be available at a regular commercial pharmacy. This could mean smaller or liquid doses for pediatric use or removing certain additives for patients with allergies or sensitivities. With flumazenil, compounding pharmacies make it possible to offer the drug in a substantially different form. The FDA website states: "Compounded drugs can serve an important medical need for patients, but they do not have the same safety, quality and effectiveness assurances as approved drugs."

THE FIRST AND ONLY LIVING WITH HYPERSOMNIA CONFERENCE

When organizers first posted notices online about the 2014 Living with Hypersomnia conference online, they thought they would just attract thirty or forty people, but the number kept growing. The managers at the airport hotel had to keep assigning them a larger room. The final count was more than 150 people.

The majority was from Georgia, but a few people came from France, Canada, and Australia. Prabhjyot Saini, Rye's "right hand" in the sleep clinic, served as cameraman. A few others besides patients and supporters were there, including Hughes and a medical science liaison from Jazz Pharmaceuticals. (Jazz will play a larger role in the IH story later, but its product Xyrem was not discussed extensively at the 2014 conference.)

The proceedings started earlier than many would have liked. Not many people dozed off in their chairs, countering organizers' expectations. The organizers sold T-shirts with the apples/oranges "seized by sleep/consumed by sleep" logo. Other shirts proclaimed "Living with Hypersomnia" or "Powered with Flumazenil."

The conference was Lloyd's day in the sun—he served as jovial moderator, while Diana was on a patient panel. He kept the group chuckling with a series of jokes. Rye reprised his "What's in a Name" talk on patient evaluation and diagnosis. Trotti methodically went through data available on both conventional treatments for hypersomnia: modafinil and stimulants such as methylphenidate and amphetamines. In a second presentation on newer treatments, she posed the question: "Should you be lobbying your doctor to give you flumazenil?" Not yet, without robust efficacy or safety data. "It may turn out to be safer than amphetamines, but we just don't know right now."

Trotti laid out a four-part checklist she and Rye would use for prescribing flumazenil for hypersomnia. These were: at least two other drugs didn't work, hypersomnia interfered substantially with job or family life, hypersomnia wasn't caused by something else, and a spinal tap revealed the presence of GABA-enhancing "sleepy stuff" in the patient's cerebrospinal fluid. At that point, Trotti said she was willing to prescribe clarithromycin without seeing results from a lumbar puncture, unlike flumazenil.

She also disclosed the results of a small clinical trial of clarithromycin.[21] The study was performed from July 2010 to September 2012 with a crossover design: two weeks of either clarithromycin or control, with a week in between. Taste and gastrointestinal side effects may have made some participants aware of what they received, she said. All had "GABA-related hypersomnia," as verified by lumbar puncture. Participants could not be taking flumazenil, but they could be taking medications such as modafinil.

With clarithromycin, participants reported an average improvement of almost four points on the Epworth Sleepiness Scale (from 14 to 10, on a scale from 0 to 24), but they did not experience an improvement in reaction time, as measured by the Psychomotor Vigilance Test. Despite the hiccups, Trotti later concluded: "The benefit observed is large enough to be clinically meaningful, and is of the same magnitude or higher than that reported with modafinil in narcolepsy and shiftwork trials."

When patients asked her about current studies, Trotti answered, "We don't have *a* study," since small-scale trials with flumazenil and clarithromycin were finished. However, the supply of compounded flumazenil she and Rye had established for the 2010 study was kept open. On paper, it was an improvised "expanded access" program without a corporate sponsor.

During an energetic talk on "GABA 101," Jenkins was asked why people with hypersomnia have more GABA-enhancing activity in their spinal fluid, after a day when several people had been speculating about the role of anesthesia. Lloyd stepped in and asked how many in the audience believed that their hypersomnia was a result of anesthesia. More than a dozen hands went up, surprising Jenkins. It was an example of how organizing could aggregate patient anecdotes. An Emory paper later documented the phenomenon of sleepiness exacerbation after anesthesia, sometimes lasting weeks or months.[22]

In a symbolic move at the end of his talk, Jenkins donned an orange "Powered by Flumazenil" T-shirt. Addressing the group, he said: "Your brains are so tough—you're fighting off the anesthesiologist in your head."

BEYOND EMORY

At the end of the afternoon, Page-Rye introduced the newly formed Hypersomnia Foundation. A foundation could act as a research charity and advocate for people with IH, going beyond Emory University, she said.

The Ryes had hosted prospective donors at their house the evening before, and earlier that day he had admitted that his lab's research funds were running low, explaining delays in testing patient samples. Rye's voice choked up while he described his patients' experiences and the skepticism of other sleep specialists. "Do I think I know everything about this? No," he said. "But one thing I do know is that these medications work."

Rye had set up a simple website for the Hypersomnia Foundation at the end of 2012.[23] At its inception, the foundation appeared to be a vehicle for Rye and his colleagues at Emory. Early versions of the foundation's website featured a prominent photo of him. Initially, the three medical and scientific advisors listed were Rye, Trotti, and Jenkins.

When the foundation was formally established in 2014, all three founding officers—Jennifer Beard, Cate Murray, and Catherine Page-Rye—had some connection to Rye.[24] Jennifer Beard's link was through Diana. Murray had worked with Rye as executive director of the Restless Legs Syndrome Foundation.[25] Murray had a personal stake, since her daughter recently had been diagnosed with IH, along with other conditions. Page-Rye was named as the first CEO.[26] While Rye's links to the Hypersomnia Foundation raised eyebrows at Emory because of past conflict-of-interest controversies at the medical school, there has been no self-dealing. The foundation has provided small-scale research awards to junior researchers at Emory but not to Rye, Jenkins, or Trotti.

Patient advocacy organizations come in a wide range of sizes, depending on their specific foci.[27] In the sleep disorders world, others such as Narcolepsy Network, the Restless Legs Syndrome Foundation, and the American Sleep Apnea Association were organized earlier, at the time of rapid growth in the sleep medicine field. Such groups also vary in terms of how much funding they accept from pharmaceutical industry sources; the Hypersomnia Foundation initially did not seek industry funding but later did for specific programs.

Speakers at a 2015 conference, the first organized by the foundation, included the neurologist Isabelle Arnulf from Paris and the psychiatrist David Plante from the University of Wisconsin. Since that time, the foundation has recruited a broader stable of medical advisors, and they don't necessarily agree with all of Rye's ideas about hypersomnia. What he launched grew into something larger.

Later conferences allowed the foundation to attract several parents of young people with IH, as well as people with IH themselves, as board members. One example was Diane Powell, a licensed clinical social worker and psychotherapist who attended the 2015 Hypersomnia Foundation conference. "Until I went to one of the conferences, I didn't really get it," Powell said. Her daughter had experienced excessive sleepiness throughout her teen years, and a physician at her college campus finally sent her daughter for a sleep study. After her daughter was diagnosed with IH, she had been hungry to find out what was known about the disorder. Her first reaction upon hearing her daughter's diagnosis was relief—"Great, she'll get treatment, she'll be fine"—followed by puzzlement and frustration. Powell joined the organization's board later that year and eventually became chair and CEO in 2017.

In additional to updates on medical research, the foundation's conferences have included motivational speakers, advice on educational accommodations, and presentations from an attorney and board member who specializes in Social Security Disability Insurance. Following Twery's suggestion, the foundation established a relationship with South Dakota–based CORDS (Coordination of Rare

Diseases at Sanford) and formed a registry for idiopathic hypersomnia patients, which both academic researchers and pharmaceutical companies have tapped.

At a foundation conference in Baltimore, I attended a session for supporters of people with IH. Several spoke with regret about how they had previously viewed their loved ones' sleepy behavior as a pattern of laziness or complaining. A few supporters seemed like they were on the verge of tears. They said they could see how difficult it was to have IH, and they were proud of their partners for facing up to it.

SNOOZE CRUISERS

The Atlanta hypersomnia support group began meeting around the same time as the Living with Hypersomnia conference. Sometimes the group assembled at a casual restaurant. The group later met on Saturdays at Emory Sleep Center, which moved to a modern building with a spacious lobby in 2015. "We realized that we had to do it every month, whether we knew people were going to show up or not," Diana said. Atlanta support group members have exchanged tips about doctors, how to deal with insurance companies, and specific medications. Diana has also organized week-long support retreats on cruise ships—known as Snooze Cruises.

Many in the Atlanta group have tried various dietary supplements and low-carbohydrate or ketogenic diets. They've talked about their struggles to have friends and relatives see what they have as real. One woman discussed whether she should disclose her diagnosis to her supervisor, since she was managing well enough with a remote work schedule. Another said she had trouble relating to those she met at a Narcolepsy Network conference because everyone seemed to have a job.

Members have discussed "spoon theory," a term used in several chronic illness communities when fatigue limits daily activities.[28] Spoons refer to units of motivation and concentration that are used up as someone makes it through the day. At one meeting, a supporter of an IHer explained that he envisioned spoons like magical energy in a fantasy role-playing game, consumed upon casting spells.

Notes of levity appear when members discuss alarm clocks that flash and shake or make someone perform math calculations before they shut off. At another meeting, members couldn't stop laughing at a phone-based alarm with an opera singer proclaiming "You must get up! You must get up!"

"Hah, I could sleep through that!" someone said.

CHAPTER 9

THE STORY OF FLUMAZENIL

Although our prime intention was to create a powerful scientific tool for future studies of the function of the BZR [benzodiazepine receptor], we did quite clearly foresee a number of therapeutic applications for a BZ antagonist. The enthusiasm of the marketing department was modest at the time.

—Walter Hunkeler and Willy Haefely, 1988

David Rye was not the first to test flumazenil in the context of sleep disorders. Working decades before its repurposing for hypersomnia, the scientists who first synthesized flumazenil cited it as "a classic example of preclinical serendipitous drug discovery."[1] While searching for an improved version of Valium, they found its opposite along the way.

In the form of a little yellow pill, Valium inspired Mick Jagger's barbs in "Mother's Little Helper" and made Roche the biggest drug company in the world in the 1960s and 1970s.[2] Riding a wave of profit, the Swiss company invested in research facilities in Europe and the United States. But since Valium-related patents were anticipated to expire, Roche was headed for trouble, and competitors were already encroaching.[3] Users' difficulties with dependence and withdrawal made Valium a target for policy makers. Senator Edward Kennedy of Massachusetts held hearings in 1979 denouncing the drug. "Millions of dollars are spent each year to convince physicians to use tranquilizers for a wide variety of things—some legitimate, and some, like the stress of everyday life, not," Kennedy said then.[4] "Our message to the American people is clear. If you require a daily dose of Valium to get through each day, you are hooked and you should seek help."

The Roche of the twenty-first century became more conservative and more focused on cancer, its largest source of revenue. An official history criticized the company's scattered approach in the 1970s: "The underlying hope that someone would somehow stumble on another money-spinner like the benzodiazepines appears to have distracted people's attention from the key issues."[5]

GRIND AND BIND

Willy Haefely, the leader of Hoffmann-La Roche's central nervous system research unit, is not as celebrated as Leo Sternbach, the chemist who first synthesized Valium and other benzodiazepines. As a colleague wrote: "He had no desire to place himself in the limelight; he did not covet titles or seek power. For all his outward robustness he was a sensitive man, easily hurt."[6]

Haefely (figure 9.1) was known for his fierce intelligence, and he deserves some credit for correctly deducing that benzodiazepines exert their effects through GABA, years before the mechanism became clear.[7] Today, we have amassed a huge amount of information about neurotransmitters, drugs, and receptors. Back then, these relationships were still being teased out. More pharmaceutical companies like Roche were conducting basic neuroscience research. Scientists didn't know much about the complex variety of receptors they were dealing with, and GABA receptors' diversity did not unfold until later.

A standard "grind and bind" experiment from the period went like this. Put rat brains in a blender, breaking apart the cells, and centrifuge the mixture to obtain membranes. Add something "hot": a radioactive drug. Then put the membranes on a filter and slurp off the liquid with a vacuum. A scientist can test how tightly the drug is sticking to its receptors by checking if a surplus of the "cold" drug or a similar compound can make the radioactivity wash away.

In 1977, grind and bind was how Hanns Möhler, a pharmacologist working for Haefely in Basel, identified what were called "benzodiazepine receptors," which are now called GABA-A receptors. In this case, the radioactive probe was Valium, whose generic name is diazepam. Möhler could see that diazepam and its relatives stuck to something on cell membranes all over the brain.[8]

A year later, Möhler was using the grind-and-bind technique to screen a host of new compounds the Roche chemist Walter Hunkeler had been synthesizing. Their overall goal was to find new benzodiazepine drugs that would have

FIGURE 9.1. Willy Haefely, circa 1980.

Source: Roche Historical Archive.

antianxiety and anticonvulsant properties without being sedating, or vice versa. One of the compounds, a parent of flumazenil with the code number Ro 14-5974, caught the attention of Möhler, Haefely, and their colleagues. This compound was capable of competing with diazepam for binding, but by itself, it had none of diazepam's classic effects in live animals, such as sedation or stopping seizures. "We checked if it entered the brain, and it clearly did," Möhler said in a phone interview. "But it was inactive in all the assays we had at the time. So the idea came up: maybe it's an antagonist, not an agonist, like the benzodiazepines."

Remember what benzodiazepines do in molecular terms: they make it easier for GABA receptors to open their gates. Haefely and Möhler thought it was possible for a molecule to occupy the same slot that diazepam did but just sit there without facilitating anything: an antagonist. They had searched for benzodiazepine antagonists in the early 1970s without success. Ro 14-5974 still had some weak activity on animals' spinal cords, but Haefely and Möhler knew they were close to having something "with as little agonistic (intrinsic) activity as possible." Grind and bind made it possible to look for antagonists more systematically.

In a "crash program," Hunkeler synthesized many variations of molecules resembling the parent, and the group tested them all over again. One with the code number Ro 15-1788, later named flumazenil, came out on top. In an initial test, Möhler had a technician give enough diazepam to a mouse to make it fall asleep, then inject the proposed antagonist. The mouse promptly woke up and began running around, as if the diazepam in its body had been wiped away.

In 1981 in *Nature*, the Roche team reported that in experiments with several types of animals, flumazenil had none of the benzodiazepines' characteristic effects.[9] At the same time, it did not sensitize animals to seizures or act as a stimulant. These results agreed with the Roche researchers' emerging view of the drug as biochemically inert.[10]

When it came time for human experimentation, a physician named Walter Ziegler insisted on testing flumazenil on himself first. On the morning of April 3, 1981, at Zürich's university hospital, Ziegler was first put to sleep with the fast-acting benzodiazepine flunitrazepam (Rohypnol, now banned in the United States). Before the entire intravenous dose of flumazenil entered his body, he turned his head and opened his eyes. Ziegler was quoted in Roche's in-house magazine as saying that he felt "as if I have been torn from sleep by an alarm clock, and want to get up."[11] He was still dizzy, but within another minute, he stood up, went to lunch as usual, and then got back to work.

The first published reports on reversing benzodiazepine-induced sedation in humans with flumazenil came from a Roche-affiliated clinical research facility in Ireland.[12] These studies were mainly aimed at demonstrating flumazenil was safe enough for one-time use. In less than a minute, intravenous flumazenil could rouse heavily drugged volunteers, all healthy young men, who otherwise would not respond to vigorous shaking, having their names called, or ten seconds of alarm bells. Separately, flumazenil—in large doses, from 200 up to 600 milligrams by mouth—did not affect volunteers' performance on a series of benzodiazepine-sensitive tests, such as reaction time and the ability to quickly copy numbers on paper.[13]

THE BRAIN'S OWN ANTIANXIETY SUBSTANCE

In the laboratory, flumazenil by itself sometimes displayed weak effects. Since the drug was thought to be inert neurochemically, this was explained by the proposed existence of endogenous benzodiazepines, or "endozepines." Speculation about endogenous ligands—something that binds to a receptor, presumably fulfilling a physiological role—had begun even before flumazenil was identified.[14] Opiate receptors, the molecules that morphine and heroin act upon in the brain, were a hot topic at the time. When opiate receptors were identified in 1973, *Newsweek* quoted Solomon Snyder—one of the flashier neuroscientists of his time—as predicting prophetically: "We can assume that nature did not put opiate receptors in the brain solely to interact with narcotics."[15]

A similar set of assumptions lay behind research on benzodiazepine receptors. Scientists rationalized that something in the body must take advantage of the sites accessible to benzodiazepines because their receptors were present in all vertebrates, and evolutionary pressure kept them there.[16] A commentary in *Nature* described the unknown entity as "the brain's own anti-anxiety substance." At Roche, Willy Haefely was skeptical of arguments based on divining nature's intent but said that the possibility that an endogenous ligand bound at the same sites as benzodiazepines "cannot be dismissed and has to be examined seriously."[17]

In the 1970s, major pharmaceutical companies such as Eli Lilly and Sandoz were developing enkephalins, endogenous pain relievers that interact with opiate receptors, as potential drugs. The hope was that these molecules might be nonaddictive, or less harmful compared with opiates, because they were naturally found in the body. However, because they were peptides, they didn't last long in the body and did not become commercial drugs.[18]

NOT TOTALLY INACTIVE

The prevailing view of flumazenil as an inert antagonist began to change as others began trying it out. While not sedating like diazepam, the drug had antiepilepsy properties, which Roche and allied investigators puzzled over.[19] In healthy people, the drug could elicit "increased discontent, increased headache and sweating" and did have the potential to make people dizzy or uncomfortable in high doses.[20]

Jean-Michel Gaillard, a psychiatrist in Geneva, was one of the first to test flumazenil's effects on sleep. Beginning in 1981, Gaillard compared flumazenil's effects on how long it takes someone to fall asleep with that of caffeine. With a small number of volunteers, he found its activity profile was similar to caffeine's, although weaker. Gaillard wrote: "To our surprise, flumazenil alone was not totally inactive, but exhibited some alerting effect. When we presented these

results to Haefely, he was somewhat skeptical, because there was not the slight-
est evidence for such an effect in animal experiments.... We continued to believe
that our results were not due to chance and this little controversy became a joke
between Haefely and us."[21]

Other investigators gave flumazenil to sleep-deprived people, reasoning that
the stimulant effect Gaillard observed would then be easier to detect. Peretz Lavie,
a sleep researcher from Israel, had volunteers stay up all night and take flumazenil in the morning.[22] Large doses (60 or 120 milligrams by mouth, every four
hours) helped them resist slumber. Here, the idea of endozepines resurfaced. Since
flumazenil was supposed to be inactive, Lavie interpreted his results as revealing
that the drug was displacing a benzodiazepine-like substance that had accumulated during his charges' sleepless nights.

Other studies from Germany and Switzerland support Lavie's finding that flumazenil can dispel drowsiness resulting from lack of sleep, although ready comparisons between these studies are challenging.[23] Alertness tests were conducted
in the morning or the evening; sometimes the drug was given orally, sometimes
intravenously, and doses vary. One study from the United Kingdom found that
flumazenil could trigger panic attacks, but this effect was limited to people who
had been taking benzodiazepines for panic attacks to begin with.[24] Generally, in
healthy study participants without a sleep disorder diagnosis, the effects of flumazenil were mild.

Research on flumazenil's effects in animals was more provocative. One of the
better-publicized examples comes from Thaddeus Marczynski at the University of
Illinois. Marczynski, funded by the Air Force, gave flumazenil to older rats for
months. Even after the drug was removed from the rats' drinking water, on memory tests, the older rats performed as well as control animals that were a year
younger.[25]

Marczynski was so enthusiastic about flumazenil that he had to promise Roche
executives not to try it himself.[26] He told *Newsday*: "I rejuvenated the rats. Exploratory behavior is extraordinarily enhanced. They want to know everything." *Science* noted: "Though Marczynski's results are provocative, that's nothing new in
a field that has seen many leads fail to pan out."[27]

By the time Rye, Parker, and Jenkins became interested in flumazenil, the
stream of publications on the cognitive or sleep-related effects of the drug had
dwindled. For his part, Rye relied mainly on a study from researchers from the
United Kingdom,[28] which concluded that "the beneficial effects of flumazenil on
cognitive performance appear limited to the reversal of benzodiazepine-induced
impairment (in humans at least)."

"All those early experiments with flumazenil were interesting, but you have to look at the effect sizes," Rye said. "They weren't that big. Not like what we saw with Anna."

A ROMANTICIZED QUEST

A full account of the search for endozepines, which some scientists have called a "romanticized quest," would take up more space than available here.[29] The quest was not resolved to the satisfaction of the person who had the largest role in driving it: Erminio Costa. A formidable neuroscientist, Costa ran a National Institute of Mental Health lab located at St. Elizabeth's Hospital, across town from the intramural campus in Bethesda. He and Haefely were both friends and allies in debates over benzodiazepines' mechanism of action, since they had both inferred the benzodiazepine-GABA connection before the receptors were fully identified.

Costa, originally from Sardinia, was a lover of opera, outgoing and imaginative. He was known for asking tough questions in public, but he was devoted to supporting the careers of his trainees from all over the world.[30] Colleagues described him as having a fiery temper and as "a truly brilliant man who simply did not tolerate fools easily."[31] "He was a very excitable man," said Alessandro Guidotti, Costa's partner in running his lab for many years. "He believed in what he was doing and would strongly defend his ideas."

Starting in the late 1970s, Guidotti and Costa were looking for proteins that would interfere with diazepam at its receptor sites in the brain. They identified a protein called "diazepam binding inhibitor," or DBI, which fit criteria for an endogenous benzodiazepine receptor ligand. DBI's purification began with the extraction of rat brains with steaming acetic acid—a pungent, old-school technique that left other proteins behind in a scrambled mass. But when DBI was injected into the brains of animals, it seemed to heighten anxiety rather than dampen it, a result that Costa rationalized. "It makes sense that the endogenous system in which the benzodiazepines operate is there to create anxiety, not to limit anxiety," he told *Science News*.[32]

Researchers looked for DBI in the cerebrospinal fluid of people with various mental illnesses, finding that its levels were elevated in depression. However, the more DBI was probed, the murkier its role became. Smaller fragments of DBI sometimes had opposite effects from the full peptide. And DBI appeared to have another identity; it was the same as another protein whose function was escorting fatty acids *inside* cells, instead of modulating GABA signals outside.

Doubts accumulated. Over time, Costa changed his ideas on DBI's function, postulating that its effects on GABA receptors might be indirect. In a 1994 review, he wrote that "the term endozepine should not be used to refer to DBI, because it is now reserved for endogenous ligands of benzodiazepine recognition sites which do not include peptidic bonds in their chemical structure."[33]

Costa's competitors at the National Institutes of Health's Bethesda campus identified several other candidates for "the body's own Valium," as the hypothetical benzodiazepine-like compounds were called. The major candidates to emerge were derivatives of progesterone and other steroid hormones, known as neuroactive steroids or neurosteroids.[34]

In 1986, a team led by Steven Paul at NIH discovered that these hormones signal through GABA-A receptors. Progesterone was known to have anesthetic properties since the 1940s, and Paul's research explained why. Paul went on to lead Eli Lilly's research in the 1990s and later cofounded a company, Sage Therapeutics, focused on developing related compounds to treat disorders such as depression. In an interview, Paul said: "After all these years, except for GABA itself, the only truly physiological ligands for GABA-A receptors appear to be neuroactive steroids."

Like benzodiazepines, neurosteroids can have antianxiety, antiseizure, and sedative effects. These compounds are formed inside the brain, rather than in the gonads or adrenal glands. Some neurosteroids can make GABA receptors more active, and others can push in the opposite direction. One of the most potent and best-studied members of the neurosteroid family is allopregnanolone, whose levels rise throughout pregnancy and fall after giving birth. Allopregnanolone levels in the blood also rise rapidly following acute stress, and deficiencies in allopregnanolone have been observed in several diseases, such as depression and Alzheimer's.[35] Neurosteroids can be found in both men and women, and their fluctuations may account for some of the effects of menstrual cycles on the nervous system—including a rare form of hypersomnia known as "menstruation-related hypersomnia."[36]

However, neurosteroids bind at separate sites on GABA-A receptors from where benzodiazepines bind, according to biophysical studies (figure 9.2).[37] Whether flumazenil interferes with a molecule like allopregnanolone depends on the experimental setup.[38] We can't cleanly explain flumazenil's effects on the nervous system—and especially, its ability to wake some people up—through the displacement of neurosteroids.

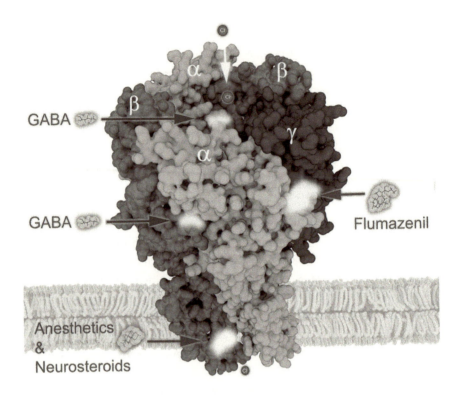

FIGURE 9.2. Diagram of GABA-A receptor assembly and ligand binding sites.

Source: Illustration by Juan Gaertner, based on data from Shaotong Zhu et al., "Structure of a Human Synaptic GABA-A Receptor," *Nature* 559 (2018): 67–72. SciencePhoto.com.

ELIMINATE THE WORMS

At Roche, management didn't jump at the prospect of a benzodiazepine antidote when flumazenil was first identified, according to Haefely and Hunkeler's account. Interest came from another arm of the company aiming to treat schistosomiasis, a waterborne parasitic infection that affects millions in developing countries. Unexpectedly, a benzodiazepine called meclonazepam could paralyze the worms that cause schistosomiasis, but the amounts needed to eliminate the worms made patients groggy.[39] Roche didn't pursue meclonazepam further, but it kept the ball rolling for flumazenil when an antidote's commercial potential was uncertain.

While flumazenil was undergoing early clinical tests, Roche had introduced midazolam as a successor to diazepam but with a more limited set of uses.

Midazolam was meant for sedation during medical procedures, not as a mass-market antianxiety pill. Around this time, Imperial Chemical Industries was developing the anesthetic propofol, which wore off more quickly than other anesthetics. With competition in mind, executives at Roche looked more favorably at flumazenil, which could make midazolam's effects dissipate in a few minutes. Accordingly, when flumazenil was introduced in European countries, it was marketed together with midazolam.[40]

Meanwhile, doctors' unfamiliarity with midazolam and its greater potency had caused problems, which first surfaced in Europe in the mid-1980s. At a congressional hearing, midazolam's allegedly inadequate labeling was blamed for overdoses and respiratory and cardiac arrest, leading to dozens of deaths.[41] Most occurred in connection with sedation during endoscopies, usually with elderly patients. These incidents made the need for a benzodiazepine antidote clearer. In 1988, flumazenil won the Prix Galien, an Academy Award for the pharmaceutical industry, in the category of neuropsychiatry. In 1991, flumazenil was approved as a benzodiazepine antidote by the U.S. Food and Drug Administration.

ENHANCING VIGILANCE AND SOBERING UP

If flumazenil's properties were intriguing, then GABA receptor "inverse agonists" represent an even stronger version of what might have been. In the 1980s, researchers at Roche's competitors, such as Schering and Ciba-Geigy, were studying flumazenil-related compounds, with some reports claiming that they displayed memory- or learning-enhancing effects in animals.[42] These so-called inverse agonists pushed GABA-A receptor sensitivity in the other direction, compared to benzodiazepines.

Since benzodiazepines seemed to embody the opposite of anxiety, neuroscientists used *their* opposites, inverse agonists, to create chemically induced models of anxiety in animals. Inverse agonists acquired an aura of danger because they could make the nervous system more sensitive to seizures. One such compound brought on "intense motor unrest" and "an impending fear of death or annihilation" when it was injected into human volunteers.[43]

A less potent but longer-lasting inverse agonist called 3-HMC attracted the attention of Wallace Mendelson, who worked with Steven Paul as director of the sleep lab at NIMH. They reported in *Science* that 3-HMC kept rats in a state of "quiet wakefulness," and it did not induce agitation, like amphetamines or caffeine did.[44] It appears that 3-HMC was never tested in human studies, although

Mendelson saw the possibilities. He told the *Washington Post*: "Here is a drug that has several effects that are opposite of the benzodiazepines. So we wondered, if the benzodiazepines put you to sleep, will this drug wake you up? . . . Maybe these chemicals would be useful drugs for people with disorders in which they get too much sleep."[45]

Around this time, Roche was investigating a relative of flumazenil called sarmazenil. The company got as far as conducting small studies with sleep-deprived volunteers, finding that the drug could reduce sleepiness and improve reaction time and performance on other cognitive tests.[46] In a 1989 report to his colleagues, Roche research manager Peter Schoch proposed that sarmazenil or a related compound could be developed as a potential "vigilance enhancer" for indications such as Alzheimer's disease. He noted that two other companies were pursuing similar compounds.[47] "It could be of clinical use in narcoleptic, depressed and/ or geriatric patients with a reduced level of arousal. . . . Evaluation of the pharmacological properties and therapeutic potentials of compounds like Ro 15-3505 [sarmazenil] has to proceed rapidly if Roche is not to lose this field to its competitors," Schoch wrote.

Sarmazenil might have become a competitor for wake-promoting medications such as modafinil, if it had progressed. But company documents reveal that Roche chief Jürgen Drews doubted that vigilance enhancement "fulfills a real medical need—Such a drug would most probably face strong regulatory resistance anyway."[48] Some volunteers reported unpleasant side effects such as dizziness, and there were hints from EEG studies that seizure risk might be elevated. Sarmazenil was registered for reversing benzodiazepine sedation, but for veterinary use only.

Another tweak of flumazenil's structure led to an unexpected effect. Roche scientists attached a light-sensitive chemical group to flumazenil, with the aim of creating a probe that could label benzodiazepine receptors in cells. They discovered that the tweaked version, called Ro 15-4513, could reverse the behavioral effects of alcohol in animals—something flumazenil can't do.[49] Ro 15-4513's properties excited some on Haefely's team. However, he thought that wide availability of an antidrunkenness pill might encourage risk taking and promote alcohol consumption, and the company decided not to develop it further.[50] Other researchers later used Ro 15-4513 as a probe to dissect how alcohol exerted its effects on GABA receptors and the nervous system.

After the early 1990s, Roche management paid little attention to flumazenil, sarmazenil, or Ro 15-4513, according to research steering committee meeting minutes in the company's archive. Several strands of research on flumazenil were clipped. For example, epilepsy studies were phased out because "the results accumulated suggest that flumazenil exhibits too weak an antiepileptic activity." Möhler returned to academic research at the University of Zürich, and Roche's central nervous system unit underwent reorganization.

During the 1990s, gene cloning techniques advanced, revealing a forest of genes encoding different GABA-A receptor subunits. Some varieties were present only in particular regions of the brain. The dominant form was sensitive to benzodiazepines, but others were not. Their receptors' distribution in the brain was more complicated than anticipated, slowing efforts to devise drugs that were specific to one receptor or only a few. "It was a complete shock," John Kemp, Haefely's successor as head of central nervous system research, told *Forbes*.[51]

Haefely's original goal of finding alternatives to Valium was achieved by competitors, through the development of drugs such as Xanax and Ambien. His last paper mentioning flumazenil, published two months after his death in 1993, stated that the drug "exhibits virtually zero intrinsic efficacy."

THE PATIENT OPENED HER EYES

A separate line of research on flumazenil in people with hepatic encephalopathy unfolded in the 1980s and 1990s. Roche sponsored some of the studies but didn't push for them especially hard. This use for flumazenil emerged unexpectedly, through risk taking in the hospital. In comparison with Anna's experience, there was less precedent, and the patients were more vulnerable.

Hepatic encephalopathy can develop as a result of alcoholic cirrhosis, viral hepatitis, acute poisoning, or surgery that bypasses the liver. Alcohol consumption, gastrointestinal bleeding, or having to digest a large amount of protein can trigger an episode. Sometimes the symptoms—sleepiness, agitation, erratic behavior, memory problems, tremors—creep up slowly or fluctuate; people with liver damage can be unaware they have it. At its most extreme, hepatic encephalopathy results in an unwakeable stupor or coma.

In the early 1980s, it was known that people with liver disease were more sensitive to benzodiazepines, and hepatologists were exploring various explanations.[52] Perhaps more GABA, produced by gut bacteria, was leaking into the nervous system, or maybe GABA receptors in the brain were becoming more sensitive? Research on rats with acute liver failure was suggestive enough for some clinicians to try flumazenil in humans with liver diseases.[53]

Giuseppe Scollo-Lavizzari, a neurologist in Basel, was the first to test flumazenil in a hospital's emergency department on people with benzodiazepine overdoses.[54] Encouraged by his experiences, Scollo-Lavizzari tried again, this time with a twenty-five-year-old woman who was infected with hepatitis B as a result of heroin addiction.[55] Her liver disease had left her in a coma, motionless and unresponsive to painful stimuli. She had not been given benzodiazepines, he noted in a 1985 letter to *Lancet*.

After intravenous flumazenil, "the patient opened her eyes, reacted to verbal commands, and moved spontaneously and in response to painful stimuli, but she did not speak." Flumazenil's effects were reproducible, but in contrast, the opiate antagonist naloxone did not affect the twenty-five-year-old, who died two weeks later. Before his experiment, Scollo-Lavizzari had asked permission from the head of the hospital's emergency department, but regulators might not look kindly on the venture today, he told me. Scollo-Lavizzari was one of David Rye's forebears as a flumazenil enthusiast. He also tested flumazenil as an anticonvulsant and in people who were intoxicated by alcohol. In addition, he obtained a patent for flumazenil in the treatment of stroke.[56]

In the same issue of *Lancet*, doctors in Zürich reported similar results with four cirrhotic patients. One of the Zürich authors was the self-experimenter and Roche employee Walter Ziegler. The *Lancet* letters led to other small-scale trials of flumazenil in people with hepatic encephalopathy. Flumazenil does not resolve underlying liver damage, so why go through the trouble and risk? It provides symptomatic relief, makes managing the patient's care easier, and gives time for other interventions.

One spectacular case report came from Newcastle, England, where a forty-one-year-old woman had been drinking a bottle of vodka daily for several weeks. Her pupils reacted to light, but she would only move her limbs in response to deep pain. In response to intravenous flumazenil, "the patient wiped her eyes and mouth, exclaimed that she was 'starving,' and was verbally abusive to the nurse." A minute later, she could answer questions and focus visually. The woman stayed awake for about two hours. Flumazenil woke her up a second time the next day. Her condition gradually improved, and she was able to leave the hospital twenty-five days later.[57]

In Vienna, using flumazenil, physicians were able to stave off coma for almost two years in a woman who had her liver bypassed after gallstone surgery.[58] Two decades before Anna Sumner's experience, this was a rare example of a patient chronically treated with flumazenil. The woman had been in and out of the hospital several times, where she had been treated with lactulose enemas and antibiotics. Her long treatment and its twists and turns make it unlikely that she was somehow deceiving her doctors.

The Viennese woman was able to avoid coma in two stretches in 1985 and 1986, interrupted by a serious intestinal infection. While taking flumazenil, she did not need to avoid large amounts of dietary protein, which would normally lead to encephalopathy. One drawback was that she consistently became anxious for about half an hour after taking the drug. She was taking 50 mg per day—more than Anna. The Viennese doctors concluded: "In contrast to when she was not taking the drug, while receiving flumazenil she was able to lead a normal life."

STORM RUNOFF

Despite some intriguing parallels, we can only learn so much about idiopathic hypersomnia from looking at people with hepatic encephalopathy, who are seriously ill and have a high risk of mortality. Given how the liver cleanses the body, studying liver dysfunction is like analyzing agricultural runoff or sewer water after a storm. High levels of ammonia are a problem, but many other metabolites build up as well. Intestinal barriers break down, allowing bacteria and their waste products access to the nervous system via a "leaky gut." How should researchers pick out which toxin is the most important?

The reports of patients with liver failure waking in response to flumazenil presented the same mystery as with Anna. They were not known to have any benzodiazepines in their systems, so why did flumazenil wake them up? One proposed explanation was that endogenous benzodiazepines were clogging up the brain's GABA receptors, and flumazenil displaced them. Researchers at NIMH led by Phil Skolnick, a colleague of Paul's, obtained brain tissue samples from people who had died of acute liver failure after acetaminophen overdose. None of the deceased had received benzodiazepines while in the hospital, the researchers reported in the *New England Journal of Medicine*. Analyzing liver and brain samples, the authors used mass spectrometry to show that one of the peaks represented diazepam, although not all of the peaks were identified.[59] Skolnick told the press: "This is the first evidence that benzodiazepines found naturally in the brain play a role in illness. This study provides a rational basis for a cause and a cure for hepatic comas."[60]

Skolnick's 1991 paper represents a peak of enthusiasm for flumazenil's use in hepatic encephalopathy. His colleagues later found hints from animal models of liver failure that intestinal bacteria could be producing the benzodiazepine-like compounds, but they were not able to identify the specific chemicals responsible.[61]

As research progressed, a less intriguing explanation loomed larger: hepatic encephalopathy patients' previous medication intake. When liver function is compromised, benzodiazepines last longer in the body because the drugs are normally broken down by the liver. To know that something generated *within* the body was causing patients' stupors, doctors needed to make sure that the patients had not previously been given synthetic benzodiazepines. In earlier studies, adequate screening tests were not performed.

The Montreal-based hepatologist Roger Butterworth supervised the first randomized, double-blind trial of flumazenil in hepatic encephalopathy. His study, published in 1994, found that a minority (40 percent) of patients in hepatic coma displayed clinical improvement in response to flumazenil.[62] A 1998 follow-up concluded: "These findings do not support a role for 'endogenous' benzodiazepines in the pathogenesis of HE in chronic liver disease, but suggest that pharmaceutic benzodiazepines administered to cirrhotic patients as sedatives or as part of endoscopic work-up could have contributed to the neurological impairment in some patients."[63]

Early studies, conducted without placebo controls, claimed that a majority of patients responded favorably to flumazenil. The numbers fell as more rigorous studies were performed, in which thorny issues emerged. Was it better to enroll comatose patients or those who were still wakeable? Patients with acute liver failure or chronic liver disease? If someone became encephalopathic because of gastrointestinal bleeding, maybe they would have recovered anyway, once the bleeding was controlled.

The largest study of this type recruited more than five hundred participants with cirrhosis in Italy and found neurological improvements in less than 20 percent of those treated with flumazenil. Others obtained similar results, and the hepatology field began to drift away from flumazenil.[64] Butterworth later reexamined the issue and proposed that neurosteroids were likely to be playing a major role in amplifying GABA signals in hepatic encephalopathy, rather than endogenous benzodiazepines.[65]

A CONCLUDING NOTE ABOUT ENDOZEPINES

Endozepines' relevance for sleep disorders remains tenuous. Diazepam binding inhibitor, the protein originally identified by Costa and Guidotti, does loosely fit the information we have about the hypersomnia somnogen, the GABA-enhancing substance in patients' spinal fluid. Peptide fragments of DBI are around the right size, and it binds to GABA-A receptors in a flumazenil-sensitive mode.

When Rye raised money in 2014 to look for the unknown somnogen (see next chapter), testing DBI fragments was the first task on his list.[66] Rye and Jenkins were encouraged when epilepsy researchers at Stanford found that DBI was capable of acting analogously to diazepam within the reticular thalamic nucleus: a region of the brain with a central role in sleep oscillations.[67] The Stanford findings confirmed the role that Costa had envisioned for DBI years before. But when Jenkins obtained some DBI peptide, it was inactive in his patch clamp experiments, he said. Based on that negative result, DBI does not account for the GABA-enhancing activity of CSF samples from hypersomnia patients.[68]

Neurosteroids could also be candidates for the hypersomnia somnogen. At first glance, they don't fit the detectives' clues from chapter 6. They are not peptides, so peptide-chewing enzymes should have kept their activity intact—and they don't bind at the same sites on GABA-A receptors as benzodiazepines. In her thesis research, Jenkins's graduate student Olivia Moody did find that hypersomnia CSF samples, like neurosteroids, can modulate the activity of GABA-A receptors that are benzodiazepine insensitive.[69] Rye and Jenkins said they had tested hypersomnia CSF samples for neurosteroids, but that her findings made them want to take a second look.

For a time, Rye was having hypersomnia patients perform a visual test called critical flicker fusion, which has been used to identify people with covert hepatic encephalopathy.[70] Benzodiazepine-like compounds synthesized by intestinal bacteria might account for some of the wake-promoting effects of the antibiotic clarithromycin. In future research, potential contributions of both neurosteroids and the intestinal microbiome to idiopathic hypersomnia should be examined more thoroughly.

CHAPTER 10

WEIRD DRUGS

Any consideration of the development of psychopharmacology makes it quite clear that good marketing of such ideas can capture a field, either before the evidence is in on an issue or even in the face of considerable contradictory evidence. Ideas, such as the dopamine hypothesis of schizophrenia or the amine hypotheses of depression, have functioned very much as brand names.

—David Healy, *The Antidepressant Era*, 1997

After the 2014 Living with Hypersomnia conference, identifying the somnogen became a rallying cry for fundraising. A development officer at Emory's medical school cooperated with a donor from Colorado to create a hypersomniaresearch.org webpage. The page, promoted on Facebook groups such as Idiopathic Hypersomnia and Major Somnolence Disorder, included photos of Jenkins's patch clamp rigs. Rye was quoted as saying: "I believe that discovering the mystery somnogen holds the key to curing hypersomnia once and for all."[1]

Online enthusiasm was strong. Someone posted a note on one of the Facebook groups saying, "I TESTED POSITIVE FOR SLEEPY JUICE!!! Never have I been so excited for a positive test result," and received forty-six comments in response. The fundraising page set out the goal of gathering about $70,000 for hypersomnia research. In a few months, that goal was fulfilled, with more than sixty donations contributing.

At that point, Rye's lab was running low on money. He called in a favor from Garcia, who had established his own lab with funding from the Department of Veterans Affairs. Garcia temporarily paid the salary of one of Rye's researchers. During this period, Rye was able to manage with additional support from a university neuroscience initiative and from private donors, such as Anna's father. In mid-2015, Anna lent her support by asking her wedding guests to give money to Emory, directed to Rye's lab. Her wedding to fellow Atlanta lawyer Nick Pieschel demonstrated her recovery and her confidence in the future.

With a high-profile publication in hand, Rye had applied to the National Institute of Neurological Disorders and Stroke (NINDS) for a large grant in early 2014. The application was well received, and after revisions, the grant was to start in August 2015. It resolved the funding problem, providing almost $400,000 per year in direct costs.[2] Jenkins estimated that up to that point, he and Rye together had submitted at least twenty unsuccessful applications since 2009.

With all the new funding, the sleepy juice idea was supposed to be tested in a more comprehensive way at Emory. Jenkins or one of his trainees would compare GABA-enhancing activity in CSF from patients with "primary hypersomnia" versus people with sleep apnea as well as nonsleepy controls. The elusive somnogen might be identified, with the help of mass spectrometry and proteomics, which had advanced enough to be able to identify minute amounts of substances in CSF. The hypersomnia team began working with biochemists at an Emory facility meant to support Alzheimer's disease research.

The scarcity of CSF, the abundance of other proteins in CSF, and the slow, low-capacity nature of patch clamping were all still obstacles. But according to preliminary data in the NINDS grant application, patient CSF samples induced sleep when directly injected into rats' brains. Potential somnogen hits were emerging from the proteomics collaboration. Another aspect of hypersomnia would be investigated by Trotti, who had obtained a smaller grant from NINDS for a brain imaging study.[3] "If our hypotheses prove correct, they would call for a sea change in scientific thought and clinical practice that has considered the brain's wake promoting regions as the principal arbiters of pathological hypersomnolence," Rye wrote.

RETROSPECTIVE ANALYSIS

In an August 2014 thank-you letter to donors, Rye noted that close to one hundred sleep disorders patients had received flumazenil in its unconventional form.

He and colleagues at Emory's sleep center were evaluating several new patients every week. In addition, doctors outside Emory were beginning to prescribe it.

At the beginning of 2015, Trotti and Rye reviewed records of 153 patients with treatment-refractory hypersomnolence, to whom they had prescribed flumazenil. People who met the MSLT requirements for IH or narcolepsy type 2 comprised less than half of the group—a larger percentage either had sleep apnea or took longer to fall asleep than eight minutes on an MSLT. On average, the entire group reported habitual sleep of seventy hours per week and had tried more than four other medications already.

This was not a randomized controlled clinical trial. Rather, it was a retrospective analysis, with patients choosing the dose, mode of delivery (lozenges or skin cream or both), and whether to stop. According to this chart review, flumazenil helped about 60 percent of the group feel more awake.[4] A smaller number (39 percent) stuck with the drug long term. A cost of hundreds of dollars per month, with few people getting reimbursed through insurance, was another reason why some people stopped. The most common side effects were dizziness and anxiety. There were a few more serious adverse events, such as a transient ischemic attack in a patient with a history of atrial fibrillation, but determining if flumazenil was the cause was not possible.

Flumazenil responders reported a large average drop, from 15 to 10, on the subjective Epworth Sleepiness Scale. Whether that corresponded to fewer hours of sleep or other measures of alertness during the day, the Emory authors could not say. In this group, testing positive for CSF "sleepy stuff" was a prerequisite. Flumazenil responders didn't have more GABA-enhancing activity than nonresponders, but one predictor of flumazenil response was the symptom of sleep inertia.

The Hypersomnia Foundation's website acknowledged where flumazenil could be obtained and what it was prescribed for. Rye said he occasionally asked pharmacists to curb efforts to signal its availability online, because of concern about "off-label promotion" (discussed in chapters 14 and 15). There was no shortage of people waiting to see Rye and Trotti. But the painstaking patch clamp test had become a bottleneck for the process of diagnosis. Some people waited months for their results or never received them at all. Jenkins said: "They must be horribly frustrated with us. But this isn't a diagnostic lab, it's a research lab."

When Trotti presented the results of the retrospective analysis at a 2016 sleep research conference, she noted that she and Rye had gained enough confidence in flumazenil to stop requiring lumbar puncture and GABA activity measurements before a prescription. "Prospective controlled studies of flumazenil for treatment of hypersomnolence are certainly needed, but in the absence of those data there's at least a rationale for people who are severely affected and have nowhere else to turn to consider flumazenil," she said at the conference.[5]

The overall conclusion was that flumazenil was another option for people with hypersomnolence who had found other drugs unsatisfactory. It wasn't clearly better than drugs that acted by other mechanisms. Many people reported contrasting results with lozenges versus skin cream. It was difficult to determine whether someone didn't respond to flumazenil because it didn't work or because their body wasn't really getting enough of the drug.

Flumazenil could perhaps be formulated into a skin patch, similarly to methylphenidate or the Parkinson's medication rotigotine. But for something like that, a pharmaceutical company would need to invest money in developing a delivery system. As it was, only a few people with hypersomnia were able to try a continuous infusion, Australian style. Diana Kimmel eventually did. "It felt amazing—but it was awkward to move around," she said.

Despite frustrations with flumazenil, enough people in this niche community depended on it that supply problems caused disruption. In mid-2015, the Atlanta compounding pharmacy Rye had engaged was temporarily unable to fill new or renewing prescriptions. The pharmacy had been receiving flumazenil from a company in Canada that imported it from China, and changes in pharmaceutical import regulations, along with purity issues, caused a delay. Members of hypersomnia interest groups posted frequent anxiety-tinged updates on Facebook until the shortage was resolved.

SUCCESS STORIES

In the summer of 2015, Romy, a woman from the French-speaking part of Switzerland, came to see Rye. She was his first hypersomnia patient from Europe. It was an indicator of how far word had spread, since she had learned about his research through a Facebook group.

Romy recalled that when she was in school, she slept for twelve hours per day. She would ask her teachers for permission to go to the bathroom—and used the opportunity to nap. She found a job in fashion design but had to stop working because sleepiness was interfering too much.

She saw several physicians, who mainly assumed that she was depressed. She was told: "Everybody is tired—you are just lazy." Blood tests for metabolic disorders were negative. In early 2015, she visited the university hospital in Lausanne, where she finally got a confirmation that her long sleep times were "neurological, not psychological." At a sleep lab, she dozed overnight for fourteen hours straight. Standard medications were unsatisfactory or ineffective. She slept for twenty hours per day after becoming tolerant to methylphenidate. "It feels like this is the last chance," she said when she came to Atlanta. "After seeing several sleep neurologists in Switzerland and trying all the different treatments available, I decided to cross the ocean to meet this professor who everyone was talking about on social media groups."

Contacted after she returned home, Romy said that the flumazenil skin cream made her feel much better. The first week was "magical," although the effect gradually decreased, and she perceived little benefit from flumazenil lozenges. Together with flumazenil, Adderall gave Romy a window of time when she felt safe enough to drive, but in the evening, she would have a headache and stomach pain. Over the next few years, Romy continued to visit Atlanta to see Rye and update her flumazenil prescription, sometimes staying with a member of the Atlanta hypersomnia support group. "It's still hard to do things, but it [flumazenil] did help," she said in 2018. "But not enough to feel like everybody else."

That same year, Anna's brother James credited a higher dose of flumazenil, combined with conventional medications, with making it possible for him to get his dream job: animator at Industrial Light and Magic in San Francisco. In competitive fields like video game design and animation, James needed to be able to work hard at crunch time. After years of feeling stagnant, he was more productive.

There were other celebrated success stories, such as Sigurjon, a young man from Iceland who came to see Rye in 2018. Formerly active and athletic, Sigurjon dropped out of college and lost touch with friends because of his ravenous need for sleep. His hypersomnolence seemed to vary with the strong seasonal light variations in Iceland; he slept almost ten days straight in the winter of 2017, and his MSLT results varied with the seasons as well. Standard medications such as modafinil and methylphendiate could keep him physically awake, but he still felt worn out inside.

Rye had Sigurjon try flumazenil for the first time on camera, as part of a Georgia Public Broadcasting television feature on hypersomnia, with both his family and Anna watching. Sigurjon's response sounded like Anna's years ago: "It's like my eyes are being lifted up . . . it feels really good. I don't even remember feeling like this." With flumazenil skin cream, he was able to wake up spontaneously for the first time in years, without the aid of an alarm clock. Rye visited Sigurjon in Iceland afterward and reported that he was working at his father's construction company and was still taking flumazenil.

THE ZOMBIE APOCALYPSE

At the same time, plenty of people with IH have tried flumazenil without finding it a satisfactory solution. One of them was Meghan Mallare, who exemplified several elements of the IH experience. Meghan was referred to Rye in 2014, after having difficulty managing her sleep schedule at a university in North Carolina. She had started feeling excessive sleepiness in high school, falling asleep in class. Diagnosed with inattentive ADHD, she started off with the conventional stimulant methylphenidate, which her mother would wake her up to deliver at four in the morning. "Stimulants got me through high school, and then by college, I could take a higher dose, and I could just sleep through it," Meghan said. "Sometimes they kind of work, but not really."

The greediness of her sleep schedule followed her to university, where Meghan set up special alarm clocks that would shake her bed in the morning. Her roommates called it the "zombie apocalypse" and called campus security a few times because they were unable to wake her. She went to see a psychiatrist through student health services, but the dominant assumption was "you're a teenager, you're depressed," she said. The psychiatrist was willing to consider narcolepsy after she described a possible hypnogogic hallucination; during class, she saw herself at a desk floating in the middle of the ocean. The psychiatrist contacted Stanford, and somebody there recommended Rye.

In a thorough examination, Rye measured her levels of iron, vitamin B12, and thyroid hormones. On an MSLT, she fell asleep in an average of 2.8 minutes and only entered REM once—clearly meeting ICSD criteria for idiopathic hypersomnia. Afterward, she returned to university with a letter from Rye to show her instructors, who laughed at the letter. She did not receive any disability accommodations, and Meghan eventually withdrew from university after sophomore year, feeling defeated. "They did not believe it [IH] was a real thing," she said. "They thought it was a code name for partying too much."

Through Rye, Meghan was able to try both of the GABA antagonist medications flumazenil and clarithromycin, with mixed results. "Flumazenil made me feel dizzy and almost drunk and didn't help with the tiredness," she said. In contrast, clarithromycin was "amazing" and provided a definite improvement in sleepiness, but the side effects were too strong.

When Meghan came to Emory as part of a grand rounds presentation, she said she doesn't drive long distances and only drives short distances in the afternoons. By necessity, she worked part-time at a doctor's office, only in the afternoons, since she was rarely able to wake up before noon. Her dog was trained to wake her up, and friends had a key to get in, just in case.

Meghan's case was noteworthy because she exhibited all the aspects of the classic long-sleep form of IH, including severe sleep inertia. She recalled being clumsy in the mornings and "confused and agitated" when others try to wake her up. Her later MSLT results also demonstrated the test's limitations. A year after her initial diagnosis, she retook the test and entered REM in all five naps, but falling asleep slower: 7.9 minutes. In 2016, her average sleep latency was 11.4 minutes, which would be considered normal—but during this period, the amount of time she spent sleeping increased from ninety up to more than one hundred hours per week.

CHALLENGE FROM FRANCE

In 2016, the French sleep neurologist Yves Dauvilliers and colleagues published an *Annals of Neurology* paper saying they had not been able to replicate the Emory results.[6] They observed no differences in GABA-enhancing activity between CSF samples from people with IH, narcolepsy, and controls. They argued that the findings did not support prescribing GABA antagonists such as flumazenil and clarithromycin to IH patients.

Dauvilliers was the respected leader of a sleep disorders center in Montpellier and lead author of several studies on narcolepsy medications. A few years before, his group had brought down another theory about hypersomnia, involving levels of the neurotransmitter histamine in CSF.[7] He was a foe of what he called "circular reasoning," or making assumptions that ensure the result one is supposedly testing.

Several factors could have explained discrepancies between the French and American research. The Montpellier group was using eggs from *Xenopus* toads as vehicles for GABA-A receptors, rather than the human kidney cells used by Jenkins's lab. They were delivering puffs of CSF onto the toad eggs rather than bathing cells by continuous flow. Also, the patient populations were different; the Montpellier authors emphasized recruiting long sleepers. The majority had slept for more than sixteen hours total in an extended overnight protocol. In comparison, Emory patients were less uniform in their diagnoses.

When the French paper was published, Jenkins was annoyed, saying that the Montpellier group had made a mistake in calibrating their assays. A ceiling effect imposed by the wrong GABA concentration was obscuring a modulating effect of the CSF, he said. "Basing therapeutic decisions on results of an in vitro assay that is methodologically flawed is a disservice to IH patients," Jenkins and his colleagues wrote in a rebuttal to Dauvilliers's *Annals of Neurology* paper.[8]

The rebuttal contained data from Jenkins's graduate student Olivia Moody and included results from a laboratory at the University of Queensland that had used a different patch clamp technique on the same CSF samples. As a bonus, the Emory group added testimonials from colleagues in Florida and Minnesota who had begun prescribing flumazenil to their patients starting in 2014.

Around this time, Rye told me he felt like he was on a "clinical treadmill," unable to devote as much dedicated time to research as he had earlier in his career. And while the number of patients who had tried flumazenil had climbed above three hundred by mid-2016, there wasn't much new biochemistry to throw back at the French. The identity of the sleepy juice factor in patients' CSF remained undetermined.

ALTERNATIVE MECHANISMS

Other mechanisms could explain the sleep-dispelling action of flumazenil, and they do not require invoking the existence of an endogenous benzodiazepine-like substance. Instead of displacing something from GABA-A receptors, flumazenil may be weakly signaling through them or reducing the receptors' abundance on brain cells' surfaces. These changes would take hours to occur, rather than minutes—an extension of changes that occur all the time. In animals, sleep deprivation causes an increased level of GABA-A receptor expression on hypocretin-producing and other midbrain neurons.[9] A subset of GABA-A receptor levels also change after acute alcohol consumption.[10]

When asked about this possibility, Rye referred to patients such as Anna and Sigurjon, who noticed the wakening influence of flumazenil quickly. While others also say they felt flumazenil's effects immediately, not everyone has the same experience. A few reported feeling dizzy or sleepy right after taking it. Some people said they felt its effect the next morning, after taking their first dose in the evening. For them, flumazenil seems to have a stronger effect on sleep inertia, the transition out of sleep, than on overall daytime alertness.

"I noticed that it basically cured my sleep inertia in the morning when I put it on before bed," said Amy Desmarais, who was diagnosed with IH in 2015 after an initial diagnosis of "probable narcolepsy" during graduate school a few years before. "I was too afraid to stop taking stimulants during the day and didn't really notice a difference during the day, so decided to only use it at night when I knew it definitely helped." She said: "I still don't like waking up in the morning, but then again I haven't met many people who do. But the difference is that when I put flumazenil on, I *can* wake up in the morning. My alarm goes off and I still want to snooze, but the fact is, I hear it and I wake up. Whereas before flumazenil, I would sleep through anything and everything."[11] Other IHers have reported that for them, flumazenil took days or a week to kick in. They gradually felt less sleep inertia and were able to reduce the number of hours of sleep they needed per week, but its wakening effect wasn't as dramatic as what others experienced.

For support for this alternative mechanism, we can look to researchers studying how flumazenil could be used to treat addiction and withdrawal. In rats exposed to chronic methamphetamine, levels of a benzodiazepine-insensitive form of the GABA-A receptor (alpha 4) increase in the hippocampus.[12] Those levels stay elevated when the rats are in withdrawal. Flumazenil can push the levels of the alpha 4 form back down, reducing withdrawal-related jumpiness in the rats. The presence of flumazenil appears to drive neurons to recycle GABA-A receptors more quickly.[13] In addition, researchers in Italy have observed that neurons respond strongly to flumazenil after the cells are exposed to alcohol. That responsiveness goes along with elevated levels of the alpha 4 GABA-A receptor.[14] Adjustments in GABA-A receptor expression are part of compensatory changes in neural sensitivity that occur during addiction and withdrawal for several drugs.

Addiction treatment is one area of medicine where a limited number of physicians have gained experience with flumazenil comparable to Rye's and Trotti's. Flumazenil found its way into a flashy addiction treatment called the Prometa protocol, which was promoted in Hollywood around the time that Anna was going through her struggles.[15] The Prometa protocol failed in a rigorous clinical trial,[16] but clinicians at the University of Verona have reported success using continuous low-dose flumazenil in treating benzodiazepine withdrawal.[17] The leader of the Verona group, Fabio Lugoboni, has published commentaries with titles such as "What Is Stopping Us from Using Flumazenil?"[18] Perhaps the sleep research and addiction medicine communities could learn more from each other, since there is no licensed treatment for benzodiazepine dependency or addiction.

A THIRD GABA OPTION

In 2014, just after the Living with Hypersomnia meeting, Rye became aware of the ongoing development of another drug acting on GABA that deserved consideration, given the limitations of flumazenil and clarithromycin. He was speaking at a sleep research conference in Texas, where he met the Stanford neuroscientist Craig Heller.[19]

Heller was part of a group investigating GABA receptor antagonists for another purpose: as cognitive enhancers for people with Down syndrome. In mouse models, the Stanford researchers had observed benefits from both flumazenil and a related drug, pentylenetetrazol, or PTZ. Those benefits were on *memory*, not sleep.

Some of the Stanford group had formed a company, Balance Therapeutics, to commercialize PTZ. When Rye met Heller, Balance had already started a clinical trial in Australia in people with Down syndrome, and company executives were looking around for other possibilities. Heller introduced Rye to Balance executives, and the company was willing to expand into sleep disorders. Balance thus became the first pharmaceutical company to sponsor clinical trials designed for people with IH. It demonstrated Diana Kimmel's and Jennifer Beard's idea from a few years before: if people with IH organized themselves, industry partners will make themselves available.

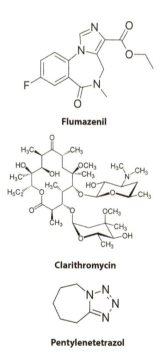

Flumazenil

Clarithromycin

Pentylenetetrazol

FIGURE 10.1. Chemical structures of flumazenil, clarithromycin, and pentylenetetrazol.

While the development was encouraging, Balance was a small company, with just a handful of employees and limited resources. Balance licensed the Parker-Rye-Jenkins flumazenil patent from Emory University but did not organize clinical trials for other drugs besides PTZ, such as flumazenil.

In addition, PTZ differs from flumazenil in a critical way. It doesn't act on the benzodiazepine site—it directly blocks GABA-A receptor function, and its action does not depend on the presence of a somnogenic ligand, peptide or otherwise.[20] The rationale for PTZ in hypersomnia was that it might push back against overactive sleep-related signals from GABA. PTZ's GABA-based pharmacology was different enough from conventional stimulants for it to be a possible alternative, but its success would not have proven Rye's "sleepy juice" theory.

FROM THE DUSTBIN OF PHARMACEUTICAL HISTORY

PTZ's long history—part notorious, part benign—may not inspire confidence. Pentylenetetrazol's first medical uses in the 1920s predate the synthesis of amphetamines.[21] Known first as Cardiazol and called Metrazol in the United States, it was manufactured by the German pharmaceutical firm Knoll.[22] In the 1930s, it was used in psychiatric hospitals in Europe and the United States to deliberately induce seizures in people with schizophrenia.[23] Decades later, it was advertised as a mild stimulant for the elderly before being removed from the market as part of FDA reforms.[24]

PTZ's revival in the twenty-first century was part of a burst of drug discovery efforts aimed at Down syndrome, which researchers had assumed for years was too complex and permanent for pharmaceutical treatment.[25] The origins of the wave came from research on Ts65DN mice, which have an extra copy of almost one hundred genes, modeling the situation in Down syndrome, in which an extra copy of chromosome 21 leads to intellectual disability. In brain regions important for learning and memory, some of the extra genes are thought to shift the balance between excitation and inhibition more toward the latter.

The neuroscientist Craig Garner, then at Stanford, and colleagues wanted to push back against GABA's inhibitory signals as a way to recalibrate the balance. Although PTZ's reputation as a seizure-induction tool put off some scientists, it did not intimidate Garner. "Every drug has a toxic range," he said. "Seizures are just toxicity. In the mice, we could see positive effects without getting anywhere near that point."[26]

Ts65DN mice that were given small doses of PTZ every day for a couple weeks performed better on memory tests.[27] Specifically, they were able to recognize new objects, when usually Ts65DN mice could not. The mice didn't perk up right after the first dose, but the memory effects lasted for up to two months after the drug was withdrawn. The observation suggested that low-dose PTZ was not acting as a conventional stimulant, since its effects lasted after the drug had left the body.[28] "We think that it's more about network adaptation," Garner said. "The circuits in the brain readjust."

After Garner unsuccessfully tried to get established pharmaceutical firms interested in his Down syndrome work, he formed Balance Therapeutics in 2009. He said his lab also had obtained positive results with flumazenil,[29] but industry advisors told him that PTZ was a better drug candidate, because flumazenil didn't last long in the body and had an inconvenient mode of delivery.

Balance sponsored a study of PTZ in Australia called Compose21, which enrolled eighty-eight young participants with Down syndrome, up until 2015.[30] The Compose21 results remain unpublished; positive results would have been reported. Similar studies in other neurodevelopmental disorders, such as fragile X syndrome, have stumbled over differences between homogeneous populations of lab mice and humans of varying ages and abilities.

Confounding matters, humans with Down syndrome tend to have more fragmented sleep than neurotypical people, and they often have obstructive sleep apnea because of their altered craniofacial structures.[31] On top of that, Ts65DN mice are not sleepier than standard lab mice.[32] However, since brain fog can be a prominent feature of IH, PTZ's effects on memory are intriguing, even if they have been demonstrated only in mice.

PTZ FOR HYPERSOMNIA

At the 2017 World Sleep Congress in Prague, Rye reported his positive initial findings with PTZ—but only with five patients, in a non-placebo-controlled format.[33] Balance Therapeutics has sponsored two clinical trials testing PTZ in people with narcolepsy type 2 and idiopathic hypersomnia.[34] The company was able to recruit participants through the Hypersomnia Foundation's conferences, for which it was listed as a sponsor. Balance also recruited through the Foundation's IH registry, showing how compiling data on people with IH could facilitate clinical trials.

Some members of the hypersomnia community, including Meghan Mallare and Diana Kimmel, have reported favorable experiences in Balance's clinical trials. Some said it was the only experimental medication that seemed to work for them. A few have gone as far as to seek unconventional access to the drug; one person obtained it as a component of the cough syrup Cardiazol Paracodina, which is available in Italy.[35] Balance was not able to respond to expanded access requests, executives told me.

When Balance evaluated the results from its second clinical trial for sleep disorders, development of PTZ was stopped, and the board of directors decided to wind down the company. In 2020, Balance was put up for sale by Gerbsman Partners, which specializes in restructuring technology and life science firms, but there were no buyers.[36] And as of 2022, the results of Balance's clinical trials of PTZ in people with sleep disorders are unpublished. Despite anecdotal success, PTZ did not provide benefits to enough people in the second clinical trial such that the company's advisors thought it would succeed.

Although PTZ did not appear to have a path forward, some valuable precedents emerged from Balance's efforts. It was the first company to obtain an orphan drug designation for IH and conduct clinical studies focused on IH. Also, for its second clinical trial, Balance developed its own outcome measure panel: a questionnaire on IH symptoms. According to Gerbsman, the FDA has agreed that the panel can be used as the basis for a marketing approval. We will discuss orphan drugs and the FDA approval process more extensively in chapters 14 and 15.

CHAPTER 11

THE HEART OF THE BRAIN

The hypothalamus does mundane things, but it does them well, and how it solves difficult challenges can tell us much about what neurons and networks are capable of.

—Gareth Leng, *The Heart of the Brain*

In this chapter, we will venture beyond Rye's "sleepy stuff" and look at other mechanisms that may explain how and why idiopathic hypersomnia occurs. In general, sleep researchers have moved beyond looking at sleepiness as being brought on by a single substance. The field has shifted to brain circuits and cellular clocks, and two proposals we will examine involve disturbances of those circuits and clocks. Experts don't yet know enough to say definitively whether IH comes from sleep being too "thin"—missing some element of its mysterious, restorative oscillations—or the reverse, abnormally deep and prolonged because of a distortion of regulatory forces. In addition, uncertainty remains regarding to what degree IH is imposed from outside by injury or stress, versus an inherited disorder that unfolds from within. Still, enough is known to sketch out two possibilities.

First, we must acknowledge that in clinical practice, the label of idiopathic hypersomnia is applied to individuals with a variety of underlying issues. More than one pathological mechanism may apply to people who fit under this umbrella. As currently defined, IH includes both people who display extra-long sleep

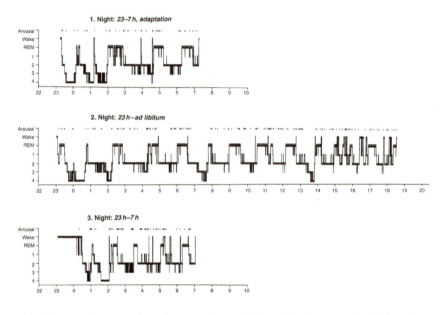

FIGURE 11.1. Sleep recordings from a patient with idiopathic hypersomnia with long sleep and sleep drunkenness.

Source: U. Voderholzer et al., "A 19-Hour Spontaneous Sleep Period in Idiopathic Central Nervous System Hypersomnia," *Journal of Sleep Research* 7, no. 2 (1998). Reprinted by permission from John Wiley & Sons ©1998.

periods—eleven hours of sleep or more on most nights (figure 11.1)—and those who do not.[1] Some with IH diagnoses may have chronic pain or other factors disturbing their nighttime sleep, enough that they fall asleep quickly in the setting of an MSLT. It is also possible that some diagnosed with IH may have other conditions that are difficult to spot right away: metabolic diseases, vitamin deficiencies, or even covert hepatic encephalopathy. One of the seven flumazenil-responsive patients described in Rye and Jenkins's 2012 paper was discovered later to have myotonic dystrophy, an inherited multisystem disorder that includes daytime sleepiness as a prominent symptom.[2]

Putting those possibilities aside, one proposed mechanism for IH is an old idea, extending back before Bedřich Roth's time. It is partly based on an analogy with narcolepsy. The basic idea is that an infection somehow damages part of the brain whose function is to regulate sleep and wake. More than one hundred years ago, this idea manifested in a way that shaped the history of neurology and produced a group of famously sleepy patients.

SMOLDERING WILDFIRES

In the middle of World War I, doctors at a hospital in Vienna began to see patients with signs of inflammation in the central nervous system: encephalitis. Amid wounded soldiers who had returned from Austria-Hungary's battles, these civilians with encephalitis were distinctive. Affected people usually had fever and headaches. They developed double vision, and the muscles controlling their eyes became paralyzed. And they were profoundly sleepy. They could be wakened and answer questions but would fall into delirious quasi-sleep if left alone and would sometimes progress to stupor and coma. More than a quarter died. Despite differences between individual cases, they had enough in common that a visionary neurologist named Constantin von Economo (figure 11.2) grouped them together and named their disease "encephalitis lethargica." "The most striking symptom of this disease is the sleepiness of the patients, which at times is associated with delirium and at times is not. This somnolence can vary from light sleep, entirely

FIGURE 11.2. Constantin von Economo was one of Austria-Hungary's first credentialed pilots. He was serving as a wartime pilot until 1916, when he returned to Vienna, where he would investigate encephalitis lethargica.

Source: Photo by Max Schneider. Public domain.

resembling physiological sleep, to the deeper stupor, independent of any fever," von Economo wrote.[3]

Independently from von Economo, doctors in the United Kingdom and France made similar observations. An epidemic of encephalitis appeared in Europe and the United States in 1918. A second wave in the 1920s was more varied, with a dominant "somnolent-ophthalmoplegic" form (sleepiness plus eye paralysis), but other patients experienced different symptoms, such as insomnia, agitation, and involuntary movements.[4]

For those that recovered, the acute state of lethargy lifted after days or weeks. However, the epidemic left thousands of people with a variety of neurological and psychiatric disorders—usually with a puzzling delay of months or years before the appearance of chronic symptoms. Encephalitis lethargica was like a series of smoldering wildfires, burning across the brain along trails that differed from person to person.[5]

Many survivors had symptoms similar to Parkinson's disease: rigidity, immobility, and tremors. Those with symptoms like these lingered in New York City hospitals long enough for Oliver Sacks to learn their stories in the 1960s for his book *Awakenings*, describing people who had been immobilized for years. Other survivors developed sleep disorders, such as severe insomnia or sleep inversion: being drowsy during the day but having disturbed sleep at night. Some of those with sleep inversion found night jobs because of their need to sleep during the day.[6] A handful of encephalitis survivors were reported to have narcolepsy, although terminology was then looser, and part of this group did not display clear cataplexy.[7] A few individuals developed a sleep disorder that resembled idiopathic hypersomnia.

NIGHTMARE LETHARGY

While living in New York City in 1923, a woman named Eleanore Carey developed severe pain in the back of her head, which migrated to her shoulder. She later became feverish and fell into a "drugged lethargy, a semiconsciousness filled with the hodge-podge thoughts of a nightmare." Years later, Carey, who started but did not complete medical school, wrote about her experiences for the *American Mercury* magazine.[8] She wrote that her semiconscious state lasted for more than three months. She only dragged herself out of it because of her need to care for her young daughter. Her description of her daily existence sounds like Anna's: "My entire day was clouded by one obsession—the wish to sleep. It was torture—this continual forcing oneself to keep conscious, and a great part of that time I was not entirely conscious—going about in a daze." Friends viewed her with condescension or aversion, Carey wrote. She lost several jobs because her employers thought she was taking recreational drugs. A landlady accused her of being a "chronic inebriate" because of her confused appearance and behavior. A doctor who had cared for her during her months of delirium told her she should be grateful to be alive.

In her article, Carey did not describe any cataplexy-like symptoms, so it was unlikely that she had the type 1 form of narcolepsy. Despite pleas to her doctor, she did not mention being given any medications; the stimulant ephedrine was first given for narcolepsy around 1930. She did recall symptoms consistent with hypnogogic hallucinations and also sleep inertia: "I find it difficult, particularly when I first rouse myself out of the lethargy, to remember directions."

Three years after the acute phase of her illness, Carey wrote that she could fall asleep in a warm bathtub and wake up hours later, the water having grown cold. She could slip into sleep while getting dressed or while waiting on a crowded subway platform. She thought of her life as divided like a soldier's: before and after the war. Carey would fit in today at a hypersomnia support group meeting, although the term *idiopathic* would not apply, because encephalitis was the precipitating event.

DEVASTATION WITHOUT A KNOWN CAUSE

The pathogen responsible for encephalitis lethargica remains a mystery. To this day, nobody knows what caused it. Many in the 1920s proposed that encephalitis lethargica was related to the worldwide influenza pandemic that killed millions during and after World War I. Confusion came because encephalitis lethargica arrived when understanding of viruses was limited. It was before scientists had developed tools, such as filters and electron microscopes, for separating and visualizing viruses. In addition, because of press attention, encephalitis lethargica was overdiagnosed, and experts today believe that the label was applied to people with more than one underlying disorder. For her part, Eleanore Carey reported having the flu in 1917, years before her disabling encounter with encephalitis.

Von Economo doubted that influenza was the cause of encephalitis lethargica, because he began seeing encephalitis cases before the flu arrived in Vienna, and he thought encephalitis lethargica was not as contagious as flu. He did believe encephalitis lethargica was infectious and tried to culture the pathogen responsible in monkeys. Several researchers unsuccessfully searched for a herpes-like virus as a potential culprit for encephalitis lethargica. Research on the origin of the disease had "run into sand" by 1930, according to the Australian medical historian Paul Foley.

In the early 1980s, William Foege, then head of the U.S. Centers for Disease Control, reignited the debate. In the *Lancet*, Foege and his colleague Reimert Ravenholt argued that influenza infection almost certainly caused encephalitis lethargica. Their research was based on analyzing health records from Seattle and comparing the epidemic's course in American and Western Samoa.[9] They wrote that von Economo's attention to extreme sleepiness "helped distinguish such cases, but distracted diagnostic and epidemiological attention from the much broader range of neurological disorders caused by influenza."

Modern molecular tools were used to look at autopsy samples from a few encephalitis lethargica cases, but no influenza virus was detectable.[10] In autopsy samples, investigators in London have glimpsed viral particles and detected genetic sequences resembling those of enteroviruses, relatives of the virus that causes polio.[11] However, this still hasn't settled the matter, given the limited number of samples available and their condition after almost a century. Other researchers have proposed that encephalitis lethargica was the result of a slow burn, not driven by the pathogen directly: an autoimmune reaction perhaps brought about by bacterial infection.[12]

NEUROANATOMICAL LEGACY

In addition to human devastation, encephalitis lethargica established a scientific legacy. Von Economo observed that while most encephalitis patients developed lethargy and hypersomnolence, some developed insomnia. The first group had damage to a region of the brain near the junction of the brainstem and forebrain, close to the nerves that control eye movements (*N. oculomot* in figure 11.3). The second group had lesions in a more anterior region.

Some of the sleepiness in acute encephalitis lethargica probably came from temporary inflammation and cytokines. Still, the anatomical distribution of sleep disturbances drove von Economo to postulate the existence of a two-part "center for regulation of sleep" in the brain. He attained his insights based on pathology; he described in detail what the destruction in the brain looked like under a microscope. He was building on ideas from others, such as the Austrian neuroanatomist Ludwig Mauthner, who made a similar proposal in the nineteenth

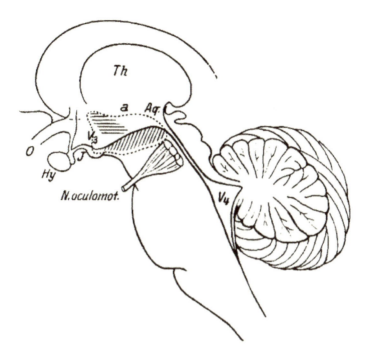

FIGURE 11.3. Diagram from von Economo. Lesions in the posterior wake-promoting region, marked by slanted lines, resulted in hypersomnolence. Injury to the anterior sleep-promoting region, marked by horizontal lines, resulted in insomnia.

Source: Lazaros C. Triarhou, "The Percipient Observations of Constantin von Economo on Encephalitis Lethargica and Sleep Disruption and Their Lasting Impact on Contemporary Sleep Research," *Brain Research Bulletin* 69, no. 3 (2006). ©2006 with permission from Elsevier.

century. Von Economo gave a nod to Pieron's hypnotoxins, suggesting that the sleep regulatory center was more sensitive to fatigue-related substances than other brain regions.

Over the long term, von Economo's ideas were influential, although they met with resistance at the time. Neuroscientists celebrate him today because his predictions approximated what is now known about the brain's circuitry. Parts of von Economo's sleep regulatory center may match up to regions of the hypothalamus.[13]

Sandwiched between the thalamus and the pituitary, the hypothalamus is a fingernail-sized structure in the middle of the brain. The hypothalamus lies at the center of a network of brain circuits that filter sensory stimulation and pain and focus attention and alertness. In addition to regulating sleep and wake, the hypothalamus is home to several bundles of neurons that regulate body temperature, heart rate, blood pressure, and appetite. It's as if a house's light switches, plumbing shutoff valve, thermostat, and alarm clock were all together on one panel.[14]

Von Economo sometimes gets credit for correctly guessing that the hypothalamus is ground zero for narcolepsy type 1. In a 1929 lecture, he predicted that narcolepsy "has its primary cause in a yet unknown disease of that region." However, he wasn't very precise. His diagram appears to include both the posterior hypothalamus and territory farther back, extending to a region known as the periaqueductal gray. Today, neuroscientists have established that several wake-promoting circuits run through that area.

Putting this together, a venerable and still viable theory for idiopathic hypersomnia is that people with IH have experienced something similar to what happened to Eleanore Carey. This idea frames IH as a regional neuronal injury, somewhat like narcolepsy type 1. Many people with IH recall that they began to feel extra sleepy after a viral infection, whether it was influenza, Epstein-Barr, or some other virus. Widespread neuroinflammation, seen in postviral chronic illnesses, can leave individuals with a long list of symptoms, sometimes including excessive sleepiness and brain fog. For IH, a case can be made that a more selective strike is occurring against the regions of the brain that regulate sleep. The site of injury could be part of the hypothalamus or another region of the brain.

Bedřich Roth considered this mechanism a possibility, since he had several patients with hypersomnia resulting from encephalitis. His French and Czech mentors had studied encephalitis lethargica, and Roth cited von Economo extensively in his books. Although Roth made a distinction between *symptomatic* hypersomnia, coming from encephalitis, and functional or idiopathic hypersomnia, he wrote in 1980: "It is therefore highly probable that the same structures are affected in the two conditions and that they have the same type of pathophysiological mechanisms—i.e. excessive facilitation of non-REM sleep."[15]

Supporting a hypothalamic lesion theory, people who undergo surgery for tumors or cysts close to the pituitary or hypothalamus often experience severe hypersomnolence, attributed to tissue injury inflicted by the tumor or by surgery. One 2002 study found that children who underwent such surgeries slept an average of more than thirteen hours per day.[16] Another case review of children with brain tumors concluded: "Children who sustained damage to the hypothalamic/pituitary region developed EDS [excessive daytime sleepiness] regardless of whether the damage was the result of the tumor, surgery, hydrocephalus, or radiation to the whole brain."[17]

While the hypothalamus is a plausible place to look for problems in both nar-colepsy type 2 and IH, we have to be careful about availability bias: the error of the drunk man looking for his keys under a bright street lamp. Also, perturbing the hypothalamus can affect several other aspects of physiology besides sleep. Injury to another area of the hypothalamus can lead to *diabetes insipidus:* daily generation of gallons of urine, which is generally not a symptom of IH!

However, many people with IH do experience symptoms of autonomic ner-vous system dysfunction.[18] Examples include feeling lightheaded or fainting upon standing up, as well as numbness in the extremities in response to cold (Rayn-aud's syndrome), both of which are related to problems regulating blood pressure or blood flow. Others include problems regulating body temperature, including heavy sweating or feeling colder or warmer than others in the same room, and gastrointestinal difficulties.

Whatever is happening in IH has to be restricted in scope. IH doesn't leave someone with tremors or paralyzed limbs, and it is uncommon for people with IH to have recognizable abnormalities on a brain MRI scan.[19] With IH, we should not focus on lesions in regions such as the parabrachial nuclei of the brainstem, the equivalent of the fuse box in the basement, where injuries can produce a coma. Instead, we are looking for what may be a subtler injury—a bruise, not a gaping wound.

THE HOME BASE OF SLEEPINESS

Learning about the hypothalamus also allows us to envision the sources of the symptoms experienced by people with IH. If sleepiness can be localized within any specific place in the brain, the hypothalamus could be considered its head-quarters. Neurologists now think that the anterior part of the hypothalamus is the most probable location of the sleep-promoting region von Economo identi-fied. He certainly knew about the hypothalamus; some of his teachers in Vienna were the first to do experiments testing its physiological function with electrical stimulation. However, in von Economo's often-cited 1929 lecture, he only men-tions the hypothalamus once, to say that the sleep-promoting region may extend into it.

Decades of experimental work in animals, beginning in the 1940s with the Dutch neuroscientist Walle Nauta, have refined our understanding of the sleep-promoting region's location and function.[20] In mice and rats, one of the critical bundles of neurons within the hypothalamus is called the VLPO (VentroLateral PreOptic area). During sleep, neurons within the VLPO and allied regions act like a source of soothing music, helping keep the rest of the brain asleep with inhibitory GABA signals.[21] The VLPO has connections to other parts of the hypothalamus and to several wake-promoting regions in the brainstem and forebrain. Along with the basal forebrain, the VLPO is thought to be a main target for both adenosine and for GABA-enhancing sedative and anesthetic drugs. Because of its role as a source of inhibition via GABA, the VLPO may be the site in the brain where flumazenil loosens the grip of sleepiness.

In the 1990s, Cliff Saper's group at Harvard defined the importance of the VLPO in regulating sleep in rats.[22] VLPO neurons are responsible for sending messages of sleepiness to the rest of the brain when time awake has lasted too long. That is, they increase firing when an animal is kept awake during a time when it would normally be asleep, even for a few hours.[23] VLPO neurons are active during both REM and non-REM sleep, and their activity ramps up as sleep deepens.[24]

While most studies of the VLPO's role in sleep were performed in animals, some evidence for its importance comes from looking at older humans. Loss of neurons in the VLPO's counterpart in humans, the intermediate nucleus of the hypothalamus, has been associated with greater sleep fragmentation, or awakenings that interrupt nighttime sleep.[25] Aging-related loss of sleep-promoting cells may partly explain why older people are less susceptible to sleep deprivation, in terms of sustained attention.[26] It's not that older people need less sleep; they are less sensitive to the molecules that prompt the transition into sleep.

Saper has described the mutually inhibitory relationship between VLPO neurons and wake-promoting regions as resembling a "flip-flop" electrical circuit. He credited one of his graduate students, trained as an electrical engineer, with introducing him to the concept. In a flip-flop circuit, when signals from one side get strong enough, they squeeze out activity on the other side. Like a see-saw, the system is stable on either side but not in the middle. The circuit's organization may explain why sleep typically dissolves quickly after someone wakes up. In people with IH who experience sleep inertia or sleep drunkenness, some of the flip-flop switches may have become stuck, so that VLPO activity continues after waking.

David Rye's proposal about IH and related sleep disorders was not mainly about *where* the problem in the nervous system occurs. It was about *what* makes people sleepy (the suspected somnogen) and *how* (GABA). However, his proposed mechanism shares something with Mahlon DeLong's insight regarding Parkinson's. Underactivity in one part of a brain circuit can result in overactivity in another. When Rye described giving stimulants to Anna as like driving a car with the parking brake on, he envisioned the sleep-promoting regions of the brain—namely, the VLPO and allied regions in the hypothalamus—as the overactive brake. On the other side, several regions of the brain activated by stimulants, both within the hypothalamus and beyond it, grind against the brake.

The hypothalamus appears to be the site where many forces—temperature, sensory stimulation, attention, extended time awake, or recent food consumption—engage in a tug of war over sleepiness (figure 11.4). Recent research has revealed that the VLPO is not a uniform bundle, and some neurons in it can be wake-promoting. Nearby bundles of neurons in the hypothalamus are also active in promoting sleep, and their function is intertwined with temperature regulation.[27] This may explain why sleepiness is so sensitive to skin temperature.[28]

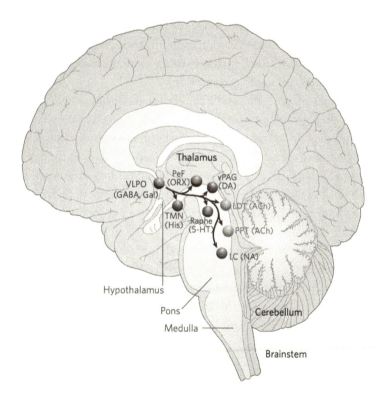

FIGURE 11.4. The VLPO (ventrolateral preoptic area) inhibits the activity of several wake-promoting regions, such as the tuberomammillary nucleus (TMN), the ventral periaqueductal gray (vPAG), the locus coeruleus (LC), and the dorsal raphe. Damage or impaired function of these regions in idiopathic hypersomnia has not been documented.

Source: C. B. Saper et al., "Hypothalamic Regulation of Sleep and Circadian Rhythms," *Nature* 437 (2005). Reprinted by permission from Nature ©2005.

A LONG BIOLOGICAL NIGHT

The second theory about IH hasn't received as much fanfare as Rye's GABA-enhancing somnogen theory, but it has caught on with others in the sleep research field. It is more concerned with the processes that are disrupted rather than with a particular region of the brain or an underlying cause. The second theory says that people with IH, especially those who require long sleep periods, may have a distortion of their circadian rhythms. Robert Thomas, a neurologist at Beth Israel Hospital in Boston, uses an evocative phrase to describe this imbalance, saying that people with IH have a "long biological night." It may explain why IHers have such difficulty waking up in the morning—their bodies and brains tell them that it's the middle of the night.

In the early 1980s, the Swiss sleep researcher Alexander Borbely developed a "two-process" model for how the body regulates sleep. It's a bit simplified, and the two-process model leaves out a lot, including the powerful effects of our conscious behavior. Still, it's a helpful way to think about the forces that may be driving excessive sleepiness in someone with IH.

The first major force is Process C: the circadian rhythm or body clock, which corrals our sleep into daily cycles of light and dark imposed by the sun. Circadian rhythm–promoted wakefulness is part of what keeps someone functioning during daylight hours, even if they have had little sleep or poor sleep the night before. Circadian rhythms drive inconvenient wakefulness in jet-lag, when someone quickly shifts to a different time zone. In addition to sleepiness, circadian rhythms modulate body temperature and processes such as metabolism, urination, and digestion.

Circadian clocks can be found in every cell in the body. They consist of oscillations in the levels of a set of core clock proteins. When enough of the core clock components accumulate, they are able to block activity of the corresponding genes that encode the same proteins. This creates a delayed negative feedback loop, making the clock proteins' abundance swing up and down, with a rhythm of about twenty-four hours. Many cellular enzymes modify the core clock proteins, offering opportunities for regulatory adjustment and *entrainment*—the influence of light.

The second force is Process S, sleep pressure: the longer someone stays awake, the more pressure builds to fall asleep. This is sometimes described as the *homeostatic* sleep drive, because when there is more need for sleep, Process S is supposed to return a human or animal to a state of balance. When someone does get to sleep after extra time awake, an abundance of slow wave or delta sleep, the deepest form of non-REM sleep, is a sign of relief of that pressure. We know that people with IH experience sleep pressure. Indeed, it makes them miserable when they are awake. They bear the imprint of one half of Process S but not the other: its release.

MEASURING MELATONIN

One reason to suspect an alteration of circadian rhythms in IH is a change in the nightly pattern of melatonin, the hormone that signals nighttime to the body and helps set our day-night schedule. Despite its common use as a sleep aid, melatonin should be considered a "hormone of darkness" rather than a sleep-induction molecule like adenosine. Nocturnal animals such as mice have higher melatonin levels at night, even though they are awake and active at that time. So changes in melatonin reflect a distortion of the underlying circadian rhythm.

In two studies of people with IH, the production of melatonin appears to be weaker and more stretched out, compared with healthy sleepers. Investigators in Prague, led by Bedřich Roth's associates, were the first to show this among people with IH in a publication from 2000.[29] The results resemble melatonin profiles in habitual long sleepers.[30] Normally, melatonin levels in saliva begin to increase at around 9 or 10 pm before bed, peak at around 2 am, and fall off as morning approaches. According to the findings from Prague, in people with IH, melatonin levels began to increase later, peaked at a level that was less than half normal, and then remained at an elevated level into the early afternoon. The rhythms of cortisol, a stress- and alertness-related hormone, were also shifted and delayed in IH patients. The Czech researchers focused on people with IH who reported long sleep times and sleep drunkenness; their results for people with narcolepsy with cataplexy and nonidiopathic hypersomnia were described only as "heterogeneous."

In Boston, Thomas has made similar observations. In his brief report published in 2020, most IHers (forty out of fifty) displayed an extended period when melatonin was elevated, compared to their awake state.[31] Thomas has proposed that a long biological night has implications for how IH could be treated: with bright light therapy or the carefully timed application of melatonin-suppressing beta blockers. While the success of his clinical approach is currently anecdotal, it makes use of existing tools that are relatively accessible.

THE SUPRACHIASMATIC NUCLEUS

The long biological night theory could potentially fit with the proposal that an injury to part of the hypothalamus causes IH. The hypothalamus contains a structure called the suprachiasmatic nucleus (SCN), which sets circadian rhythms and transmits information about environmental light to the rest of the body. Its firing rate peaks in the middle of the day and reaches a minimum in the middle of the night. When stimulated by light, through a series of connections involving the retina and the spinal cord, the SCN inhibits production of melatonin. Beyond inhibiting melatonin, the SCN acts as a master regulator of circadian rhythms, sending signals to the rest of the hypothalamus and other parts of the brain. The SCN does not connect with most of the brain's sleep-wake circuits directly, and its influence is exerted through other parts of the hypothalamus.

The SCN's importance first emerged in the 1970s, when researchers showed that destroying the SCN in rats eliminated the rats' circadian rhythms: they drank water and ran around without regard to the time of day.[32] Some studies of the SCN in animals showed that lesions in the SCN dramatically increase the total amount for sleep. In squirrel monkeys, animals that display consolidated sleep and wake periods similar to humans, SCN lesions resulted in them sleeping four hours more every day, and their average time awake without naps lasted only fifteen minutes.[33] However, the same type of effect has not been observed consistently in rodents; this discrepancy may come from neuroanatomical differences, imprecise experimental surgeries, or the environments the animals were kept in.

It is possible that defects in SCN function weaken circadian rhythms in people with idiopathic hypersomnia. Degeneration of the SCN is also thought to lie behind disrupted circadian rhythms in Parkinson's disease and other neurodegenerative diseases.[34] Sleep researchers at Northwestern have observed a weakened, extended nighttime profile of melatonin in people with Parkinson's, particularly in those who had stronger symptoms of excessive daytime sleepiness.[35]

CIRCADIAN RHYTHM GENES

The 2017 Nobel Prize in Medicine or Physiology was awarded for the discovery of circadian clock genes in fruit flies, and similar genes exist in humans too. Mutations in genes encoding parts of the circadian rhythm's cellular clock have been found in "morning lark" humans, who have an advanced sleep phase. Other mutations appear in their "night owl" opposites, who have a delayed sleep phase.[36] The mutations compress or extend the clock's timing, producing a circadian clock that would fit better on a planet with a twenty-three- or twenty-five-hour day, respectively. These mutations' effects go beyond what we see in older people, who tend to go to sleep early and wake up early, or in many teenagers, who display the reverse. The mutations shift the time when people carrying them go to bed and wake up compared to typical sleepers, but their total amount of sleep need does not change.

Additional support for altered circadian rhythms in IH has come from studies conducted at the University of Münster in Germany. Researchers asked people with IH to donate skin cells, which were cultured and artificially synchronized to follow the oscillations of their circadian rhythm genes. In IHers' cells, the activity of core clock genes swung up and down markedly less, compared to controls.[37] In addition, the overall activity of the core clock gene *BMAL1* was reduced. A follow-up study showed that circadian period length was slightly longer in IHers' skin cells.[38]

These were intriguing results because it looked like the alterations in circadian rhythms were independent of the nervous system and present in every cell in the body. However, signals carried in the blood are continually corralling the clocks of peripheral tissues, so weakened circadian rhythms in IH may not be "cell-intrinsic."[39] In addition, the Münster authors left out information on study participants' circadian rhythms and sleep habits. Did some exhibit abnormally long sleep periods at home? What about their melatonin profiles? All we know is that they were diagnosed with IH according to current criteria and thus fell asleep quickly during an MSLT. Altered circadian rhythms have been detected in several other disorders, ranging from diabetes type 2 to schizophrenia.[40] Side-by-side comparisons are necessary to assess whether circadian alterations are more pronounced in IH compared with other conditions.

The reports from Prague, Boston, and Münster highlight the need to more thoroughly examine circadian rhythms in IH patients, including measurements of melatonin, cortisol, and body temperature, along with the activity of their circadian rhythm genes. Measurements of melatonin are inherently tricky because of dim light requirements, but a major focus of sleep researchers' efforts over the last few years has been to develop robust blood biomarkers for circadian phase.[41] Future studies of IHers could incorporate measurements of such biomarkers, as well as examining the effects of medications on their circadian rhythms.

TOWARD GENETIC STUDIES OF IH

Along with probing circadian rhythms, future studies of IH should include genetic sequencing of "multiplex" families. For IH, musing about a genetic link extends back to the 1950s. Four members of a family with similar symptoms occupied a central place in Bedřich Roth's initial report. Later sleep researchers encountered families having several members with IH, suggesting strong inheritance effects.[42] For example, Michel Billiard and Yves Dauvilliers reported a patient whose sister, mother, and maternal uncle all had IH.[43] Most clinicians who have studied IH in detail have found that a large fraction of IH patients report having relatives with a history of excessive daytime sleepiness, if not outright IH. Examples in this book include Anna Sumner Pieschel and her brother James.

Then why has progress been so meager? Partly, the lack of scientific agreement about how to define IH clinically and the absence of research funding. Those obstacles are beginning to change, at a time when genomics could make progress more rapid. Today, modern genetics should allow the historic strands connecting narcolepsy and idiopathic hypersomnia to be unraveled.[44]

So far, no mutations closely analogous to those in Yanagisawa's *sleepy* mice have been found in humans, but technology has advanced to a point that might astonish Bedřich Roth. It is now possible to sequence an entire human genome within a week, for less than $1,000.[45] Whole exome sequencing—a limited survey of the protein-coding regions of the genome—is even more accessible. If a few members of a family are affected by the same disease and donate DNA samples, researchers can compare patients' genomes to unaffected relatives. By sorting through a list of genetic changes present in the patients' genomes, it is possible to identify mutations that may be responsible for their disease. This approach is now used to diagnose cases of autism spectrum disorder and early-onset epilepsy, although the diagnostic yield (the fraction of cases in which a cause can be assigned) is generally below 40 percent.[46] Similar studies on multiplex IH families could be critical for establishing IH's biological basis.

JAPANESE GENETIC STUDIES

Before whole exome or genome sequencing became widespread, another approach was in common use at the beginning of the twenty-first century. A GWAS (genome-wide association study) makes use of single nucleotide polymorphisms (SNPs), places where one letter of the genetic code varies in a fraction of the population. In a GWAS, researchers examine hundreds or thousands of human genomes in people with a particular disease—reading SNPs but not the entire genome. The SNPs don't necessarily cause the disease in question, but they offer hints of where to dig deeper. GWAS doesn't require a strong inheritance pattern; it can pick up weaker influences.

The only published attempt to find genetic risk factors for IH using GWAS techniques provided limited results. In this study, Japanese researchers used

GWAS techniques to study 408 patients with "essential hypersomnia," a category that corresponds to narcolepsy without cataplexy and IH without long sleep combined. What emerged was just one marker, close to a gene encoding an enzyme called carnitine O-acetyltransferase (CRAT).[47] Carnitine helps cells metabolize fatty acids as fuel, but how CRAT fit into the IH puzzle was not clear, and the CRAT mutations' effect was not strong.[48]

As this book was being revised for publication, the same Japanese team published the first study of IH to identify a mutation with a strong link to the disorder.[49] The mutation may account for only a sliver of IH cases in Japan, since only twenty out of 598 IH-diagnosed patients had the identified mutations. Still, the study was significant because it was the first to find anything potentially causative. The mutation was in the *prepro-orexin* gene, encoding the neuropeptide orexin/hypocretin (discussed in the next chapter). The mutation did not appear to induce sleep-onset REM, even though the overall effect of an orexin/hypocretin mutation should be like a limited version of narcolepsy type 1.

LONG SLEEP IH MAY BE DISTINCT

Future genetic studies may need to separate out patients with the long sleep form of IH. Supporting this point, a 2021 paper from Nevšímalová and her colleague Karel Šonka argued that IH with long sleep should be considered a distinct clinical entity, based on stability of the symptoms and clinical course of the disorder. In their paper, the Czech authors reexamined IH patients they had seen over the last twenty years.[50] Those in the long sleep duration group could sleep for an average of more than twelve hours if given the chance to do so. They reported a higher frequency of sleep inertia and autonomic nervous system symptoms, along with a poorer response to conventional medications. Other investigators, such as Isabelle Arnulf, have also found that people with long sleep IH tend to display severe sleep inertia or sleep drunkenness more than IH without long sleep.

Longitudinal and genetic studies could resolve the issue. Individuals and especially families with long sleep IH have not been studied in a way that would allow researchers to say that someone who habitually sleeps seventy or more hours per week needed the same amount one or five years ago. Most previous studies of IHers have relied on patients' recall of their habitual sleep duration. Also, IHers' attempts to manage their symptoms with stimulants may obscure shifts in their underlying sleep need.

WIDE VERSUS DEEP

Studies on families and twins indicate that at the population level, a large part of habitual sleep duration is genetically controlled. From an evolutionary point of view, humans sleep far less per twenty-four hours than expected based on comparisons with other primates, so the genetic changes that facilitated that difference may be relatively recent.[51] Based on recent studies, it looks like dozens of genes can influence sleep duration, each one having a small effect. In the general population, excessive daytime sleepiness is more likely to come from insomnia or sleep apnea, not from an inborn increased need for sleep.[52]

One of the largest studies to probe genetic influences on sleep duration comes from the Broad Institute, where investigators were able to tap into the UK Biobank project, which has collected DNA from close to half a million people.[53] In this analysis, the 5 percent of people who carried the most sleep-increasing genetic variants reported sleeping about twenty minutes more than those carrying the least. The largest effect that any single variant had was to increase sleep duration by 2.4 minutes. Genetics explained a small part of the total variation in sleep duration; other factors—age, job, or previous health history—influenced sleep duration more.

Genome-wide association studies like this rely on tracking common variants, which usually don't make a large difference in gene function. GWAS can spot interesting gene variants that influence a universal trait—such as sleep duration—or a common disease.[54] GWAS is less applicable toward getting to the bottom of a rare sleep disorder. Mutations driving IH are unlikely to be present at a sufficient frequency in the general population. For rare disorders such as IH, going deep—looking for rare variants with a large effect on sleep duration—may be better than going wide. The long sleep form of IH would be more unambiguous and better suited for this type of study.

SHORT SLEEP FAMILIES

The type of study I am proposing has already been performed with families who exhibit the opposite phenotype from IH. The geneticists Louis Ptacek and Ying-Hui Fu, a husband-and-wife team at University of California–San Francisco, have had extraordinary success finding natural short sleep families and isolating the mutations responsible. If this is possible for short sleep, it should be possible for long sleep IH as well.

Ptacek and Fu came to this topic through their work on advanced sleep phase syndrome: people who both wake up and go to bed several hours earlier than others, without sleeping more overall. Their investigation of a Utah family, sparked by a colleague's encounter with a woman with pronounced advanced sleep phase, began before the Human Genome Project and other advances made such studies less burdensome.[55] Until they were close to their quarry, their effort was independent of knowledge of circadian rhythm genes in fruit flies. Discovery of the responsible mutation was a confirmation that clock genes isolated through fly genetics were relevant for human biology.[56] Since then, DNA sequencing has become easier and less expensive, enabling the investigation of smaller families with just a few affected individuals.

Fu and Ptacek's initial reports on the Utah advanced sleep phase family had led to a stream of people contacting them. In their initial investigations of other advanced sleep phase families, the mutations were easier to find because of clues about where to look.[57] Instead of having to search across the entire genome, the scientists took an educated guess about what might be affected: the other components of the circadian clock. Their guesses were correct, at least for some families. However, the alterations in sleep did not manifest in the same way in each family—in some families, affected individuals displayed signs of seasonal depression or migraine headache.[58]

Other families who contacted Fu and Ptacek were different, in that affected members woke up early but didn't go to bed early to compensate. These individuals needed less sleep than other people in their families—usually less than six hours every night, and they didn't seem to suffer negative health consequences.[59] In Fu and Ptacek's first paper on a short sleep family, the "candidate gene" (or educated guess) approach led them to a mutation in a component of the circadian clock called *DEC2*.[60] But the other short sleep families they identified did not carry mutations in circadian clock components. Instead, affected family members have other mutations affecting neuronal signaling. Fu and Ptacek have identified more than fifty short sleep families and have published genetics on only a few so far, and they have not described the characteristics of those families' sleep in detail.[61]

From research elsewhere, there is evidence that habitual short sleepers display a higher proportion of slow-wave sleep; they may be packing the same level of restorative sleep into fewer hours spent in bed.[62] When they're awake, genetic short sleepers appear to be more efficient and optimistic than other people. Fu told *Scientific American*: "They like to keep busy. They don't sit around wasting time."[63]

It is striking how some genetic mutations in humans shift sleep cycles forward or back, while others decrease the need for sleep, but none have been found that markedly increase sleep need—yet. Fu and Ptacek said their team had only begun to collect long sleepers and had not begun to screen for mutations. I asked Ptacek whether it would be possible for a mutation in a circadian rhythm clock gene to lengthen the nighttime phase, producing long sleep IH. This hypothetical mutation would be analogous to the *DEC2* mutation but would have the opposite effect. "I believe the answer is yes, but we have not found a mutation that does it," Ptacek said.

LESSONS FOR IH

As investigations of short sleep families suggest, it is unlikely for long sleep IH to be caused by only one type of mutation, such as a mutation in a circadian clock component. When German or Japanese research teams used the candidate gene approach to look for circadian clock mutations in IH, they did find a few associated variants.[64] However, the associations were weak, similar to those reported in the GWAS study of essential hypersomnia, and individuals with long sleep duration were not separated out. Genetic studies of patients diagnosed with IH also have the potential to detect inherited conditions with overlapping symptoms, such as myotonic dystrophy.

Other genes related to IH may include those responsible for sensitivity to sleep deprivation or sedatives.[65] As stories in this book show, idiopathic hypersomnia has emerged as a clinical topic of interest partly because caffeine and conventional stimulants were unsatisfactory options. Thus, a reduced response to caffeine or stimulants or an inability to tolerate them may be a factor appearing in genetic analysis of IH because of altered drug metabolism or heightened aversive effects. For a large number of people with IH, their increased need for sleep was not present at birth but emerged around adolescence, with triggering events such as a viral infection or anesthesia. Thus, the underlying biology may be sensitivity of neural sleep-wake circuits to injury, possibly bearing some similarity to narcolepsy type 1.

CHAPTER 12

IMMOBILIZED BY HAPPINESS

One of my objects here is to rescue the word narcolepsy from the confusion that now surrounds it, and to reinstate it as the name of a highly remarkable and by no means very rare disease, with peculiar and unmistakable features that distinguish it from epilepsy and all other morbid conditions in which excessive or untimely sleep is an occasional symptom.

—William Adie, 1926

This book focuses on idiopathic hypersomnia, but we've spent a lot of time comparing IH to narcolepsy. Narcolepsy has a longer history, more scientific information is available about it, and sleep medicine specialists have been trained to recognize it. Current understanding of the type 1 form of narcolepsy represents a scientific success story—one that carries both clues and distractions for those interested in IH.

A central part of narcolepsy type 1 is the symptom of cataplexy. Let's hear from someone who lives with it: Ann, whom I met through a narcolepsy support group organized by her mother. Ann distinctly remembers the first time she experienced cataplexy. She was in her junior year of high school. She was at her church in north Atlanta, and it was Youth Sunday, when she and her peers would be playing lead roles in the service. Ann was feeling excited and happy. She had to fetch something in an area of the church where not many people were. "Suddenly, my legs just didn't work anymore," she recalled. "And my body didn't want to sit up. It was very uncomfortable."

At church, she could still call her mother with her cell phone. Another adult tried to help her onto a bench and then outside, to her mother's car. The odd paralysis of her legs went away after a few minutes. Over the next few weeks, it started happening more often. Sometimes her legs would kick involuntarily, and her shoes would go flying. "For me, cataplexy always occurs legs first," she said. "Like someone poking me in the back of the knee."

One day at school, she was late for science class, and the teacher shut the door and locked her out. She was looking in the window, saying "Let me in, please." Her legs buckled. Unfortunately, she was standing on a concrete floor, and she hit her head on the concrete and experienced a concussion. As a result, teachers and administrators became concerned about her cataplexy events. They didn't know exactly what was happening, but they didn't want any more concussions or potential legal liability. Seeking a temporary remedy, her father—a mechanic—took a racing car harness and attached it to a wheelchair. During class, she was supposed to stay in the restraints. "I hated that wheelchair," she said. "Half of the time, I used it as a book cart and would push it from class to class."

For most of her junior year, Ann was experiencing cataplexy several times per day—although she didn't know what to call it. The collapses tended to occur when she was excited. At church youth group gatherings, a refrain was: "Don't make Ann happy." She would make sure to sit down if someone had good news. She didn't avoid church or her friends ("I did avoid concrete"), but she did tend to remove herself from social situations and become more passive, zoning out watching television. A girl she considered a friend began talking behind her back, saying that she was faking to attract attention from boys.

At one point, Ann was swimming at a neighborhood pool. She saw a family friend's toddler behaving in an amusing way—and she suddenly couldn't move enough to stay above the water. Her mother had to jump in and enlist others to lift her out. "It was the scariest moment of my life," she said. "To this day, I won't go to that pool by myself. I didn't want to go to the beach for a long time."

During this period, Ann thought of herself as "a sleepy teenager." She took naps at school—sometimes a few times a day, other times once a week. A few times she slept almost the entire weekend: going to bed Saturday and getting up on Monday, waking up just long enough to change clothes and use the bathroom. However, in terms of disrupting her life, cataplexy was at the forefront.

She and her mother visited many doctors trying to understand what was behind these events. Their interpretations were colored by her previous medical history. In high school, Ann was already seeing a therapist and taking an anti-anxiety drug (buspirone), following her parents' divorce. Initially, doctors thought that her sudden collapses were anxiety related. One diagnosed her with conversion disorder—what was called hysteria in the nineteenth century—and prescribed hypnotherapy.

The next proposed diagnosis was epilepsy. After a difficult premature birth and a febrile seizure she experienced as a small child, Ann had taken an antiseizure medication for a while. Perhaps the seizures had returned? The idea was reasonable. Epilepsy was much more common than narcolepsy.

Ann visited the epilepsy monitoring unit at an Atlanta-area hospital and underwent a seventy-two-hour EEG examination, during which she waited for a putative seizure to occur. For support, her mother invited someone she liked and trusted—the youth director from her church—to come to the hospital. "He walked in the door, and down I went," she said.

Even then, doctors and others at the epilepsy unit didn't recognize what had happened, and she left without a definite diagnosis. As another explanation for her falls, a pediatric cardiologist proposed dysautonomia, or difficulty regulating blood pressure, causing dizziness. After a collapse, her mother was supposed to place a bulky heart monitor on her chest and record its signals—but this revealed that her heart's rhythms were normal.

Around this time, her mother had been Googling extensively and had come across the term "drop attacks"—or sudden falls without a loss of consciousness. A neighbor who was a doctor had mentioned another term, cataplexy, but it hadn't caught on with other doctors they saw. "We had this word," she said. "We told the doctors, but they didn't necessarily agree."

In the end, a neurologist Ann and her mother found confirmed what they had been suspecting, saying: "You fall down when you get excited—that's cataplexy." Ann had been taking a low dose of an antidepressant—an SSRI (selective serotonin reuptake inhibitor). At the neurologist's suggestion, she doubled the dose. The frequency of her falls decreased from several per day to every other day or even once a week.

Ann was formally diagnosed with narcolepsy with cataplexy in September, when she was starting her senior year of high school. She and her mother had already figured it out anyway. Her MSLT was inconclusive because construction noise prevented her from falling asleep quickly. But in retrospect, the puzzle pieces fit together: her sleepiness, along with what she came to recognize was cataplexy and its response to antidepressants.

Assembling a complex regimen of medications helped Ann continue her education. She only considered schools close to home, opting for the Atlanta campus of Savannah College of Art and Design. There, employees at the office that organized accommodations for students with medical needs were helpful, having dealt with another student with narcolepsy the year before. At her SCAD graduation ceremony, a celebrity (Oprah Winfrey) came to give a surprise speech. "As I was walking away, my legs were shaking like crazy," she recalled. But she didn't fall.

Ann credited the expensive but effective nighttime medication Xyrem with controlling her cataplexy and making the episodes rare. Despite the rigid daily routine imposed by Xyrem, she recommended it to others, saying: "It keeps me vertical." In 2014, she went to Disney World—full of exciting rides—without experiencing any cataplexy events.

Although Ann's symptoms were relatively well controlled, dealing with narcolepsy had drained her confidence. She was living at home, even after finishing college and finding a job at a museum. When her mother would travel outside Atlanta, she'd ask a friend to come stay with her. Although she had obtained her drivers' license just before her cataplexy appeared, she avoided driving. She still wore a bracelet that instructed others *not* to call an ambulance if they found her in a state of temporary paralysis, although that didn't happen very often.

While every person with narcolepsy and cataplexy is different, Ann's situation does represent aspects of what many experience: a glass half full. Confusion at the beginning but support and recognition available from a well-established community after diagnosis. Since the 1980s, the national organization Narcolepsy Network has held annual meetings, where people with narcolepsy can meet others with similar concerns. In the United States, there are local in-person support groups, supplemented in recent years by online groups.

Support is necessary because narcolepsy is a major detour on life's road. In a 2013 survey of more than 1,400 people with narcolepsy in the United States, most said that the disorder limits their performance at work or school. It constrained their ability to express emotion, drive safely, and interact with family and friends. Medications can provide substantial improvement of daytime sleepiness and control of cataplexy, but most people with narcolepsy still experience residual sleepiness.[1] A substantial fraction do not take any medications, possibly because of unpleasant side effects.[2]

In the past, narcolepsy advocates used bleak terms to discuss how their conditions affected their lives. As recently as 1990, the secretary of the American Narcolepsy Association (the predecessor of Narcolepsy Network) submitted testimony to a Senate committee, saying: "Most victims of narcolepsy spend their lives moving from job to job after being fired because they are considered lazy. However, many narcolepsy victims are totally disabled and do not lead much of a life at all."

It would be disturbing to hear that kind of pessimistic "victim" language today. Part of the shift comes from real change. Compared with thirty years ago, more medications are available that help people with narcolepsy manage their symptoms, work full time, and generally participate in life. There has also been a brightening of tone; the narcolepsy community now celebrates those who are leading lives of connection and accomplishment, aiming to encourage people with narcolepsy and show them that they are not alone. Still, when it comes to awareness and diagnosis, there is room for improvement.

CONFUSION AND DELAY

It took more than a year between the onset of Ann's symptoms and a correct diagnosis. This is typical. In the 2013 survey of people with narcolepsy, most reported a delay of three or more years between symptom onset and diagnosis, and this overall picture of delay has persisted.[3] A minority of primary care physicians in the United States could name both excess daytime sleepiness and cataplexy as the most prominent symptoms of narcolepsy.[4]

A smaller Swiss study had similar findings: more than five years between onset and diagnosis, usually because of insufficient knowledge of narcolepsy in the treating physician. On average, at least three doctors were seen before a correct diagnosis.[5] Common misdiagnoses were sleep apnea, depression, blood pressure dysregulation, iron deficiency, and more subjective judgments: "burnout," "psychosomatic," and "puberty related." A Yale study found a striking years-long delay in narcolepsy diagnosis among women compared with men.[6]

In the U.S. survey of people with narcolepsy, two factors predicting a longer diagnostic delay were pediatric onset and the absence of cataplexy. Without recognition of cataplexy, doctors are more likely to interpret excess daytime sleepiness as a symptom of depression or sleep apnea, which are more common conditions. The other symptoms of narcolepsy can be misleading. Substantial weight gain at disease onset can lead doctors to investigate a possible metabolic or endocrine disorder. Hypnogogic hallucinations can be interpreted as a sign of schizophrenia.[7]

Exceptions do occur, when a few patients accurately diagnose themselves and come into a doctor's office with a description of cataplexy printed out. And in both North America and Europe, diagnostic delays do seem to be getting shorter, possibly reflecting increased awareness. Still, as a matter of necessity, these studies leave out people with narcolepsy or IH who have not been able to access a specialist or are assumed to have depression, sleep apnea, or another condition.

INTIMATE DISRUPTOR

One factor complicating and delaying narcolepsy diagnosis is that cataplexy is difficult to observe in a hospital or clinic setting because a certain intimacy or comfort level is necessary to evoke it. Even an experienced sleep specialist may have witnessed cataplexy in person only a handful of times. Doctors depend on asking patients or their families to describe what they've experienced or seen. Home videos can help.

The precise emotional triggers that bring on cataplexy vary from person to person. For most, it's not stress or social anxiety. While laughter is the strongest, telling a joke can be just as powerful as passive amusement. Julie Flygare, author of a memoir about narcolepsy, told me she could visit and enjoy a comedy club without trouble. Although she was effectively medicated like Ann, she still might experience cataplexy if she unexpectedly saw a friend or an animal near her house.

Severe cataplexy can lead someone to suppress their emotions and push them away from social interactions with friends and loved ones. One example is the Oregon neuroscientist Matt Frerking, profiled in 2010 by *This American Life*. Speaking in a flat tone of voice, Frerking told an interviewer that he tried to enjoy things less and thought of himself as a robot. The radio program recounted stories about him falling into a flower bed and being unable to pet a puppy or attend a toddler's birthday party without collapsing.[8] His wife, Trish, wrote on her blog: "He's missed weddings, birthdays, and pretty much all of our family gatherings. He has to completely avoid interacting with our grandkids, spending any time he is in the same house with them off by himself in a quiet room, either working on his laptop or napping."[9]

Frerking's story inspired the 2019 film *Ode to Joy*, which framed cataplexy as an obstacle to romance. This movie was a more humane, empathetic vision of cataplexy than crude comedies such as *Deuce Bigalow*, but it still contained some inaccuracies. While some people with narcolepsy do take steps to try to push off cataplexy, such as tensing their muscles to avoid laughing, the techniques portrayed in the movie, such as listening to funeral music on the way to work or jabbing one's foot with a tack, are not likely to work.[10]

In some people with narcolepsy and especially in children, cataplexy can appear more subtly.[11] It may affect only part of the body, such as the face or arms, or it can appear as disturbed gait or repetitive movements of the fingers or tongue, without an obvious emotional trigger. In a video shown at a narcolepsy conference, a group of children from the Netherlands were at a birthday party, clapping their hands and singing a song. One girl kept clapping, but her face was distorted and her mouth was agape, the only indicators that she was experiencing partial muscle weakness. Cataplexy often changes as a child becomes older, shifting to the form more typical for adults.

The age-related differences can lead to problems when doctors impose assumptions about narcolepsy in adults onto children. In her memoir *Waking Mathilda*, Claire Wylds-Wright describes missed chances to diagnose her daughter, then three years old.[12] Wylds-Wright tells how her daughter experienced disturbed nighttime sleep and vivid hallucinations. Doctors first thought Mathilda had a brain tumor because of her wobbling gait and then referred both mother and daughter to psychiatrists. Narcolepsy was erroneously ruled out because her daughter didn't collapse after a quick surprising tickle.

CONDUCTORS OF THE ORCHESTRA

Both Ann and Anna Sumner were young women in the Atlanta area with neurological sleep disorders. Their first names may be similar, but there is a critical difference. Although Ann's diagnostic journey had more twists and turns, scientists now know much more about what lies behind her condition: narcolepsy type 1, also known as narcolepsy with cataplexy. For Ann's sleep disorder, a wealth of research has shown that the problem lies with part of the hypothalamus.

In narcolepsy type 1, an overreaction by the immune system appears to rub out a group of neurons in the hypothalamus. This selective process leaves neighboring neurons intact. We will discuss in more detail how this happens in the next chapter; for now, let's focus on what these neurons do. They produce a neuropeptide called hypocretin, which plays a critical role in stabilizing the "flip-flop" switches we discussed in the previous chapter. These are the connections between the sleep-promoting VLPO and other wake-promoting areas in and near the hypothalamus. It explains why people with narcolepsy type 1 have disturbed nighttime sleep as well as trouble staying awake—in the absence of hypocretin, sleep-wake transitions take place more readily.

Located in just one small region of the brain, hypocretin neurons have wirelike projections connecting to areas that promote alertness and regulate sleep/wake transitions. The receptors for hypocretin are spread around the brain, and engagement stimulates a network of arousal neurons. The network resembles an orchestra, with each section dominantly producing a different neurotransmitter: dopamine, norepinephrine, serotonin, or histamine. When someone is awake, each section is playing in concert with the others. Peptides such as hypocretin act more slowly and stick around longer, compared with classic small-molecule neurotransmitters—so we can view hypocretin as raising or lowering a curtain for the performance, while the neurotransmitters make up individual notes of the music.

WIZARDS OF LIGHT

Two separate labs identified hypocretin, also known as orexin, in the 1990s.[13] Both groups were interested in the hypothalamus, even though they weren't looking for something related to narcolepsy. The term *hypocretin* comes from the hypothalamus and from its similarity to another hormone called secretin, while *orexin* refers to how it can stimulate appetite when injected into a mouse's brain.

Before hypocretin was discovered, some information was available from studying dogs with inherited narcolepsy. Like humans, the dogs experience cataplexy when they feel excitement—when they are given food or toys. With the dogs, scientists were able to test various drugs and gather information about which regions of the brain were altered. But after the identification of hypocretin, researchers could more finely dissect the relevant brain circuitry. Electrical recordings in rats have shown that hypocretin neurons are most active when the animal is awake and exploring, less if awake and quiet, and inactive during sleep. They increase firing several seconds before animals wake up.

However, hypocretin neurons are not only "wake up" neurons; they also respond to novelty or excitement. With the optogenetics technique, by engineering light-sensitive proteins from algae into an animal's brain, scientists can use light from a fiber optics cable to activate or suppress preselected populations of neurons. Experiments from the lab of hypocretin's co-discoverer, the Stanford neuroscientist Luis de Lecea, show that hypocretin neurons start firing when a mouse is presented with a new toy, another mouse, or a snack—or is grabbed by a human hand or smells a potential predator.[14] De Lecea has quipped that in mice, hypocretin-producing neurons are active during the four Fs: feeding, fighting, fleeing, and mating.[15] "Hypocretin neurons' activity correlates with all of those activities," he said. "The key aspect is how they're responsible for integrating inputs of arousal and motivation."

It is now possible to create a mouse that experiences cataplexy by deleting the hypocretin gene; the animals are immobilized when presented with something enticing, such as chocolate. Since mice missing the hypocretin gene are thought to retain the neurons that produce the peptide, hypocretin is probably what's important, even though the neurons produce other neuropeptides. And despite their central roles, if hypocretin or the neurons that make it are taken away, mice can still sleep, eat, and mate—even if cataplexy may get in the way.

In addition to regulating sleep and wake, several research teams have shown how hypocretin neurons are connected to cravings for drugs of abuse.[16] At UCLA, Jerry Siegel and his colleagues have shown that in the human hypothalamus, chronic exposure to opioids appears to increase the number of hypocretin-producing cells—the opposite of narcolepsy.[17] Opiate exposure does not appear to be causing hypocretin neurons to multiply, but it is somehow activating hypocretin production in cells that had the potential for it.

The Rutgers neuroscientist Gary Aston-Jones, whose lab observed similar effects in rats exposed to cocaine,[18] said he viewed hypocretin as "a molecule for translating motivation into action." Both Siegel and Aston-Jones have suggested that the findings may offer a way to pharmacologically blunt drug cravings. By temporarily inducing a partial narcolepsy-like state, drugs that diminish hypocretin signals, such as the sleep aid suvorexant, may be helpful in treating addiction.

SOCIAL CONNECTIONS

Nobody has tried optogenetic techniques on people, but neuroscientists can use the presence of hypocretin in cerebrospinal fluid to infer the kinds of roles it plays in humans. Once released from the brain cells where they are made, peptides such as hypocretin diffuse into the ventricles, the reservoirs for cerebrospinal fluid.

Recall chapter 6, in which epilepsy patients came to the hospital for seizure diagnosis, and researchers took the opportunity to probe their brains for changes in adenosine. A similar set of experiments was performed to look at hypocretin levels' responses to different activities.[19] When Siegel's group at UCLA did this, they saw that hypocretin levels increased markedly when epilepsy patients were visited by family members or interacted with physicians or hospital staff (figure 12.1).

However, if we want to understand how emotions trigger cataplexy, the pattern of hypocretin production during social interactions is suggestive but not specific enough. In de Lecea's studies on mice, hypocretin neurons became activated during many situations reflecting novelty or stress, but not all of these situations trigger cataplexy. We need to integrate what is known about neurotransmitters with additional information about the regions of the brain involved.

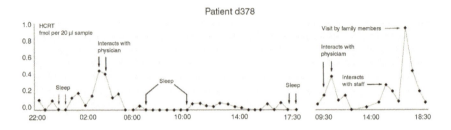

FIGURE 12.1. In people undergoing intracranial epilepsy diagnosis, hypocretin production appeared to increase during interactions with family or medical staff.

Source: Ashley M. Blouin et al., "Human Hypocretin and Melanin Concentrating Hormone Levels Are Linked to Emotion and Social Interaction," *Nature Communications* 4 (2013). Reprinted with permission ©2013.

IS CATAPLEXY UNBRIDLED REM SLEEP?

For decades, the leading theory for explaining cataplexy was as an uncontrolled manifestation of the loss of muscle tone that accompanies REM sleep. In the late 1950s, Michel Jouvet discovered that cats with lesions in a dorsal region of the pons, part of the brainstem, appeared to act out their dreams. This area, called the sublaterodorsal nucleus, orchestrates immobility during REM sleep. Spurred by the sublaterodorsal nucleus, neurons in the medulla and spinal cord send inhibitory signals to motor neurons throughout the body. Cataplexy was thought to represent a short circuit, so that the process gets activated while someone is awake.

Just as an aside, hypnogogic hallucinations are not as well studied, probably because they are less predictable, and sleep paralysis is known to occur occasionally in neurotypical people, outside the context of narcolepsy. Both are similarly thought to involve the aberrant activation of REM-like programs.

Current research is coming close to substantiating the connection between cataplexy and REM sleep.[20] However, the relationship becomes more complicated the closer one looks. Mehdi Tafti, a sleep researcher at the University of Lausanne, said that cataplexy and REM sleep may share the same neural pathways for loss of muscle tone, but beyond that, they don't necessarily have the same mechanisms. Cataplexy is not a uniform state and has its own distinct EEG features, he said.[21]

Some pharmacological evidence supports the cataplexy–REM sleep relationship. As Ann's experience demonstrates, various antidepressants reduce the frequency and severity of cataplexy, and this is attributed to their ability to suppress REM sleep. The correlation seems to work in the other direction as well.

Some medications, such as the blood pressure medication prazosin, can exacerbate cataplexy. In one severe case from Michigan, a woman in her forties began taking prazosin for hypertension about a year after she began to display signs of cataplexy.[22] The medication worsened her cataplexy to the point that she had "virtually continuous brief episodes of weakness of the face, neck, trunk, or legs." She could barely walk across a room without cataplexy occurring. When prazosin was stopped, her cataplexy ratcheted back to face and limb weakness at times of strong emotion. Prazosin works by blocking one type of receptors for the neurotransmitter norepinephrine, so researchers could infer that signals from norepinephrine help increase muscle tone and stave off cataplexy.

Current theories propose that signals that make people go weak in the knees with laughter are activated *all the time* during emotional experiences or social interactions. While it may seem odd to imagine this happening constantly, neurologists have observed that H-reflexes, or muscles' responses to electrical nerve stimulation, weaken during laughter—even in neurotypical people.[23] Hypocretin neurons appear to be part of a loop connecting several regions of the brain, with an offshoot leading to the brainstem. Stabilization by hypocretin usually prevents incoming emotional signals from causing widespread muscle weakness, but if hypocretin is not present, the circuit is imbalanced.

A team of neurologists at the University of Bologna has been gathering information on regions of the brain activated during cataplexy.[24] To set the scene, they asked children and teenagers with narcolepsy, newly diagnosed and unmedicated, to choose short movie clips—Roadrunner and Coyote cartoons or YouTube cat videos, for example. For each young study participant, a video that worked well in evoking cataplexy was shown while they were wearing EEG electrodes and inside a magnetic resonance imaging scanner.

During cataplexy events, imaging detected increases in blood flow in several brain regions, including the amygdala and part of the prefrontal cortex associated with pleasurable stimuli. The amygdala is known for its role in processing fear and fear-associated memories, but neuroscientists believe it is just as important as an "intensity detector" for positive emotions.[25]

If a study participant simply laughed at a cartoon, the same regions were not activated. The Italian researchers concluded that during laughter-induced cataplexy, the activated regions of the brain are the same regions that physiologically process positive emotion and amusement or reward.

Some additional evidence supports the idea that the amygdala's function is impaired in people with narcolepsy. Usually humans reflexively blink more quickly in response to unpleasant or disgusting stimuli, a process controlled by the amygdala, but people with narcolepsy do not display this response.[26] A model of how cataplexy occurs is that the amygdala sends signals that activate a REM-like paralysis program within the spinal cord, and those signals are successful because they are unopposed by hypocretin neurons.

Recent research suggests an important role for the peptide oxytocin, known for its roles in both childbirth and in promoting social bonding.[27] Oxytocin is supposed to drive us to seek out social situations and draw our attention to social cues. Preliminary findings indicate that in mice that model narcolepsy, oxytocin-sensitive neurons in the amygdala are active just before the onset of cataplexy. The involvement of oxytocin may begin to explain why in humans, laughter with friends or loved ones is so often a trigger for cataplexy. Oxytocin could be the spark of the emotional short circuit.

TURNING A KEY IN A LOCK

Neuroscientists have made considerable progress in understanding what hypocretin does and how its absence leads to cataplexy and disturbance of sleep and wake. However, all that research hasn't provided people with narcolepsy with new ways to manage their symptoms—yet.

A proposal for a nasal spray containing hypocretin generated excitement—but not a viable product. It took time for pharmaceutical companies to begin developing hypocretin receptor agonists, which mimic hypocretin's effects on its receptors. Because the potential market for *insomnia* was much larger, drugs that antagonize the same receptors such as suvorexant (FDA approved in 2014) came first. But another hurdle was pharmacological: finding candidate compounds that cross the blood-brain barrier and activate the receptors. De Lecea said it has been a hundred times more difficult to find an agonist, something that activates the receptor mechanism like turning a key in a lock, than an antagonist, which obstructs the same lock.

In 2017, Takeda Pharmaceutical Company began testing hypocretin receptor agonists in clinical trials. Other firms are developing similar drugs. These compounds are supposed to imitate the signals from hypocretin the brain is missing in narcolepsy type 1. Eventually, these could compete with the array of stimulants, antidepressants, and other medications that many people with narcolepsy now use. Whether they will be better than previous medications remains to be seen. Like conventional stimulants, hypocretin receptor agonists may also carry the risk of abuse or unpleasant side effects.[28]

CHAPTER 13

FRUSTRATING AND MOSTLY FRUITLESS

This is something like the finding that everyone who had cataplexy had red hair. While it might not cause narcolepsy, it might indicate who could develop it.

—American Narcolepsy Association newsletter, 1984

This chapter takes its name from a 2006 commentary by the Harvard sleep neurologist Thomas Scammell. At the time, researchers studying narcolepsy had clues to how hypocretin was being eliminated from the brain but were not able to discern the specific mechanism.[1] A few years after publication, the 2009–2010 H1N1 flu pandemic pushed the field forward. In the aftermath, researchers were able to obtain evidence that narcolepsy type 1 arises through the immune system making a disastrous mistake, coming from a combination of genetic susceptibility plus an environmental trigger. It is possible that a similar mechanism can explain other sleep disorders, such as narcolepsy type 2 or IH, but the full picture is not yet available. Plus, because the definitive laboratory test for narcolepsy type 1 has not been widely available until recently, many people currently diagnosed with narcolepsy do not know which category they actually fall into.

OLD-SCHOOL DEFINITIONS

Beginning in the nineteenth century, clinicians recognized that narcolepsy could run in families. One of the original descriptions of narcolepsy, from the German psychiatrist Karl Westphal, was of a mother and son who both experienced cataplexy and sleep attacks. Familial links helped strengthen neurologists' case for narcolepsy being a biological disorder.

In the 1950s, Yoss and Daly at the Mayo Clinic studied a Minnesota family in which seven out of sixteen siblings—twelve out of twenty-four people in the family they interviewed—had what they labeled as narcolepsy.[2] Their descriptions sound familiar. Taken to the clinic by her insistent daughter, one forty-three-year-old woman said that she had been "fighting sleep all of my life." That woman's brother denied abnormal sleepiness but said he had three times driven into a ditch on the way home from work, and his family said that he sometimes fell asleep at the dinner table.

It was just before the discovery of sleep-onset REM in daytime naps; all Yoss and Daly could really do was ask their patients questions. For most patients they studied, the family links were not as obvious. Yoss and Daly's survey of four hundred people with narcolepsy found that about a third had a family member with the disorder.[3] Other investigators found that this figure varied anywhere from 10 to 50 percent. The inheritance pattern for narcolepsy was more sporadic than for other diseases.

When advances in genetics in the 1980s made it possible to search more systematically, Japanese researchers included narcolepsy in a program investigating several complex disorders. A driving force behind this effort was the psychiatrist Yutaka Honda at the University of Tokyo. In the 1960s, Honda had helped organize Japan's first narcolepsy support group, and he was a pioneer in the use of antidepressants to control cataplexy.

Honda, working with Takeo Juji, a specialist in transfusion medicine, observed an association between narcolepsy and one form of an HLA gene called *HLA-DR2*.[4] The discovery was the first clue that narcolepsy had a connection with the immune system. HLA (human leukocyte antigen) genes were known to shape whether a transplant is perceived by the host's immune system as foreign and subject to attack. When someone needs a bone marrow transplant, HLA type is the main factor determining whether a donor is a match. The HLA locus is the most polymorphic and diverse in the human genome, probably because a variety of HLA genes provides an evolutionary advantage in the arms race against pathogens.[5] By that time, some HLA gene variants had been linked to other diseases such as rheumatoid arthritis and multiple sclerosis. The *DR2* variant doesn't sensitize people to autoimmune disorders in general—it is protective against type 1 diabetes.

The link between narcolepsy and *HLA-DR2* was very tight—all the Japanese narcolepsy patients were *DR2* positive. But around a quarter of the general population is also *DR2* positive. HLA genes may set the stage, but genetics do not determine whether someone is going to develop narcolepsy. They only indicate whether someone is susceptible.

Later studies of narcolepsy genetics including African Americans and European Americans revealed that another HLA genetic variant called *DQB1*0602* was a better marker than *DR2* for population groups outside Japan.[6] Again, most people who have this risk factor—still the strongest identified—do not develop narcolepsy. With identical twins, usually just one will have narcolepsy. Some other factor has to propel the disease to appear.

Throughout the 1980s, some tension existed between Honda and sleep researchers at Stanford over the proper definition of narcolepsy.[7] Among sleep specialists in the United States, the MSLT was widely adopted. In contrast, for Honda and his colleagues, displaying sleep-onset REM during an MSLT was not specific enough, and only patients with a history of cataplexy should be diagnosed with narcolepsy.[8] To emphasize the distinction, the Japanese group used the term "essential hypersomnia" instead of "narcolepsy without cataplexy." Some specialists outside Japan, such as the United Kingdom's David Parkes, reportedly preferred Honda's strict, old-school definitions. Parkes was known to tell trainees, "A good sleep centre has far more need of a psychiatrist than an EEG machine."[9]

As the dispute proceeded, genetics was beginning to show the differences between the two types of narcolepsy. When Stanford investigators surveyed people with narcolepsy, those with severe cataplexy were almost always *DQB1*0602* positive.[10] The proportion of people with narcolepsy *without* cataplexy who had the same HLA risk factor was 40 to 60 percent: higher than the general population but lower than those with cataplexy. These findings bolstered the view of narcolepsy without cataplexy as a separate, possibly nonuniform category.

A LONG MARCH, AIDED BY DOGS

Although many sleep researchers had a part, one who was central to defining hypocretin's relationship to narcolepsy was Emmanuel Mignot, director of the Center for Narcolepsy at Stanford (figure 13.1). Mignot served as chair of Narcolepsy Network's medical advisory board for many years. David Rye viewed him

FIGURE 13.1. Emmanuel Mignot and his dog Watson, who has narcolepsy.

Source: Lenny Gonzalez.

as a friendly rival; they've exchanged patient samples and reviewed each others' papers. Mignot's ambition to solve the puzzle of narcolepsy was tempered by empathy for his patients and a willingness to tell self-deprecating jokes.

One of Mignot's signature achievements was identifying the genetic basis for narcolepsy in dogs, after Dement had established a colony of the occasionally floppy animals at Stanford in the 1970s. When he first arrived from France, Mignot spent a great deal of time in close contact with the dogs, playing with them and bottle-feeding puppies if they were rejected by their mothers. "Even after I took a shower, you could still smell the dogs," he told the magazine *The Scientist*. "It was a strange part of my life."[11]

In dogs, the inheritance pattern for narcolepsy was simpler than in humans: there appeared to be a mutation in one gene. However, this project played out before the human or dog genomes had been sequenced. There was no link to the canine version of HLA. It took a decade of painstaking work and exploring dead ends to find the mutation.

The long march through canine chromosomes was nearly finished in the 1990s when the work of Mignot's team converged with independent research on hypocretin in mice. Masashi Yanagisawa's laboratory, originally focused on appetite rather than sleep, had generated mice missing the hypocretin gene. For months, it looked like the mice didn't have altered behavior, because the Texas researchers were monitoring them during the day. At night, when mice are more active, their behavior was different. The mice would be running around burrowing and grooming themselves but would suddenly become immobile.

Yanagisawa's group first thought the mice were having seizures, but EEG and muscle activity measurements revealed that what they were observing was a mouse version of cataplexy. The two labs sniffed each other out at a sleep research meeting and published papers in 1999 two weeks apart.[12] It turned out that canine narcolepsy came from a mutation in a hypocretin *receptor* gene.

While this did not explain human narcolepsy, Mignot and his colleague Seiji Nishino then guessed that it would be possible to see changes in hypocretin levels in cerebrospinal fluid.[13] This was a confident leap, given the disappointing track record of using CSF to reveal neurochemical alterations in narcolepsy. However, using an antibody-based laboratory test for the hypocretin peptide, it worked. A bank of postmortem brain tissue established at the University of Michigan enabled other investigators to see directly that a large number of hypocretin-producing neurons (more than 90 percent) were missing in people with narcolepsy *and* cataplexy.[14] While someone was alive, hypocretin CSF measurement was a way to see into the hypothalamus.

Since then, surveys of hundreds of people with narcolepsy and other sleep disorders found that CSF hypocretin levels are generally low or undetectable in the classic form of narcolepsy *with* cataplexy.[15] Low was defined as less than 30 percent of the average in healthy controls, and some people with traumatic brain injury or encephalitis had in-between levels. Most people with narcolepsy *without* cataplexy have normal hypocretin CSF levels, an indication that something else may be going on in their brains that does not involve elimination of hypocretin neurons. There are a few exceptions to this general rule. Low CSF hypocretin can precede the appearance of cataplexy by years, and a drop in an already low hypocretin level can follow development of cataplexy.[16]

Overall, hypocretin's discovery resulted in an incomplete redefinition of narcolepsy. Previously, the disorder was defined by disturbances in REM sleep, but with hypocretin, there was something more tangible: a molecule. Experts such as Mignot expressed confidence that their discoveries would lead quickly to new treatments. They also said they thought an autoimmune mechanism for narcolepsy was likely. What remained was to provide evidence of the immune system's treachery.

XENOPHOBIC POLICE

In a sense, autoimmune diseases come in two groups. In the first, the immune system attacks the body but never succeeds in eliminating the source of its irritation. Symptoms come from inflammation and the debris of battle left behind. Examples include systemic lupus erythematosus and rheumatoid arthritis. In the second, such as type 1 diabetes, the immune system "wins" by removing the cells it objects to. Narcolepsy type 1 appeared to be a stealthy example of the second type because of the lack of obvious inflammation in the brain. Hypocretin-producing cells were apparently eliminated while their neighbors in the hypothalamus were left intact.

With the HLA genetic link in mind, several narcolepsy research groups went looking for autoantibodies, a common feature of both types of autoimmune disorders. Antibodies are blood proteins whose usual function is to grab onto invaders, and autoantibodies are reactive toward components of our own cells. They specify what the immune system is going to mistakenly attack. Despite several high-profile papers on the topic, it is still debated whether autoantibodies play a role in narcolepsy.[17]

A 2009 *Nature Genetics* paper from Mignot's group pointed in a different direction than antibodies. The study of more than eight hundred people with narcolepsy, all with *DQB1*0602* and cataplexy, showed that a variant of one T cell receptor gene was a risk factor. This was a substantial clue but one whose significance did not become clear until later.

T cells represent the other major arm of the adaptive immune system, complementary to antibody-producing B cells. When both B and T cells develop, each goes through a set of mix-and-match genetic changes called V(D)J recombination. In T cells, the process starts with a large array of T cell receptor genes, rearranging and gluing together just two of them and leaving the rest intact. As a result, each mature T cell is capable of recognizing a different set of protein fragments, displayed within a specific frame provided by HLA proteins. The HLA proteins appear on a partner cell called an antigen-presenting cell, which holds recycled protein fragments on the outer membrane as signals that may excite T cells.

If pathogens are present, unusual fragments of digested protein appear on the antigen-presenting cell's surface. This is how "killer" T cells recognize cells that are infected by a virus. It's like having security police sort through food leftovers in a garbage can outside a house, to figure out whether there were foreign travelers inside. Another group of "helper" T cells looks through extracellular waste. T cells that recognize protein fragments from the body itself are supposed to be edited out during development, but this process is incomplete, and T cells can usually recognize more than one sequence. The capability of launching an autoimmune attack lies within many people, but the checks and balances of the immune system usually prevent such an attack from gaining enough momentum to do any real damage.

Mignot's group was also able to show that people whose narcolepsy symptoms had appeared in the last three years were more likely to have antibodies against streptococcal bacteria.[18] This made narcolepsy appear like a neurological version of rheumatic fever, a delayed inflammatory complication of strep throat infection that can damage heart valves and joints. An environmental trigger of some kind of infection seemed likely; immune cells can migrate into the nervous system via the upper respiratory tract and olfactory nerves.

FORCED BY THE FLU

What removed uncertainty about an autoimmune mechanism was the H1N1 flu outbreak, which began in mid-2009. Public health officials feared that the impact could be similar to the 1918 Spanish flu, killing millions of people around the world. Because of the anticipated impact of H1N1, many countries organized vaccination campaigns, and vaccines were rushed into service. Supply was limited because vaccine production was lower than expected.[19] To extend that supply, some vaccines contained an adjuvant: a mixture of ingredients that boost the immune response. Adjuvants, previously used with other vaccines but not with flu vaccines, would allow a reduced dose and faster deployment.

In Helsinki, the neurologist Markku Partinen was the first to detect a possible association between one particular flu vaccine called Pandemrix and new cases of narcolepsy in children.[20] Partinen saw several children in the spring of 2010 in whom symptoms had developed rapidly, some with severe psychiatric symptoms. Almost all had cataplexy, and all those tested were *DQB1*0602* positive. Fourteen were diagnosed before August, when public health officials in Finland suspended the flu vaccination program and launched an investigation. Similar findings were reported from Sweden and later from France and other European countries.[21]

It remains in doubt whether Pandemrix's adjuvant contributed to the problem, since a similar vaccine, Arepanrix, did not lead to a significant increase in narcolepsy cases in Canada.[22] The differences between Arepanrix and Pandemrix, made by GlaxoSmithKline in different locations, may point to manufacturing or processing issues.[23] GlaxoSmithKline sponsored research to examine the connection, bringing in narcolepsy experts such as Mignot. He worked with narcolepsy patient groups such as SOUND (Sufferers of Unique Narcolepsy Disorder) in Ireland to find those who had recently developed narcolepsy after being vaccinated.

Of the roughly seven hundred cases of Pandemrix-related narcolepsy reported in Europe at that time, more than half were found in Sweden—but more than five million Swedes were vaccinated with Pandemrix.[24] Given that risk ratio, it would have been difficult to catch the side effect in small-scale vaccine safety studies beforehand. Eventually, studies from several European countries showed that the apparent risk of developing narcolepsy for children immunized with Pandemrix was between five and fourteen times higher than the baseline in previous years.

After the H1N1 pandemic swept through China, a spike in narcolepsy was reported from a large pediatric clinic in Beijing. The effect was independent of vaccination, but it strengthened the argument that an immune response to flu was involved.[25] A network of pediatricians in the United States has observed a similar but weaker seasonal pattern of new narcolepsy cases.[26] Neither Arepanrix or Pandemrix was used in the United States.

VACCINE SKEPTICISM

Partinen said that after making his concerns about Pandemrix known, other scientists ridiculed or avoided him and even raised doubts about his mental stability.[27] He and his colleagues encountered resistance when trying to publish their findings in medical journals. The caution grew out of concern that antivaccine activists might exploit the information to erode public confidence in vaccines.[28] Mignot told Reuters that "No one wants to be the next Wakefield." He was referring to Andrew Wakefield, who was discredited for claiming a connection between the MMR vaccine and autism.

A decade later, when I conducted an informal survey of immunologists outside the narcolepsy field, several said they were skeptical of the proposed mechanism. Past precedent, such as a link between the 1976 swine flu vaccine and Guillain-Barre syndrome, did exist. Public health officials have said that media attention in Europe made it more difficult to sort out the epidemiology, because greater awareness accelerated narcolepsy diagnoses that might have occurred later.[29] An international study supported by the CDC found evidence for increased narcolepsy risk related to Pandemrix in Sweden but not in several other countries.[30]

While 2009 H1N1 flu vaccination programs are estimated to have prevented hospitalizations and saved lives in the United States,[31] several confounding issues make it difficult to tally risks versus benefits. These include increased awareness of narcolepsy coming from media attention, viral infections occurring in the same timeframe, and apparently greater risk for vaccine-induced narcolepsy among children compared with adults. The H1N1 flu pandemic's relative mildness was welcome, but it also complicates retrospective analysis.

THE IMPACT OF DISCUSSION

Limited research exists on the impact that public discussion of the flu vaccine–narcolepsy link has had on vaccine hesitancy or antivaccine activism. In France, a country where vaccine skepticism is relatively high, a survey of general practitioners found that flu/narcolepsy represented one of several vaccine-related controversies and not the most prominent one.[32] In the United Kingdom and Ireland, a steady stream of news articles has appeared about lawsuits and compensation claims against government agencies because of the vaccine-narcolepsy link. However, those driving them have said they're not against vaccines but are only advocating for people with narcolepsy. The Narcolepsy UK website simply says: "We believe that the causal link between the Pandemrix vaccine and a rise in narcolepsy cases has been established, and that our society has an obligation to look after individuals who develop narcolepsy as a result." Similarly, an in-depth study of the group Narkolepsiföreningen in Sweden, which was established specifically for people who developed narcolepsy after H1N1 vaccination, concluded: "None of them opposed vaccination in general, and none would place themselves in an anti-vaccination movement."[33]

Before the uptake of COVID-19 vaccines became politically charged in the United States, a few people at Atlanta narcolepsy support group meetings expressed worry about flu vaccines or others such as the human papillomavirus vaccine. Non-flu vaccines would be unlikely to have the same effect as Pandemrix. Still, given the known genetic risk, we can understand individuals' concerns regarding possible exacerbation of existing narcolepsy or triggering of narcolepsy in their relatives.

RETRACTION AND VINDICATION

An initial fumble by Mignot's group probably delayed wider scientific acceptance of the flu-narcolepsy connection. Researchers in his lab and another at Stanford had been looking at whether fragments of hypocretin could fit into the DQB1*0602 protein and activate T cells. In blood samples from Pandemrix-vaccinated children from Ireland, they could detect T cells that became activated in response to hypocretin. When comparing those children with their siblings, they could only find hypocretin-reactive T cells in those with narcolepsy. In addition, they noticed a sequence similarity between a fragment of hypocretin and part of the H1N1 flu virus. The idea was that through "molecular mimicry," flu virus fragments that resembled hypocretin were arousing the immune system to begin targeting hypocretin-producing neurons. The paper didn't specify whether vaccination or infection was the trigger—both could accelerate the immune confusion.

The Stanford team published its results in 2013, receiving acclaim from other narcolepsy researchers. Unfortunately, after a scientist who performed several experiments in the paper left his lab, Mignot's group was unable to reproduce their results. He and his colleagues decided to retract the paper in 2014. The NIH's grant reviewers took note of the retraction, and one of Mignot's major narcolepsy grants was not renewed. Some of his research staff had to leave to work elsewhere. "It was really painful and the worst time in my career," Mignot said.[34]

Four years later, researchers in Switzerland published more solid evidence for the T cell–mediated mechanism. The Swiss team, led by the neurologist Claudio Bassetti and immunologist Federica Sallusto, used a sensitive "T cell library" approach to detect rare hypocretin-sensitive cells in narcolepsy patients' blood that others had trouble spotting before.[35] In the intervening time, Mignot's lab had redone its work. By that point, they could identify troublesome T cells directly, pulling out the T cell receptors' genetic sequences. They published their follow-up a few months after the Bassetti/Sallusto paper.[36]

The Stanford and Swiss publications diverged in a few ways. Unlike some of the narcolepsy cases in the Stanford paper, none from the Swiss paper were connected with Pandemrix. In addition, the Stanford paper showed T cells recognizing only a couple sections of hypocretin. In contrast, the Swiss paper had T cells recognizing fragments spread out along the entire protein.

Despite differences between the papers, when both groups presented their work at the 2018 International Symposium on Narcolepsy in Massachusetts, there was a feeling that the field had turned a corner. The findings pointed toward a possible blood test for T cells recognizing hypocretin, which could facilitate narcolepsy type 1 diagnosis.

A hot topic of discussion at the symposium was early intervention. Part of the excitement about narcolepsy type 1's emergence as an autoimmune disorder is that there are many "off-the-shelf" options to try. Over the previous fifteen years, physicians have tried several times to arrest new-onset narcolepsy with immune-calming medications, such as a blend of antibodies (intravenous immunoglobulin) or corticosteroids.[37] The thinking behind this was that in patients who had started to experience narcolepsy symptoms, it might be possible to quench the immune destruction. However, past attempts at early intervention had occurred on a small scale and were inconclusive.

At the symposium, Michel Lecendreux, a child psychiatrist and sleep researcher from Paris, reviewed his and others' past efforts at early intervention and suggested "the clock is ticking" for recently diagnosed narcolepsy patients. Still, there was a lack of consensus over whether a randomized clinical trial or even open-label studies would be appropriate.

The discoveries did spur clinicians in Boston to begin treatment of one young woman's early-onset narcolepsy with natalizumab (Tysabri), which is approved for multiple sclerosis and is thought to stall the advance of T cells into the nervous system.[38] The patient herself wrote a blog post saying: "I still don't know if Tysabri is helping me and whether my spinal fluid hypocretin levels might have stabilized or even increased with this treatment, but it feels good to know that I might have kick-started research into a new treatment approach focused on reversing this disease early in its course, rather than just managing a lifetime of symptoms."[39]

After the Swiss and Stanford papers were published, others jumped in with findings supporting the autoimmune mechanism.[40] Precisely how immune destruction began was still unanswered. To attack hypocretin neurons, killer T cells would be needed in the hypothalamus—but in general, DQB1*0602 and other "class II" HLA proteins direct the activation of helper T cells, not killers.

Some investigators argued that the autoimmune mechanism was not proven, because killer T cells that could be activated by hypocretin can be found in the blood of people without narcolepsy and autoreactive T cells have not been observed in the nervous system or CSF.[41] Mignot's most recent paper on the topic downplayed the concept of molecular mimicry, after an examination of autoreactive T cells did not find any activated by both influenza and hypocretin peptides.[42]

For comparison, multiple sclerosis is well established as an autoimmune disorder, and the basics of the disease's destruction of myelin—the insulating material surrounding neurons' axons—have been known for years. But extensive research has not suggested that a single decisive autoantigen drives pathology, despite accumulating evidence for Epstein-Barr virus infection as a trigger.[43] In narcolepsy, the available evidence suggests a scenario where someone has preexisting inflammation in the upper respiratory tract, followed by additional events—in some cases, vaccination—prompting T cells to enter the nervous system and begin hunting for hypocretin-producing cells.

LIMITED EVIDENCE FOR PARTIAL HYPOCRETIN LOSS IN TYPE 2 NARCOLEPSY

Neurologists have been attracted to the idea that an incomplete loss of hypocretin-producing cells could result in sleepiness without cataplexy. Perhaps a higher threshold of hypocretin loss is required for cataplexy to appear. With an intermediate amount of hypocretin production, someone might be sleepy but still have enough hypocretin to trickle out into the CSF. The trouble is that visualizing hypocretin neurons while someone is alive, without poking into the hypothalamus, isn't possible at this point. For people with Parkinson's disease, doctors can look for the loss of dopamine-generating neurons via a PET scan. Something analogous for narcolepsy doesn't really exist.[44]

Preserved brains from people with narcolepsy *without* cataplexy are scarce, so direct evidence supporting the "partial hypocretin loss" idea has rested on only a few examples. One was a man diagnosed with narcolepsy whose brain was studied by Siegel and colleagues at UCLA.[45] They interviewed two of the man's friends after his death from colon cancer to check that he had never experienced cataplexy. His MSLT results were consistent: in each of three tests, he fell asleep quickly and entered REM sleep twice. In the man's brain, almost complete hypocretin loss was visible in the posterior part of the hypothalamus, but the anterior was relatively intact.

More indirect evidence comes from a case report from the University of Bologna.[46] A woman first became sleepy at age thirty-nine but fit MSLT criteria for IH. At age forty-three, she began to experience disrupted nighttime sleep and entered REM sleep three times on a second MSLT. Cataplexy appeared at age forty-five—and at that point the level of hypocretin in her CSF was lower than before. Thus, she gradually moved across all three diagnoses: IH, narcolepsy type 2, and, finally, narcolepsy type 1, presumably as hypocretin neurons were eliminated.

Additional supportive evidence comes from some people with Parkinson's disease, who have both excess daytime sleepiness and degeneration of hypocretin neurons, without low CSF hypocretin levels. Also, mice engineered with a partial loss of hypocretin-producing cells have mild narcolepsy-like symptoms.[47]

UNFINISHED BUSINESS

Most people diagnosed with narcolepsy or IH in the United States have reached that status by completing an MSLT and have never undergone a CSF hypocretin measurement.[48] Without a specific test, people who fit a narcolepsy diagnosis and don't have cataplexy may have hypocretin deficiency—or they might not.

"I just want to know," said Stacy Erickson Edwards, a woman living near Seattle who was diagnosed first with sleep apnea and then IH but suspected that she was developing cataplexy. "I've been trying to convince my doctors. We have good insurance, but they tell me the procedure is too invasive, and there's no point. And I'm not crazy about my IH diagnosis, because to some people it sounds fake."

A commercial test for CSF hypocretin, performed by a federally certified lab, did not become available in the United States until recently. Testing for the *DQB1*0602* risk factor was more accessible but could not confirm a diagnosis. Mignot tried to convince several companies to develop a commercial test but was repeatedly told the market for narcolepsy diagnosis was too small. His lab at Stanford provided CSF hypocretin measurements as a service to researchers and sleep specialists around the world but had to stop in 2017. Mayo Clinic Laboratories filled the gap in 2019.[49] "It's important for patients to be objectively tested, rather than for physicians just to rely on clinical history," said the Mayo neurologist Michael Silber, in a statement explaining the value of the test. "Because it's a lifelong disorder, we treat it with potentially harmful medications, and there's a big differential diagnosis of other causes of sleepiness that could be confused with narcolepsy."

In some European countries, CSF hypocretin is measured more often as part of narcolepsy diagnosis than in the United States. But because lumbar puncture carries risks, CSF hypocretin is not like insulin in the blood, something that can be checked routinely.

ANOTHER AUTOIMMUNE TARGET?

The "incomplete hypocretin loss" idea is plausible for some people with narcolepsy type 2 or IH, but not enough sleepy human brains have been sliced and stained to fully evaluate it. Another mechanism might be reduced neuronal sensitivity to hypocretin, analogous to type 2 diabetes.[50] Damage to the hypothalamus outside of hypocretin neurons may be a cause, but to produce something like narcolepsy type 2, the injury has to perturb the regulation of sleep while avoiding other vital functions. Also, for IH, unlike narcolepsy type 2, there is no known connection with HLA genes. That doesn't mean an autoimmune mechanism is impossible, but it would have to happen in a different way.

Speculatively, the immune system might be provoked to excise another group of neurons in the hypothalamus, in areas such as the paraventricular hypothalamic nucleus[51] or the tuberomammilary nucleus (TMN), which produces the wake-promoting neurotransmitter histamine. Neurons in the TMN increase in number in people with narcolepsy type 1, possibly to compensate for the loss of hypocretin neurons.[52] Adult mice with an engineered deficiency in histamine have trouble staying awake, spending more time asleep during both light and dark periods than wild-type mice.[53] While these similarities are suggestive, more research is necessary to identify the neurobiological basis for both narcolepsy type 2 and IH.

CHAPTER 14

EVERYTHING OFF LABEL

After all, we have all needed to write letters of medical necessity for patients with narcolepsy who were told that our prescribed treatment was not approved. . . . These issues are just as important as determining how many SOREMPs are on the patient's MSLT.

—Alon Avidan, 2012

Until Balance Therapeutics started clinical trials testing PTZ in 2015, no company had ever pursued approval for the IH indication—but twenty years earlier, a similar vacuum had existed for narcolepsy. As a guide to the obstacles pharmaceutical companies face with IH, we can learn a lot by examining the past behavior of both industry and the FDA with respect to narcolepsy. Most of this chapter will explore the history of modafinil, the first medication specifically approved for narcolepsy.

First, we should review why the lack of drugs approved by the FDA for idiopathic hypersomnia has been a major concern of the U.S. hypersomnia community. Conventional stimulants are available and familiar to many IHers, but they have their limitations and drawbacks. One IHer wrote on a Facebook thread: "I have never actually *liked* Adderall because it really does make my head feel like a jumbled mess with the unpleasant physical sensation of too much caffeine, but it has stayed my drug of choice because as long as I stick to my schedule, it's the only thing that keeps me out of bed and off the couch."[1]

For reference, a survey of the Hypersomnia Foundation's registry found that a majority of IHers on medication were taking conventional stimulants such as amphetamines or methylphenidate. The second-most-popular options were modafinil or armodafinil, used by 38 percent of IHers.[2]

If someone has an IH diagnosis, any prescription written by their doctor was considered "off label." It is illegal for pharmaceutical companies to promote their products for off-label purposes: indications for which they did not have FDA approval. Doctors do have the professional capacity to prescribe a medicine for a patient if they think it is appropriate, and off-label prescriptions are common in oncology, psychiatry, and pediatric medicine.[3]

While physicians can practice medicine independently, insurance companies in the United States generally take their cues from the FDA. In 2007, the AASM (American Academy of Sleep Medicine) provided advice for the treatment of narcolepsy and "other hypersomnias of central origin," but it took years for insurers' policies to reflect that guidance.[4] An AASM panel later recommended modafinil as a first-line treatment for IH because clinicians had experience with it and there was much evidence for its efficacy and safety—but that's not the same as an official FDA stamp of approval.

Until recently, IH was considered too rare and possibly too poorly defined for a pharmaceutical company to bother seeking approval for one of its medications. To use a legal analogy, IH lacked standing with the FDA. IH's lack of standing played out in the bureaucratic battles IHers waged against their insurers. In 2018, a member of the Atlanta hypersomnia support group invited me to review what they and their spouse dealt with. It typically began when they tried to get a prescription filled for modafinil: FDA approved for excessive daytime sleepiness associated with narcolepsy, sleep apnea, and shift work—but not for IH. The first obstacle the couple would routinely face was a requirement for prior authorization: a filter for expensive, risky, or possibly unnecessary medications. Their doctor, Lynn Marie Trotti, would write a support letter, but the request would still be denied. The IHer's spouse showed me one of the insurance company's letters, which cited the FDA-approved indications for modafinil, with IH not on the list. Usually at least one additional level of appeal was necessary. "Most people would just stop at the first appeal denial," they said.

The IHer had unfulfilled education and career plans that IH and other conditions had eroded. The spouse had worked at a large company, where the couple had been able to eventually get coverage for both modafinil and flumazenil after a few months. But when the spouse joined a startup and changed health insurance, they had to start all over again. Sometimes a paperwork delay would knock them back down a rung on the ladder of approvals. The Sisyphean process forced the couple to pay for both modafinil and flumazenil out of pocket, although modafinil was relatively affordable at a discount pharmacy. Financially, they were just surviving and treading water, since a significant amount of the spouse's salary went to support the IHer's medications.

In their letters to insurance companies, they'd tried citing research on the fuzzy distinction between IH and narcolepsy type 2. That didn't seem to work; for effective was persistence—and making a personal connection with someone working for the insurer. Getting coverage for flumazenil, which required a special order from a compounding pharmacy, was even more of a challenge. Citing past adverse reactions to other medications such as methylphenidate and clarithromycin was helpful.

BUREAUCRATIC BARRIERS

Doctors in the United States have reported that prior authorization causes delays that can hurt patient care. In a 2021 American Medical Association survey, one-third of physicians reported that such delays have caused at least one of their patients to experience a serious problem, such as hospitalization, the development of a birth defect, disability, or death.[5] In that same survey, more than 80 percent of surveyed doctors said their patients sometimes abandon recommended treatment because of prior authorization obstacles.

People with narcolepsy have prior authorization challenges too, but the extra difficulty associated with IH has been a reason why IH has been undercounted. Physicians sometimes assign patients with IH a narcolepsy diagnosis instead, to make it easier for their patients to obtain prescribed medications.

A 2019 survey conducted by the Hypersomnia Foundation makes the burden clear. It found that more than 70 percent of people with IH reported having prescribed medications initially denied by their insurance company. Appeals were successful about half the time, and a third never started the appeals process.[6]

The foundation's survey also found that 30 percent of people with IH have chosen not to fill a prescription because of out-of-pocket costs. Half were facing costs of more than $250 per month, substantial for a group of people with a chronic disorder that can interfere with employment. The survey indicated IHers' out-of-pocket costs were comparable to those of people with multiple sclerosis, the most expensive in a 2016 survey of common neurological conditions.[7]

For IH and prior authorization, health insurers' policies varied. Those who obtained coverage through Medicare and Medicaid were subject to state-by-state variations. Openings have come incrementally as a result of pressure from physicians and patients. In 2020, the largest, United Healthcare, stated that coverage for IH could be allowed for modafinil, armodafinil, and conventional stimulants, while others' policies did not mention IH.[8] (This picture did change after 2021.)

GATEWAY TO BILLIONS

Before the 1990s, no medication was specifically FDA approved for narcolepsy. Narcolepsy was generally treated with amphetamines and REM-suppressing tricyclic antidepressants, which both have a long list of unpleasant side effects. In 1982, the American Narcolepsy Association complained to Congress that "almost nothing is being done to find better medications for treating the symptoms of narcolepsy."[9] A rare disease such as narcolepsy offered little incentive for a company to conduct clinical trials and develop alternatives.

The 1983 Orphan Drug Act was designed to address this problem, which applies to rare diseases in general. It set out incentives such as seven years of market exclusivity after approval, grants for exploratory studies, and tax credits for up to half the cost of clinical trials. Decades later, the Orphan Drug Act is widely regarded as a success. In fact, critics have argued that it has gone too far, pushing companies to develop expensive therapies for rare diseases. Companies could then turn around and repurpose their medications for more widespread conditions.[10] As of 2019, almost a third of drugs in the industry pipeline targeted rare diseases, and rare diseases represented the majority of new drug approvals.[11]

Modafinil, previously known by its brand name of Provigil, is an example of the "orphan-to-blockbuster" phenomenon. It was the first medication explicitly approved for excessive daytime sleepiness associated with narcolepsy in 1998. But modafinil's use by people who didn't have narcolepsy quickly grew dominant commercially. It demonstrates how billions of dollars in revenue lie in the space between caffeine and amphetamines and how an aggressive pharmaceutical company used narcolepsy as a gateway to more lucrative opportunities.

Surgeons and pilots took the drug in order to stay awake longer and complete their duties safely. So did college students studying for exams and computer programmers looking to maximize their productivity.[12] In 2013, *New York* magazine proclaimed modafinil "Wall Street's New Drug of Choice." In 2018, White House physician Ronny Jackson was reported to distribute modafinil freely to presidential aides on overseas flights, a practice considered mildly scandalous.[13]

Modafinil has represented a kind of forbidden fruit for many people with chronic illnesses. The novelist M. J. Hyland, who has multiple sclerosis, has written about difficulties convincing the United Kingdom's National Health Service to allow her a prescription: "The problem isn't the drug, but the curse of its image:

the stigma of its 'recreational' use; the idea that the benefit of cognitive endurance is inherently greedy and frivolous, in the same dirty class as steroids, Botox and Viagra."[14]

ON ITS FEET AND FIGHTING

Modafinil's early development was driven by the French sleep research giant Michel Jouvet. Starting in the early 1980s, Jouvet and his colleague Helene Bastuji in Lyon prescribed modafinil on a limited scale to both narcolepsy and IH patients.[15] They reported that it was safe and effective at combating sleepiness, although it did not suppress cataplexy.

In the 1970s, modafinil's manufacturer, Lafon Laboratories, had discovered a related drug called adrafinil and marketed it as a cognitive enhancer for the elderly. Still, Lafon was reluctant to develop modafinil for narcolepsy because of the limited market size. Jouvet convinced the company to start clinical trials and promoted potential military applications for the drug, leading to its use by the French military in the 1991 Gulf War. He told a Paris defense conference that modafinil "could keep an army on its feet and fighting for three days and nights with no major side-effects."

According to Bastuji, Jouvet himself experienced excessive daytime sleepiness and fatigue in the 1980s but could not be treated with modafinil, because of strict conditions on its use.[16] In France, the drug became commercially available in 1994. Initially, it could only be prescribed by a neurologist and was only available through hospital pharmacies.[17]

Jouvet's research established an apparent distinction between modafinil and conventional stimulants.[18] Giving modafinil to animals kept them awake longer but did not lead to agitation or rebound sleep afterward. Jouvet found the drug did not activate the same neurons as amphetamines and did not seem to act via dopamine. These would become key selling points.

SELECTIVE ACTIVATION

The Pennsylvania-based company Cephalon licensed the rights to modafinil from Lafon in 1993 and organized the clinical trials necessary for FDA approval. These studies set the standards for other wake-promoting medications. The sleep medicine field had established objective or physiological measures of sleepiness, and these came into play for modafinil's clinical trials. FDA officials have occasionally expressed a wish for "real-world" sleepiness outcomes, such as the frequency of unplanned naps, but clinical trials have generally not relied on those types of measures for approval.[19]

For modafinil, the primary measures of efficacy were the Maintenance of Wakefulness Test (MWT), which was more responsive to treatment than the MSLT, along with the subjective Epworth Sleepiness Scale (ESS).[20] Headache and nausea were the most common side effects. Although modafinil did not eliminate daytime sleepiness, most people in these studies saw a significant decline in symptoms. In these studies, most of the more than five hundred participants had narcolepsy with cataplexy: more than 85 percent in one, 70 percent in the other.

In addition to demonstrating efficacy in narcolepsy, Cephalon was required by the FDA to assess modafinil's abuse potential. For people accustomed to stimulants, it could weakly substitute for more potent drugs; it produced more inhibition of sleep and evoked less pleasure.[21] Modafinil's physical properties also made it difficult to inject or smoke. Taken together, these factors explain why it received a Schedule IV designation from the Drug Enforcement Administration—with fewer restrictions on phone-in prescriptions and refills, in comparison with methylphenidate and amphetamines.[22]

Modafinil became commercially available in the United States in 1999. Its emergence excited researchers who were probing which parts of the brain were activated by the drug. "This is the first drug in history to selectively promote wakefulness in a way analogous to how the brain normally wakes up," the neuroscientist Dale Edgar told the *New York Times*.[23]

Contradicting Jouvet's previous research, Edgar and others at Stanford revealed that modafinil weakly increased dopamine levels in the brain.[24] The findings continued to show a distinction between modafinil and conventional stimulants. In mice, modafinil appeared to selectively activate cells in the hypothalamus, including those that produce hypocretin, while avoiding cells in reward centers activated by amphetamines.[25] Pharmacologists currently regard modafinil as an "exceptionally weak, but apparently very selective" dopamine transporter inhibitor.[26]

FORK IN THE ROAD

With their eyes on a larger market, Cephalon executives began exploring what would be necessary to expand the drug's indication to include other conditions, such as sleep apnea and neurological disorders. At one point, FDA officials suggested that they might be willing to grant "a general claim for the treatment of EDS [excessive daytime sleepiness] if it could be shown that PROVIGIL had an effect on this symptom regardless of the clinical setting in which it occurred."[27]

This was a fork in the road, with significant implications. Being able to treat EDS without regard to underlying cause would include being able to treat EDS with an unknown neurological cause. Also, it highlights the distinction between symptom and disease. Drugs for transient symptoms such as headache, joint pain, or mild insomnia are available over the counter, and several prescription medications are approved for insomnia without specifying that a cause must be identified.

However, for sleepiness, the regulators' stance was that anything beyond caffeine or dietary supplements should require a prescription, and the FDA decided a more specific diagnosis than EDS would be required as well. The agency told Cephalon that it would need to demonstrate efficacy in well-defined groups of people. Some officials could see what was coming. The documents prepared for the FDA's advisory committee contain this comment from Paul Leber, director of Neuropharmacological Drug Products: "One concern, however, is that 'hype' about modafinil's capacity to promote arousal and vigilance without the untoward stimulant-like effects of amphetamine may promote its use in a number of non-orphan conditions, notably in children with ADHD. If this were to occur, the major use of modafinil might well become an off label use."[28]

Cephalon was gearing up to promote Provigil more broadly. A January 2002 warning letter from the FDA said that the company's brochures and advertisements in medical journals were misleading, giving the impression that "Provigil can be used to improve wakefulness in all patients presenting with symptoms of daytime sleepiness, characteristic of generalized sleep disorders, whether or not they have narcolepsy."[29]

Critics have labeled Cephalon's marketing as a classic example of disease mongering: conflating the symptom of excessive daytime sleepiness, which many people experience, into a disease requiring a remedy. One Arizona doctor complained in 2009 that "excessive sleepiness was a little-known term to primary care physicians" until modafinil's introduction.[30]

A LARGE, NATIONAL EXPERIMENT

By 2001, sales of modafinil had shot up to $150 million, mostly going to people who didn't have narcolepsy.[31] Articles appeared in magazines such as the *New*

Yorker about modafinil's potential, suggesting that it may be possible to drastically limit someone's need for sleep.[32] "There are no warts on this drug," Frank Baldino Jr., Cephalon's hard-driving CEO—who reportedly used it for jet-lag—told the *New York Times*. "The only question is how big we can make it."[33]

At the time, Cephalon was conducting studies of modafinil in both obstructive sleep apnea and shift work sleep disorder. The company donated money for an endowed professorship at Harvard for one of the leaders of the shift work study, Charles Czeisler.[34] Because standard CPAP treatment of sleep apnea can leave people with residual sleepiness, sleep apnea was an obvious choice, but shift work sleep disorder was more curious as an indication. It was circadian misalignment imposed by someone's job. Do truck drivers or night-shift nurses necessarily have a sleep disorder?

For shift work, a case for symptom management could be made based on public safety. At night, participants in the shift work studies had the same level of objective sleepiness as people with narcolepsy or IH do during the day. That is, they fell asleep in an average of two minutes during a nighttime version of the MSLT. By this measure, modafinil made them somewhat less sleepy, nudging them up to four minutes. It cut in half the number who reported accidents or near-accidents while commuting home. Yet the authors noted: "Although modafinil improves the measured levels of performance, it is far from what is needed for these patients to function at a normal level."[35]

Several in the sleep medicine community expressed caution about modafinil's promotion. "What I worry about more is the issue of somebody who wants to succeed in business and work 100-hour weeks, and the way to do that is to pop a modafinil," said the University of Pittsburgh sleep researcher Timothy Monk, who was studying the drug's temporary use by astronauts. "That is a cause for concern, given that we don't know the effects of only having four hours of sleep and then avoiding the negative sleepiness with a pill."[36]

A series of studies conducted by U.S. Army researchers concluded that modafinil does not appear to offer advantages over caffeine for improving performance and alertness in healthy adults lacking sufficient sleep.[37] In a similar vein, a 2003 editorial from the University of Pennsylvania sleep researcher Allan Pack argued against the prospect of a broad FDA approval for excessive sleepiness.[38] Simply giving someone a pill for sleep apnea's symptoms without treating their respiratory problems would put them at higher risk of heart attack and stroke, he wrote. He added that in healthy people, encouraging them to sleep less with the help of modafinil also had risks. "If modafinil is used for this purpose, we will be engaged in a large, national experiment without data to support it," Pack wrote.

That fall, Cephalon almost got what its executives wanted. An FDA advisory committee split evenly on whether there was enough evidence for a broad label for excessive sleepiness, and the agency chose the narrower path. (Emmanuel Mignot was among the "Yes" votes on this question.)[39] "We're talking about treating a symptom without understanding the many possibilities that lead to that symptom," said Advisory Committee Chair Claudia Kawas, who voted "No."

This episode showed what regulators and some doctors were not quite ready for. If Cephalon had succeeded, that would have made it easier for people with idiopathic hypersomnia—and multiple sclerosis and other conditions—to obtain modafinil prescriptions and have them covered by insurance.

Another boundary was set in 2010 when the FDA declined to approve armodafinil (Nuvigil), a variation of modafinil, for the acute, transient condition of jet-lag.[40] Cephalon brought armodafinil forward to get around modafinil's expiring patent. Modafinil consists of a mixture of two mirror-image chemicals, while armodafinil is just one of the two, and armodafinil is supposed to last longer in the body. Neither drug shifts sleep-wake cycles—they only help someone stay awake.

SKIN IN THE GAME

Starting in the 1990s, whistleblowers and later the federal government pursued several cases against pharmaceutical companies for off-label promotion, extracting large settlements and fines. Several major pharmaceutical companies—including Pfizer, Eli Lilly, Bristol Myers Squibb, AstraZeneca, Abbott, and Johnson & Johnson—ran into this type of trouble, indicating how widespread the practice of off-label promotion was.

While Cephalon's promotion of an opioid lollipop received more press attention, modafinil was part of a set of whistleblower complaints alleging that the company was courting off-label prescriptions. Its sales representatives allegedly used exploratory studies of modafinil in several conditions, such as depression, multiple sclerosis, and Parkinson's, as promotional material and aggressively pushed for modafinil prescriptions, especially among psychiatrists.[41] The complaints were eventually settled in 2008 for more than $425 million.[42]

Along the way, Cephalon's efforts to have modafinil approved for children and adolescents with ADHD (attention deficit hyperactivity disorder) ran aground in 2006 because of reports of serious skin rashes. Three cases of severe skin rashes occurred among 1,236 children and teens participating in clinical trials, according to FDA briefing documents.[43] These were severe enough to notify a doctor, but they did not require medical intervention or hospitalization. At least one was considered by dermatologists to be Stevens-Johnson syndrome, a rare life-threatening skin disease involving widespread red or itchy blisters. Eight other patients discontinued modafinil because of rash or hypersensitivity. Cautiously, FDA advisors voted that modafinil should not be approved for pediatric use.[44]

Several other drugs have been linked to Stevens-Johnson syndrome, including NSAIDs (nonsteroid anti-inflammatory drugs), antibiotics, and antiepilepsy medications. It is suspected to be like an allergy—the immune system overreacts to the drug or one of its metabolites. For some drugs, the condition is linked to certain HLA alleles, suggesting an analogous mechanism to narcolepsy type 1 and other autoimmune diseases.[45]

In 2007, the FDA issued a safety alert because of reports of serious skin rashes developing in reaction to modafinil in adults. In postmarketing monitoring for modafinil, six cases of severe skin reactions were reported to the FDA.[46] The skin rash reactions, together with concern about adverse reactions related to psychiatric disorders, prompted the United Kingdom to request a European Medicines Agency review.[47] The data subsequently compiled by the EMA was more serious than what had been reported to the FDA. The EMA cited sixteen cases of Stevens-Johnson and related skin reactions occurring in a postmarketing setting, three of which were fatal.[48]

BENEFIT-RISK RATIO

In 2010, the European Medicines Agency withdrew approval for modafinil for sleep apnea, shift work sleep disorder, and idiopathic hypersomnia. The decision meant doctors in European countries could only prescribe modafinil for IH off label, and it would not be reimbursed by national health insurance programs. Previously modafinil was explicitly approved for IH in four European countries: France, Poland, Sweden, and Norway. Sleep specialists in Europe quietly disagreed with the EMA's decision and continued to consider modafinil a first-line treatment for IH.[49] The decision was also criticized with regard to sleep apnea: "Therefore, the removal of modafinil for this indication appears to have been an unfortunate decision by the EMA, particularly when the major pharmaceutical alternatives are amphetamine derivatives, which do not have a better safety profile and are well-known potential drugs of abuse."[50]

One aspect of the EMA's review deserves emphasis. When discussing clinical studies, the reviewers prioritized *objective* measurements of patients' sleepiness, based on tests like the MWT, rather than subjective questionnaire-type data. When the objective effects were small, such as in sleep apnea, the benefits were considered not "clinically relevant" and not worth the risks of psychiatric adverse reactions and skin rashes. Only in narcolepsy was the benefit/risk ratio acceptable, the agency said.

The studies submitted to the EMA only included a handful of IH patients, so the reviewers concluded that the condition was very rare. The scarcity of information spurred clinical researchers in France and Germany to perform two small clinical trials of modafinil that included people with idiopathic hypersomnia. In both of these studies, several people were already being treated with modafinil or methylphenidate before going through a washout period. The French study included a driving test, when participants drove a 230-kilometer highway trip, accompanied by a professional driving instructor.[51] It showed that modafinil could improve driving performance in people with narcolepsy type 1 and IH, cutting in half the number of inappropriate line crossings.

The study in Germany was the first placebo-controlled trial anywhere to focus on IH, although it recruited people with IH diagnoses who did not require long sleep. A few of the thirty-three participants slept only five or six hours on weeknights, while sleeping longer on weekends.[52] Modafinil improved participants' subjective sleepiness, as measured by the Epworth Sleepiness Scale, but there was no significant difference in sleep latency, as measured by the MWT. The study's organizer, Geert Mayer, said he sent the results to the EMA and never received a response. Reversing the EMA's decision on modafinil for IH may require a larger number of patients—or a group with more uniform sleep habits. A larger Japanese study of modafinil published in 2021, also focusing on people with IH without long sleep times, did show a significant effect on participants' sleep latency.[53]

While modafinil has been studied extensively for its ability to enhance cognitive performance in healthy adults, there is very limited data on its cognitive or performance-related effects for people with IH, such as the French driving study. One recent review concluded: "Data with these stimulants is far from positive if we consider that effects are small, in experiments that do not accurately reflect their actual use in the wider population. There is a user perception that these drugs are effective cognitive enhancers, but this is not supported by the evidence so far."[54]

NEWER OPTIONS

Even as newer drugs are approved for IH, modafinil will continue to be considered as a first-line option, because neurologists and sleep specialists have the most experience with it and because modafinil and armodafinil are currently available as inexpensive generics. That hierarchy could change as prescribers gain more experience with other medications (table 14.1). Two newer wake-promoting medications became available in the United States in 2019: Jazz Pharmaceuticals' Sunosi (solriamfetol) and Harmony Biosciences's Wakix (pitolisant). Their applications have not raised grand policy questions at the FDA in the way that modafinil did—so far. That's because modafinil had established a precedent, and neither manufacturer sought to expand the boundaries of what their medications could be prescribed for.[55] The FDA approved Sunosi for the indications of narcolepsy type 1 or type 2 and obstructive sleep apnea, while Wakix was approved only for narcolepsy.

Sleep specialists seemed to view solriamfetol as an extension of drugs they were already familiar with, since it works by enhancing signals from dopamine and norepinephrine.[56] Like modafinil, solriamfetol did appear to increase blood pressure slightly, a potential drawback for cardiovascular health. The Korean company SK Biopharmaceutical had originally developed solriamfetol as an antidepressant and licensed it to Johnson & Johnson, but an initial clinical trial in depression was not successful.[57] After reporting a side effect of insomnia, the Korean firm reframed solriamfetol for sleep disorders and sought out Jazz as a partner, leading to the drug's eventual approval.

In contrast, pitolisant was distinct from many previous wake-promoting medications, because it did not act through dopamine. Instead, it increases the availability of the neurotransmitter histamine, representing the opposite of the drowsiness induced by antihistamines. Pitolisant is a histamine H3 receptor antagonist, and its main targets are the histamine-producing neurons of the tuberomammillary nucleus. Other groups of neurons are stimulated too, because the drug weakens an inhibitory receptor present on neurons that do not make histamine themselves. The effect on other functions of histamine—in allergic reactions and in the stomach and heart—is limited, because the H3 receptor is mostly found in the brain.[58]

Originally developed by Bioprojet in France, pitolisant was approved in the European Union in 2016. Because of its different mode of action, pitolisant was thought to have less abuse potential, and it was the first wake-promoting medication to avoid Drug Enforcement Agency restrictions. For narcolepsy type 1 symptom management, another attractive property of pitolisant was its ability, demonstrated in some clinical trials, to reduce the frequency of cataplexy.[59]

TABLE 14.1 Wake-promoting medications investigated for idiopathic hypersomnia

Medication	Neurotransmitter	Mechanism	Clinical studies for people with IH?
Modafinil (Provigil)	Dopamine	Inhibits reuptake	Germany 2009–2011, Japan/Alfresa 2014–2015, Emory head-to-head 2019–2023
Armodafinil (Nuvigil)	Dopamine	Same as modafinil	No
Solriamfetol (Sunosi)	Dopamine, NE	Inhibits reuptake	No
Methylphenidate (many)	Dopamine, NE	Inhibits reuptake, promotes release	Commonly prescribed
Amphetamine (many)	Dopamine, NE	Inhibits reuptake, promotes release	Commonly prescribedEmory head-to-head 2019–2023
Mazindol (Quilience)	Dopamine, NE	Inhibits reuptake, promotes release	No; expanded access program in Europe
Pitolisant (Wakix)	Histamine	H3 receptor antagonist	Harmony: 2022–
GHB/Oxybate (Xyrem/Xywav)	GABA	GABA-B receptor agonist (see chapter 15)	Jazz: 2018–2020
Flumazenil (Anexate/Romazicon)	GABA	GABA-A receptor antagonist	Emory: 2010–2012
Clarithromycin (Biaxin)	GABA	GABA-A receptor inhibitor	Emory: 2010–2012, 2019–ongoing
Pentylenetetrazol (Cardiazol, Metrazol)	GABA	GABA-A receptor inhibitor	Balance: 2015–2018, 2018–2020
GR3027	GABA	GABA-A receptor inverse agonist (neurosteroid)	Umecrine: 2017–2019
TAK-925	Hypocretin	Hypocretin receptor agonist	Takeda: 2020

A study of pitolisant from Isabelle Arnulf's sleep disorders clinic in Paris was relevant for IHers seeking additional options. Arnulf's group had people with IH resistant to conventional stimulants try pitolisant in an open-ended fashion.[60] Arnulf's retrospective analysis was similar to how the Emory group assessed the efficacy of flumazenil. About a third of those whose records were reviewed reported an improvement on the Epworth Sleepiness Scale more or equal to 3, a significant alleviation of daytime sleepiness. A majority stopped taking the drug, mainly because of lack of efficacy and more rarely because of side effects, but the proportion who decided to keep taking pitolisant was about the same as with flumazenil.

PLAYING POOL: OBJECTIVE VERSUS SUBJECTIVE

Observers of the pharmaceutical industry have compared clinical trials to playing pool. Like committing to sinking a ball in the corner pocket, study organizers need to declare a primary endpoint ahead of time to get credit for it. When researchers are designing clinical trials for people with IH, what should be the measures of success? This decision cuts to the core of what IH is and how it is defined.

Going back to Bedřich Roth, we have seen that the sleepiness of hypersomnia needs to be evaluated and measured differently, in comparison with narcolepsy. For IHers, sleepiness is multifaceted, including an inability to stay alert, the total hours of sleep someone needs or can't push away, and sleep inertia. Measuring how quickly someone falls asleep only captures one part of IH, and sleep latency is often not the strongest aspect of sleepiness. Thus, to measure sleepiness in IH patients, the field may need to make a shift away from narcolepsy-oriented measurements.

At the same time, the EMA's safety review of modafinil does indicate regulators' preference for objective measures of patient benefit, given risks of adverse side effects. We can also see this preference in the FDA's clinical review history for pitolisant: "DNP [Division of Neurology Products] advised that efficacy for EDS [excessive daytime sleepiness] should be supported by positive findings in two adequate and well-controlled studies on both an objective and subjective measure.... The Agency does have reservations about the use of the ESS, as it requires patients to answer hypothetical questions and is also subject to recall bias."[61]

In the clarithromycin study organized at Emory, participants reported a strong effect on the subjective ESS but did not experience improvements in their PVT reaction time, the declared primary outcome. That result was not a problem, given how the antibiotic was being used clinically. Trotti and Rye regarded it as a temporary stopgap, and clarithromycin was not destined for commercial approval as a sleepiness remedy anyway.

Regarding outcome measures, the two trials for PTZ sponsored by Balance Therapeutics have taken a mixed approach. The first study, conducted from 2015 to 2018, included people with both narcolepsy type 2 and IH diagnoses, and any use of conventional stimulants had to be halted for the trial.[62] The primary outcome was the subjective Epworth Sleepiness Scale, with the MWT a secondary outcome. The second study lasted from 2018 to 2020 and allowed participants to be taking other medications, including stimulants, at the same time.[63] For this study, which included only people with IH diagnoses, Balance engaged the Michigan-based sleep researcher Tom Roth, who played a key part in establishing the MWT and MSLT.

For the second study, Roth and colleagues interviewed IHers and developed a symptom diary as the primary outcome measure. It includes questions on brain fog, fatigue, and memory impairment. Before any results were announced, Balance presented analyses of patient data at the 2019 World Sleep Congress in Vancouver. The study team observed that the IH patients they recruited had high subjective sleepiness on the Epworth Sleepiness Scale but could stay awake during a MWT for a normal length of time. Without breaking the blind of the placebo-controlled study, the analysis revealed that some participants' subjective levels of sleepiness and brain fog improved but their objective sleepiness did not. The investigators' conclusion was that patients' reports of sleepiness in IH related more to mental fatigue or daytime impairment rather than physiological sleepiness, as measured by tests like the MWT or PVT. This does not mean that study participants' sleepiness was less real, only that the physiological tests currently used to measure it were inadequate.

The question of subjective versus objective endpoints has also come up for other neurological disorders in which excessive daytime sleepiness and fatigue are prominent symptoms, such as multiple sclerosis and Parkinson's disease. These are conditions for which modafinil is widely accepted but not FDA approved. Studies in Parkinson's have shown stronger effects from modafinil on subjective versus objective sleepiness, and in both Parkinson's and multiple sclerosis, modafinil has stronger reported effects on sleepiness than fatigue.[64] Looking ahead, it highlights the need to develop standardized, well-accepted measurements for other aspects of hypersomnia, such as weekly total sleep time and symptoms such as brain fog and sleep inertia.

CHAPTER 15

KNOCK YOURSELF OUT

Disease states will also define the essential scientific and medical audiences that must be persuaded if the drug is to gain legitimacy, whether through FDA approval or through other means. . . . Whenever a drug is developed for a certain disease, the various communities organized about this category quickly become the therapy's most important constituencies.

—Daniel Carpenter, *Reputation and Power*, 2010

In 2018, Jazz Pharmaceuticals began organizing a clinical trial of a drug called JZP-258 for idiopathic hypersomnia.[1] Jazz's interest was a sign that part of the pharmaceutical industry was taking IH more seriously than before. Although small compared with pharmaceutical giants such as Roche, Jazz was a bigger player than Balance Therapeutics.

Jazz's drug was a low-sodium formulation of the sedative gamma-hydroxybutyrate (GHB), eventually named Xywav. It was a new version of a drug, GHB, which had been used to treat narcolepsy since the 1970s. Jazz had already begun clinical trials of JZP-258/Xywav for narcolepsy before initiating studies on IH.[2]

In the narcolepsy community, Xywav's predecessor Xyrem was well known. Xyrem had the distinction of being the first drug approved by the FDA specifically for the treatment of cataplexy in 2002. In the scientific literature, it was known as sodium oxybate or by its active ingredient: gamma-hydroxybutyrate.

According to Jazz, around 15,000 patients were regularly taking Xyrem in the United States by 2020.[3] Some, like Ann in chapter 12, swore by it. Others with narcolepsy were intimidated by the drug and vowed not to take it. One woman in Atlanta told me she cried herself to sleep her first night because she was so afraid of its side effects, yet she acknowledged its positive effects on her narcolepsy symptoms.

The drug, mixed in water and consumed as a clear or opalescent liquid, was supposed to enforce the continuous, unfragmented sleep that people with narcolepsy tend to have trouble achieving. When taken in the evening, Xyrem quickly plunged recipients into deep, nonphysiological sleep,[4] a state that could prevent them from hearing a fire alarm or getting up to go to the bathroom. Its effects were not sustained for more than a few hours, so they were supposed to take another dose in the middle of the night. The drug was only available from one central pharmacy in the United States, which shipped it directly to patients because of federal regulations. It came with rules that can constrain someone's schedule: no food two hours before bed, and no combining it with alcohol.[5]

For people with narcolepsy or IH, regularly quaffing a strong sedative may seem counterintuitive. Yet if someone lacks the feeling of restorative sleep, then it may make sense to push them into a state where some components of sleep occur, even if that state is artificial. We might compare the drug to a cast allowing a broken bone to heal.[6] "I really thought for a long time that idiopathic hypersomnia patients should never try Xyrem," Lynn Marie Trotti said in 2018.[7] "People with idiopathic hypersomnia already sleep really well—like arguably, much much too well, and certainly much too long."

What changed Trotti's mind—and was generally credited with opening her colleagues' eyes to the possibility—was a 2016 publication from Isabelle Arnulf's group in Paris. Arnulf found that her patients with IH responded to Xyrem as well as patients with narcolepsy type 1, often managing with a smaller dose.[8] More than a third experienced side effects such as nausea and dizziness, and more than half stopped taking it after fifteen months. Still, for more than 70 percent of the patients with IH who tried it, Xyrem improved their severe sleep inertia, a debilitating symptom of IH that had been poorly studied at that point.

A full dose of Xyrem (9 grams, spread out into two quaffs) contains more than 1,600 milligrams of sodium, more than the American Heart Association's ideal limit for daily food intake. Some sleep medicine specialists had been concerned about cardiovascular health in people with narcolepsy, since they often rely on blood pressure–increasing stimulants during the day.

Among those who had tried it, the drug was known for its soapy, salty taste. When the FDA gathered information on narcolepsy for its 2013 Patient-Focused

Drug Development program, Xyrem was mentioned hundreds of times—often positively, because it controlled commenters' symptoms. Others reported that the saltiness led to nausea and loss of appetite. One person wrote that the drug had "the taste of the bottom of the inside of a shoe after it has been worn all day." A high school student wrote an entire essay on how "it tastes so bad any attempt to describe it would not do it justice."

Thus, a low-sodium version of Xyrem could be a more palatable and cardiovascular-healthy option to offer patients. But Xywav's predecessor, like pentylenetetrazol, was another drug whose past cast a shadow over its present.

THE LONG-SOUGHT SLEEP HUMOR

The first to study GHB's clinical utility was Henri Laborit, a polymathic French physician. One of his primary interests was anesthesia. In the 1950s, Laborit shared a Lasker Award for his part in developing chlorpromazine, the first antipsychotic drug. Laborit was interested in GHB because he thought it might function as a GABA precursor and facilitate GABA's synthesis in the brain.[9] GHB has a close chemical relationship to GABA (figure 15.1). Laborit favored GHB as a sedative because it induced hypothermia and because he thought it did not suppress respiration, unlike other sedatives or anesthetics. It also had a "spectacular action on the dilation of the cervix." In the 1960s, GHB was used as a sedative during childbirth in France and Italy.[10] GHB is not an analgesic—it has to be combined with something else to block pain signals.

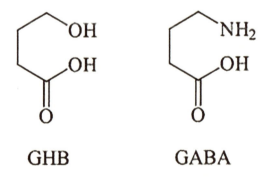

FIGURE 15.1. The chemical structures of GHB (gamma-hydroxybutyrate) and GABA (gamma-aminobutyric acid) are similar.

The Toronto psychiatrist Mortimer Mamelak brought GHB to the attention of the sleep research community in the 1970s. In people with histories of insomnia and depression, the drug promoted both deep slow-wave sleep and REM sleep. In the *Lancet*, Mamelak speculated about GHB as "an attractive candidate for the long-sought sleep humour." That is, the compound might be one of the elusive endogenous sleep substances researchers had been hunting for years.[11]

Mamelak and his Canadian colleague Roger Broughton gave a report on GHB at the landmark 1975 narcolepsy meeting in France.[12] Following Laborit's reasoning, they proposed that GHB could stoke GABA production and alleviate fragmentation of nighttime sleep. They had tested GHB in four women, all with narcolepsy and cataplexy, adding banana flavoring to mask the unpleasant taste.

With pie charts, Mamelak and Broughton showed how before treatment, the women's sleep was spread around a twenty-four-hour clock, but by the fifth day, it was confined to the hours between 11 pm and 7 am. They reported dryly: "Daily irresistible sleep attacks and cataplexy disappeared, the patients were better able to cope with daily chores, and their mood improved." At the time, Mamelak and Broughton were uncertain about how GHB worked. It might have to do with increasing GABA levels in the brain, but it might also stimulate dopamine synthesis or act through acetylcholine. They wrote: "The mechanisms of the apparent therapeutic effect of GHB in human narcolepsy-cataplexy remain obscure."

The 1975 conference proceedings contain an exchange between Broughton and Stanford's Christian Guilleminault, who expressed skepticism about a GABA-based mechanism.[13] More than forty years later, the field has not resolved the question of how GHB works.

EXERCISING LESS RESTRAINT

Mamelak and Broughton published several papers on their clinical experience with GHB over the next decade. A few other sleep specialists in Canada and France tried the drug with their narcolepsy patients, but uptake elsewhere was slow.[14] A few people in the United States traveled to Toronto to obtain access to it.[15]

In 1983, a sleep specialist in Cincinnati, Martin Scharf, obtained permission from the FDA to prescribe GHB for narcolepsy patients. He began doing so under a treatment IND (Investigational New Drug) program, which the FDA describes as "a mechanism for providing eligible subjects with investigational drugs for the treatment of serious and life-threatening illnesses for which there are no satisfactory alternative treatments."

What Scharf observed with the first thirty participants was not dramatic, when it came to measures of sleepiness. They were allowed to continue taking stimulants—and they still needed to, although not as much. However, GHB had a stronger effect on the other symptoms of narcolepsy: cataplexy, sleep paralysis, and hypnogogic hallucinations.[16] It was tricky, since most had been taking anti-depressants in an effort to control those symptoms, so withdrawal tended to make them rebound.

When Scharf began experimental treatment of patients with GHB, he admitted them to the hospital for safety reasons. He recalled that some were emotional when he came to visit them in the morning.[17] They surprised both Scharf and one another by not experiencing cataplexy. Some began testing their limits, seeking out situations that normally induced cataplexy by "exercising less emotional restraint." For one person, playing the video game Pac-Man had been problematic, Scharf said. "In every single case, we've seen improvement," he told *Cincinnati* magazine. "People who haven't been able to work, who've been on disability for years, are going back to their jobs."[18]

Without a corporate sponsor or a government grant, Scharf ran an open-label study of GHB for the treatment of narcolepsy, which eventually included about 140 people and lasted for more than a decade.

AN ORPHAN DRUG WITH SEVERAL PARENTS

When Congress was holding hearings for the Orphan Drug Act in the early 1980s, the American Narcolepsy Association had mentioned GHB as a promising drug that was not being tested.[19] A few years later, the FDA's Office of Orphan Products—created by the Orphan Drug Act—provided important support for GHB's development as a narcolepsy treatment.

The Office of Orphan Products provided a grant to researchers at the University of Arkansas, who performed a small placebo-controlled study of GHB.[20] Also, FDA officials approached several pharmaceutical companies about developing GHB. One abandoned the project after being acquired by a larger firm. The Minnesota-based firm Orphan Medical began developing the drug in the late 1990s. "We're the guys who give a damn when no one else does," Patti Engel, Orphan's director of sales and marketing, said in a Minnesota newspaper's profile of the company.[21]

While research on narcolepsy advanced, other communities had discovered the drug too. GHB was taken up by some body builders for its reported ability to stimulate the release of growth hormone.[22] Nightclub dancers found that "GHB induces a pleasant state of relaxation and tranquility and enhances one's libido."[23]

A recreational dose of GHB is not far from a dose that quickly knocks someone out, and the dose-response curve bends upward, meaning that twice the dose can have three or more times the effect. That's dangerous when a recreational user is not aware of precisely how much is being taken or is consuming alcohol as well. Consequences of overdose can include nausea, vomiting, incontinence, muscle spasms, coma, and even death.

In 1990, public health officials in several states had become concerned by reports of GHB poisoning, often in combination with other drugs.[24] That year, the FDA issued an advisory warning that GHB was unsafe and illegal outside of its use in narcolepsy research. Until that point, the substance was available in stores that sell vitamins and dietary supplements. In 1993, the rock star Billy Idol collapsed in convulsions outside a Los Angeles nightclub, reportedly because of GHB use.[25]

The drug acquired a list of colorful street nicknames, such as Grievous Bodily Harm or Georgia Home Boy; outside the United States, Fantasy or Liquid Ecstasy. GHB was also relatively easy to make at home, because its chemical relatives GBL (gamma-butyrolactone) and 1,4-butanediol were used as industrial solvents.

GHB attracted more attention when it was linked to incidents of drug-facilitated sexual assault and became known as a "date rape drug." In 1997, a member of Congress from Texas introduced a bill that would have the Drug Enforcement Administration designate GHB as a Schedule I drug—the same as LSD and heroin. This would have stopped all research. A complicated dance took place, with Orphan Medical executives meeting with prosecutors, police, and advocates for victims of sexual assault. The company argued for tight controls on GHB in order to forestall an outright ban.

In testimony before Congress in 1998, leaders of the sleep research community including William Dement joined together with Orphan Medical and the National Organization for Rare Disorders. They all pleaded with lawmakers, saying: don't ban this drug, let research on narcolepsy play out.[26] Mali Einen, a participant in one of Orphan's studies and later a narcolepsy research coordinator at Stanford, said that GHB had significant effects in reducing her cataplexy: "I feel like I have gotten a chance to have a life. One where I don't need to be afraid that I will fall and hurt myself or simply make someone else uncomfortable or afraid."

A compromise bill, signed into law in 2000, designated GHB as Schedule I but the prescription drug product would be Schedule III, allowing clinical research on narcolepsy to continue.

CLOSED DISTRIBUTION

When the FDA was evaluating GHB for narcolepsy, a large part of the safety data—in terms of total years of exposure—was from Martin Scharf's open-label study, dating back to 1983. However, Scharf was penalized by FDA inspectors because he did not keep records that were detailed enough for the FDA's standards.[27] "I took an enormous beating from the FDA, being reprimanded for violating procedures which didn't even exist at the time I was working on it," Scharf commented later. "It was a very bitter experience."[28]

When the FDA's advisory committee met to consider Orphan's application for GHB in 2001, several people with narcolepsy spoke about how much it had changed their lives. But alarm over the drug's non-narcolepsy uses meant that much of the discussion was about how to safely manage its distribution.

An emergency medicine researcher in Minnesota, Deborah Zvosec, described her studies on people who had become dependent on GHB. She called GHB "perhaps the most addictive drug ever abused" and warned that Xyrem might be attractive to people wanting to avoid the risks of illicitly produced GHB.[29]

In response to concerns about inappropriate use and diversion, Orphan Medical designed a closed distribution system for Xyrem, bypassing consumer pharmacies. There would be one manufacturing plant and one central pharmacy for the entire country. This is what people with narcolepsy who take the drug deal with today. Only a month's supply is delivered at a time, and the patient has to sign for deliveries; redelivery attempts are limited to minimize the amount of time the drug sits around in a warehouse or delivery truck.

ATTEMPT TO GO WIDER

For the first few years, Xyrem was unprofitable. But Orphan Medical executives seemed to have a strategy in mind for Xyrem similar to Cephalon's for modafinil/Provigil: obtain FDA approval with narcolepsy, then move on to other conditions that are more prevalent. "Xyrem is critical to Orphan Medical's future," the company's CEO said in 2000. "It will not only transform us into a profitable company, it will make us a recognizable emerging drug company."

Orphan's 2005 acquirer, Jazz, had fibromyalgia as a priority. Scharf had already begun research on GHB for fibromyalgia, which was estimated to affect millions of Americans.[30] Like other chronic illnesses, fibromyalgia has had a societal "legitimacy deficit" because it predominantly affects women and has subjective symptoms: fatigue and muscle or joint pain.

Unrefreshing sleep is a feature of fibromyalgia, and in the sleep lab, some people with fibromyalgia displayed "alpha intrusion": alpha EEG waves from wakefulness appearing during non-REM sleep. Scharf pushed for more studies on fibromyalgia, proposing alpha intrusion as a more objective measure of the disorder's severity. His study showed that treatment with GHB could reduce both measures of pain and fatigue.

When Jazz Pharmaceuticals bought Orphan in 2005, Orphan had already begun a clinical trial of Xyrem in fibromyalgia.[31] Since it represented many more potential customers than narcolepsy, the condition was mentioned in Jazz's annual reports as a market opportunity. Around this time, a few sleep specialists who worked with Jazz suggested that the company conduct studies with Xyrem on IH. Two, who were not willing to be identified, told me that this initial effort didn't go far, because FDA officials then regarded IH as too poorly defined, and company executives were focused on the larger prize of fibromyalgia.

During this period, Jazz was under financial strain. In 2007, after a public stock offering, the company had accumulated a deficit of more than $300 million since its founding.[32] In 2008 and 2009, the company missed more than one interest payment on its bonds, and its annual reports warned of the possible need to scale back research or declare bankruptcy.[33]

In 2010, the FDA rejected Jazz's bid to have GHB approved for fibromyalgia.[34] Although people with fibromyalgia testified to its benefits, efficacy was not the primary problem. Some of the FDA advisory committee thought scaling up Jazz's distribution and risk management system for Xyrem would be dangerous. They said it would result in too many opportunities for diversion and abuse, and GHB might be mixed harmfully with other medications such as benzodiazepines and opiates. At the advisory meeting, the Baylor College of Medicine psychiatrist Thomas Kosten was forceful in voicing his concerns about GHB.[35] "It's going to have to be very tight and otherwise you're going to have the street flooded with this [version] that is supplied by the manufacturer," said Kosten, who had studied GHB in people with opioid dependence.[36] "Right now it's viewed as a drug for a very unusual condition [narcolepsy] that doesn't get prescribed very much. That does not describe what fibromyalgia is."

GHB's inconvenience and cost, along with the existence of other options for people with fibromyalgia, made it a long shot anyway. But the FDA's negative decision showed the limits of what mainstream medicine would allow—at least in the United States. Xyrem was acceptable for people with narcolepsy but not for those with something more common and less well defined.

OFF-LABEL PROSECUTION

At the time when Orphan was applying to the FDA for approval of Xyrem, FDA officials asked the company to consider having doctors agree to prescribe the drug for patients with narcolepsy *with cataplexy only*.[37] The FDA's medical reviewer agreed, writing: "It is also to be expected that if Xyrem is approved without any restrictions on off-label use it is likely that it will be prescribed not just for the daytime sleepiness of narcolepsy (for which there is inadequate evidence for efficacy at present), but for daytime sleepiness of other causes and even for daytime fatigue."

The FDA's advisory committee voted in favor of restricting Xyrem prescriptions to "on-label use." However, implementing this restriction would have gone against policies on interfering with physicians' judgment and was something that the FDA counsel didn't feel the agency could legally do. Making the restriction explicit might have avoided trouble later. Orphan and Jazz became enmeshed in an "off-label promotion" case that eventually had far-reaching legal implications.

In 2006, federal prosecutors indicted a Maryland psychiatrist, Peter Gleason, for starring in Orphan and Jazz's promotional efforts. According to his indictment, he advocated Xyrem for a wide array of neurological and psychiatric disorders beyond narcolepsy. He gave hundreds of promotional talks to doctors around the country, for which Orphan paid him more than $100,000 in 2005. Gleason was recorded telling other doctors that "table salt was more dangerous" and that Xyrem was safe for children. In a press release, an FBI official compared his conduct to that of "a carnival snake-oil salesman." Martin Scharf, who had championed GHB for years, told the *New York Times*: "He is a very smart man, and I believe he is extremely well intentioned. . . . But this is not candy. It's not a cure-all."

It was the first time the federal government had prosecuted a physician for off-label promotion, rather than pharmaceutical companies. Gleason told the press that he was charged because he did not help the government build a case against Jazz. After pleading guilty to a misdemeanor, Gleason ran into trouble with multiple state medical boards and committed suicide in 2011, which relatives blamed on overzealous prosecution.[38]

Jazz settled its off-label promotion cases, civil and criminal, for $20 million. The company entered into corporate probation: a five-year agreement providing for monitoring of its marketing practices. Prosecutors also obtained convictions of two Jazz employees who worked with Gleason. One former sales representative, Alfred Caronia, successfully appealed his conviction on free speech grounds, with possible implications for how off-label drug promotion can be regulated.[39] If researchers can publish papers on off-label uses of a drug, why can't a pharmaceutical company tell doctors or even consumers about that research?

PRICE INCREASES

Xyrem was Jazz's main source of revenue. After the company was unable to expand the size of Xyrem's customer base beyond narcolepsy, it increased the price. According to data published by Bloomberg, no drug maker made a larger price increase for the period between 2007 and 2013 than Jazz did with Xyrem.[40] By 2019, Xyrem sales in the United States were $1.6 billion, with fewer than 15,000 active patients. On average, that's more than $100,000 per person in a year (figure 15.2). This parallels a similar increase in the cost of disease-modifying therapies for multiple sclerosis.[41] The six-figure yearly cost of Xyrem

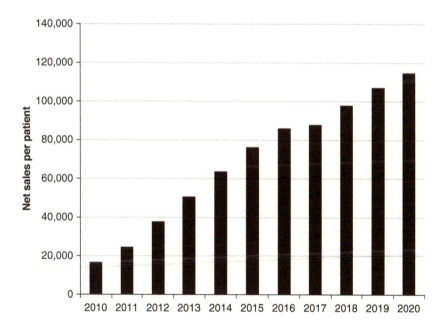

FIGURE 15.2. Estimates for the wholesale price of Xyrem, obtained by dividing total U.S. sales by active patients that year.

Source: Data obtained from Jazz Pharmaceuticals annual reports.

was considerably higher than that of other commonly prescribed wake-promoting medications, but it was more typical for drugs for other rare or orphan indications.[42]

To be clear, not everyone paid full price. Often an insurance company paid most of the cost, and Xyrem's price made it a fat target for prior-authorization screens. Some health insurers' policies stated that Xyrem should only be considered after a narcolepsy patient has already tried conventional stimulants or was unable to tolerate them.[43] Insurance coverage for off-label Xyrem was difficult to obtain, although it was possible.[44] Jazz has established support programs for patients who are uninsured or lack insurance coverage. The company was investigated for allegedly influencing an industry-funded charity, Caring Voice Coalition, which appeared to prioritize Xyrem patients over less expensive cases.[45] In 2019, Jazz agreed to pay $57 million in a settlement of the investigation with the Justice Department.

Unable to patent GHB itself, Jazz had patented Xyrem's risk management system, which it contended was an essential part of distributing its product.[46] It used that patent to fend off challenges from generic manufacturers. When Xyrem's patent exclusivity was set to expire at the end of 2022, FDA-approved generic versions and an extended-release formulation were waiting as competition—an explanation for the company's drive to develop the new low-sodium version.

In Europe, Xyrem was distributed by UCB, a Belgian company focused on neurological disorders. It cost several times less than it did in the United States, but it was still relatively expensive.[47] Until 2016, in parts of the United Kingdom, filling prescriptions for children with narcolepsy required a special request from a physician and a claim that a patient's need was "exceptional."[48]

Separate from its use for narcolepsy, GHB became available in the 1990s as a treatment for alcohol withdrawal in Italy, because of interest from addiction specialists there. Known as Alcover and formulated as a syrup, it has been prescribed to about twenty thousand people per year in Italy and Austria—more than the number of people who take Xyrem in the United States.[49] For alcohol withdrawal, the drug is given several times per day, in smaller doses than when used as a sleep aid for narcolepsy.[50]

In 2014, the French firm D&A Pharma applied for marketing approval across the European Union for Alcover. It ran into the same type of benefit-risk calculation that limited modafinil's approval. Regulators in several countries raised concerns about safety and misuse. D&A submitted studies on Alcover, some claiming its superiority to alternatives for managing alcohol withdrawal. Despite D&A offering to limit treatment to inpatient use, the application was denied in 2017.[51]

JAZZ'S INFLUENCE

Xyrem's high prices suggest consideration of Jazz's influence in the narcolepsy and hypersomnia communities. It was difficult to separate the company's role as a stakeholder from its status as a for-profit entity. Several prominent sleep neurologists were paid consultants for Jazz.[52] The company also employed people trusted by the narcolepsy community, such as the former Stanford research coordinator Mali Einen. Jazz has supported programs and events held by patient advocacy groups such as Narcolepsy Network, Wake Up Narcolepsy, and the Hypersomnia Foundation.[53]

From Jazz's point of view, expanding from narcolepsy to idiopathic hypersomnia was natural, given that the cohort of sleep neurologists who prescribed Xyrem for narcolepsy (about 1,300 nationwide) was already familiar with its peculiarities.[54] In response to questions, a Jazz spokesperson sent a statement, which said in part: "Since Jazz's inception, we have focused on identifying and developing the therapies people with serious sleep disorders need. . . . As one of the first and only companies to invest in and continuously commit to sleep medicine, we are uniquely positioned to influence the future of sleep disorder treatment, and we take this responsibility seriously."

Others may take a more critical look at Jazz's activities—which are typical for a pharmaceutical company—with the argument that commercial promotion distorts priorities. However, for the hypersomnia community, one goal of banding together was to get industry players interested—"to treat this like it's a real disease." Companies like Jazz have the muscle to get something through the FDA, so that drugs don't languish in limbo, like flumazenil had.

For other neurological disorders, patient advocacy groups such as the National Multiple Sclerosis Society have brought forward constituents' concerns about insurance coverage and out-of-pocket costs.[55] Narcolepsy advocacy groups have not done so to the same degree—possibly because of Jazz's dominance of the field. For multiple sclerosis, more companies are involved, and advocacy groups are less dependent on one single company or product.

Jazz's interests do sometimes coincide with the goals of narcolepsy patient advocates. For example, the company has sponsored awareness campaigns, and its "More Than Tired" advertising has contributed to greater public awareness of narcolepsy. In addition, Jazz has sponsored the only published studies of the cost of medical services for narcolepsy patients in the United States. In a 2018 paper, the authors' published conclusion was that timely diagnosis and treatment reduces other health care costs. Their data show that average pharmacy costs for people newly diagnosed with narcolepsy were around $6,000 per year and that claims for stimulant medications tended to decline after the first year.[56]

SAFETY ISSUES

People with narcolepsy sometimes report reluctance by physicians to prescribe Xyrem because of its reputation as a "date rape drug." However, it is unfair to confuse Xyrem's use for a legitimate medical purpose with crimes that have been committed with the aid of illicit GHB. Jazz's initial review of Xyrem's safety record did reveal some examples of abuse or dependence and two confirmed reports of sexual assault.[57] But a 2010 review concluded that other drugs were more likely to be connected to sexual assaults, and GHB's reputation was "counterproductive and misleading."[58]

More concerning for patients are the hazards Xyrem poses to the people who are supposed to benefit from it. The drug's safety has been a topic of debate in medical journals, some of it propelled by the Minnesota emergency medicine researcher Deborah Zvosec. For sleep specialists, the valued place Xyrem holds in the management of narcolepsy became clear when its safety was questioned.

In 2009, Zvosec and colleagues published a report in *Sleep Medicine* on three people who died after taking Xyrem, not illicit GHB. All three had taken other drugs, and two did not have a confirmed history of narcolepsy with cataplexy. The authors called attention to the drug's respiratory depressant effects and risks for people who were obese or taking other medications.[59] This prompted a letter to the editor of *Sleep Medicine* signed by twenty-four physicians, representing the European and North American sleep neurology establishment, defending Xyrem as "an effective and safe treatment for patients suffering from narcolepsy." Zvosec also criticized a Jazz-sponsored study of Xyrem in people with sleep apnea, which concluded that the drug "might increase central apneas and cause oxygen desaturation in some individuals and should be used with caution."[60]

The FDA eventually came to a similar conclusion, but a full picture of Xyrem's safety record was not available until several years after its approval. The close contact with patients built into Xyrem's distribution system may have provided an opportunity for thorough monitoring, yet it appears that adverse events were underreported until 2011 (figure 15.3).

At the time of an April 2011 visit by FDA inspectors, the company did not have adequate procedures for adverse event reporting, according to a warning letter released later that year.[61] As a result, serious adverse events, including

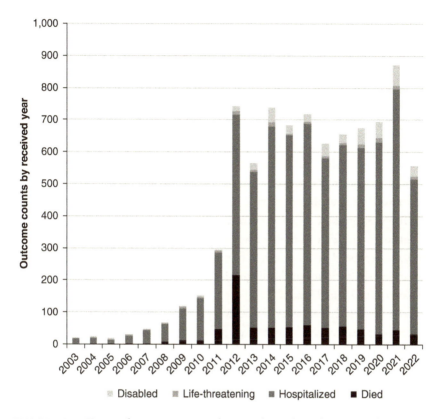

FIGURE 15.3. Xyrem adverse event reports by year, obtained via the FDA's Adverse Event Reporting System Public Dashboard. Events appear in the database according to the date received, not when they occurred. Most Xyrem-associated deaths involved other drugs, and assigning causality is not possible based on this data alone.

eighty-two deaths, had not been recorded by the central pharmacy and reported to the agency. [62] Most reported deaths associated with Xyrem until 2011 had been missed.

In 2012, the FDA found that many of the deaths came when Xyrem was combined with other drugs such as alcohol, benzodiazepines, or opioids.[63] It appeared that a majority of adverse events and deaths occurred in patients who were prescribed Xyrem for off-label uses such as fibromyalgia, insomnia, or migraine. The agency recommended that Xyrem should be used cautiously, if at all, in patients with respiratory issues such as sleep apnea or COPD (chronic obstructive pulmonary disease) because of the risk of respiratory depression. The agency also added a contraindication against alcohol use to Xyrem's label. A report from a nonprofit group, the Institute for Safe Medication Practices, concluded: "For almost a decade, the true adverse event profile of sodium oxybate was not known to the company, the medical community, or the FDA."[64]

STABILIZING THE DEFINITION OF NARCOLEPSY

Jazz-sponsored studies have shown what many experts on narcolepsy often do not acknowledge. In the United States, diagnoses of narcolepsy *without* cataplexy have become several times more prevalent than diagnoses of narcolepsy with cataplexy, contradicting an assumption expressed in older publications on narcolepsy.[65]

The Stanford-based authors of the Burden of Narcolepsy Disease study, which scanned a nationwide insurance database for narcolepsy-related claims from 2006 to 2010, expressed surprise at their findings. They wrote that it "invites suspicion regarding the accuracy of the diagnosis of narcolepsy without cataplexy by physicians in the United States."[66]

The numerical dominance of narcolepsy without cataplexy in the United States does not appear to have been the case before the twenty-first century. For example, earlier studies in Minnesota and the Seattle area found that narcolepsy with cataplexy was more common.[67] It also contrasts with recent studies from other countries such as Switzerland and Spain.[68] Thus, the increase in narcolepsy without cataplexy in the United States may be partially because of differences in diagnostic practices between countries, rather than differences between their populations. Put more simply, physicians in the United States may have been diagnosing narcolepsy less rigidly than in other countries. This could mean not checking for insufficient sleep before an MSLT or not investigating other potential reasons for excessive daytime sleepiness. In support of this point, U.S. military clinicians reexamined twenty-three service members diagnosed with narcolepsy at civilian sleep centers, but they were able to confirm the initial diagnosis in only two.[69]

Stabilizing expert opinion on narcolepsy's looser definition was in Jazz's interest, even if the company was not responsible for the expansion in narcolepsy diagnoses. Initially Xyrem was approved specifically for the treatment of cataplexy but not for treating daytime sleepiness. The FDA had concluded that Xyrem was effective at reducing the frequency of cataplexy, but the data on daytime sleepiness was "very weak at best."

In 2005, the FDA approved Xyrem for excessive daytime sleepiness in the context of narcolepsy, based on two studies. One showed its subjective benefits in combination with stimulants.[70] The other study did not require participants to have narcolepsy *with* cataplexy and showed that Xyrem could partly substitute for modafinil.[71] The approval meant that the overall diagnosis of narcolepsy was relevant for insurance coverage, rather than the specific symptom of cataplexy.

Later, Jazz supported a June 2012 conference that sought to establish a consensus on diagnosis of narcolepsy type 2. A report from the conference noted that some participants thought narcolepsy type 2 might lie in the borderland between narcolepsy type 1 and IH, while others perceived it as a distinct entity.[72] For his part, David Rye came back from the conference grumbling about his peers' willingness to rely on the MSLT.[73] The conference report emphasized the limitations of the MSLT yet said that it remained the most important test for narcolepsy type 2 diagnosis.

THE FLOOD OR THE TRICKLE?

Despite Xyrem's demonstrated efficacy in managing narcolepsy symptoms, sleep researchers remain uncertain about the drug's mechanism of action. The history of its active ingredient, GHB, stretches back decades before scientists knew much about the neurological basis of narcolepsy.

Biochemically, the drug resembles GABA. For its acute knockout effects, GHB is thought to act through GABA-B receptors, which are separate from the GABA-A receptors for benzodiazepines. GHB binds to GABA-B receptors with less affinity than GABA itself. Like GABA-A receptors, GABA-B receptors are inhibitory, but they trigger a different set of signals inside neurons, and their distribution in the brain is also different.[74]

As far back as the 1960s, neuroscientists detected GHB as a natural metabolite in the mammalian brain, leading to speculation about its possible physiological role as a neurotransmitter.[75] However, when someone takes the drug as a fast-acting sedative, the brain is flooded with GHB at a concentration hundreds of times higher than the amounts naturally found there. Do the drug's benefits come primarily from the flood or the trickle in its aftermath? The drug could also be affecting brain cells' metabolism, because GHB is quickly converted into an energy source.

One 2016 review offered this explanation: "The mechanism of the majority of GHB pharmacological effects are thought to interfere primitively with the brain GABA system," but "it seems difficult to consider GHB pharmacological effects mediated only and purely via GABA-B receptor stimulation."[76] In addition to GABA-B receptors, several research groups have identified "high affinity GHB receptors" bound by GHB in the brain—tight enough to hold on even after the flood washes away. However, the biochemical identity of these receptors has been elusive. In 2021, Danish scientists found an enzyme that appears to be the long-sought target, but investigating that enzyme's relevance for narcolepsy remains to be done.[77]

When used for management of narcolepsy, GHB's effects are not immediate. With respect to daytime sleepiness, it can take days or weeks for patients to feel the benefits.[78] The drug appears to be gradually altering how the brain's sleep and wake circuits function, even after it has washed out of the body. "Chronic treatment is doing something to reset the circuits," said Tom Kilduff, a neuroscientist in California whose laboratory has studied GHB extensively in animal models. "We don't understand the pathway that makes GHB effective therapeutically," Kilduff added. "That makes it a very attractive scientific question."

GABA-B receptors are present on several groups of neurons in the brain that regulate sleep. Because the drug's therapeutic benefits are clearest in narcolepsy, it is reasonable to ask whether it bolsters or substitutes for the function of hypocretin neurons so central to narcolepsy type 1. Which neurons are important for GHB's effects remains an unanswered question, Kilduff said. "Hypocretin neurons are part of the story, but not the whole story," he said.

When other researchers gave rats GHB, they were surprised to find that it didn't activate brain regions connected with reward and addiction, such as the nucleus accumbens.[79] Some experiments indicate that GHB acts on other neurons that connect the thalamus and cortex, which drive slow wave oscillations during non-REM sleep.

GHB appears to be an exception to the general rule that artificial sedation does not substitute for natural sleep.[80] GHB is known to increase slow waves, which are thought to be important for memory consolidation and possibly for the other aspects of restorative sleep. However, slow wave oscillations provoked by GHB don't look the same as physiological slow wave sleep.[81]

Remember the Michigan woman whose cataplexy was made worse by the blood pressure medication prazosin? GHB might be accomplishing the reverse, by stabilizing the locus coeruleus. The locus coeruleus is a group of neurons in the brainstem that produce norepinephrine, whose lack of activity is critical during cataplexy. Beyond cataplexy, it is possible that the neurons needed for GHB's alleviation of daytime sleepiness are separate. The puzzle highlights how sleep and sleepiness are orchestrated by networks of neurons, not by discrete centers in the brain.

BACLOFEN: POOR MAN'S XYREM

If acting through GABA-B receptors was the only way GHB was exerting its effects, then the muscle relaxant baclofen, less expensive and less prone to abuse, might be able to do the job. Sleep researchers have wondered since the 1970s whether baclofen—a canonical GABA-B agonist—might be helpful for people with narcolepsy. A few members of the Atlanta hypersomnia community have tried baclofen for themselves, describing it as "the poor man's Xyrem."

Because of Xyrem's side effects, cost, and restrictions around its use, clinicians keep proposing baclofen as a substitute for GHB, despite the negative experiences of influential figures such as Stanford's Christian Guilleminault.[82] Some clinicians have reported that their narcolepsy patients have had a positive response to baclofen.[83] Kilduff's work on mouse models has also suggested that R-baclofen could be effective for narcolepsy.[84] (R-baclofen is one of two mirror-image forms of the molecule; the R-form has higher affinity for GABA-B receptors.) Baclofen's lack of interaction with the shadowy high-affinity GHB receptor suggests that baclofen's and GHB's shared mechanism must be related to GABA-B receptors. However, since narcolepsy type 1 and IH are different disorders, the factors that determine efficacy for GHB or baclofen may also differ between the two groups of patients.

CHAPTER 16

BIOMARKERS OF SLEEPINESS—AND IH

In our busy world, natural long sleepers attract less interest, although their capacity of spending many hours asleep could be regarded as no mean accomplishment.

—Alexander Borbely and colleagues, 1996

For more than a century, scientists have been looking for the stuff in the brain that makes someone feel sleepy. It's becoming clear that sleepiness is better thought of as a pattern embedded within our brain cells, rather than a substance that can be extracted from the body.

Practically speaking, it's much easier to detect physiological signs of sleepiness, like droopy eyelids or impaired reaction time, than it is to isolate sleepiness itself. The signs don't *cause* sleepiness but are tightly associated with it. Scientists who study sleep have long considered it a top goal to establish "biomarkers of sleepiness," and millions of dollars of investment have gone into developing them.

We can easily see the need when considering public safety. A driver who has not slept in twenty-four hours is as impaired as someone who is legally drunk, according to measurements of their driving performance.[1] The federal government estimates that thousands of crashes and hundreds of fatalities occur every year nationwide as a result of drowsy drivers. Also, a sizable percentage of commercial truck drivers have obstructive sleep apnea, putting them at higher risk of a crash.

Clinicians know how to screen for and detect sleep apnea, but someone can be drowsy for other reasons, including simply not getting enough sleep. But some people don't recognize that they are pathologically sleepy until it overwhelms them. A broader test for drowsiness, not only one identifying breathing disruptions, could catch people who present potential risks. Professional groups whose performance depends on their ability to stay awake for long periods, such as truck drivers, pilots, or surgeons, could be asked to undergo some kind of alertness test or, alternatively, spit into a cup or even give a blood sample.

Like people with sleep apnea, people with narcolepsy or IH are assumed to have a higher risk of car crashes, but just a few studies have looked directly at their risk.[2] The most recent one from France found that patients with narcolepsy or IH had double the risk of having a car crash in the last five years, compared to controls.[3] Most studies say that the MWT can predict driving performance. But one paper by Dutch sleep researchers compared the MWT, along with the psychomotor vigilance test and a sustained attention test, to an actual driving performance test for evaluating someone's fitness to drive. The authors concluded that "none of the tests had adequate ability to predict impaired driving, questioning their use for clinical driving fitness evaluation in narcolepsy and IH."[4]

Both researchers and policy makers have lamented the lack of a broadly applicable "breathalyzer for sleepiness."[5] It has been a challenge to get to the point of having good biological markers. Scientists are close to being able to spot acute sleep loss, but chronic sleep insufficiency—getting fewer hours of sleep than ideal for a few weeks—may be harder to discern. A 2015 workshop on the topic concluded: "Laboratory measurements such as EEG-based assessment of sleepiness or objective performance measures are expensive and complex, not amenable to general use and there is no conclusive evidence that they provide gold standard measurements by which assessment of the state of chronic sleep loss can be determined."[6]

More recently, there have been signs that the field is getting closer to its longstanding goal. The Australian company Optalert currently produces infrared goggles that measure how fast and how far the eyelids open after a blink, generating a drowsiness score. The goggles are a scientifically rigorous way of measuring droopy eyelids.[7] In a different mode, researchers at the University of Surrey in the United Kingdom have been closing in on a blood test that could reliably identify otherwise healthy people who have skipped a night of sleep. Their still-experimental test looks for changes in activity in a panel of genes, tracking cellular stress.[8]

The extent of the effort to identify biomarkers of sleepiness is beyond the scope of this book, but some of this research might eventually be applied to people with narcolepsy or idiopathic hypersomnia or those suspected of having these conditions. Currently, these types of measures are not used in the sleep clinic. The standard ways to gauge someone's sleepiness are subjective, like the Epworth Sleepiness Scale, or operational, like the MSLT or MWT, which serve as imperfect measures of daytime sleepiness.

Looking ahead, it could be revealing to apply newer biological tests of sleepiness to people with IH. IHers often say that they feel like they've stayed up all night, despite spending an amount of time sleeping that most people would consider more than enough. Having an easily measurable biomarker of sleepiness might provide an explanation for how they feel.

We don't know how IHers would perform on the newer biological tests. Perhaps some IHers will have droopy eyelids or a slower reaction time for more of the day, or perhaps patterns of gene activity in their blood might match the effect of sleep deprivation. Others may not display the same similarities. Would that mean their feelings are less valid? This points to how sleepiness can be viewed both objectively or subjectively.

In this chapter, we will turn to two types of studies that do reveal something about IH and how it affects the brain. The researchers were looking for what IHers have in common, as well as what distinguishes them from other groups of sleepy patients, such as people with narcolepsy or obstructive sleep apnea. These studies are examples of insights that the increased attention to IH has made possible.

PLANTE'S PSYCHIATRIC PERSPECTIVE

The first study we will examine was conducted at the University of Wisconsin's sleep clinic. Its director, David Plante, comes to IH research with the perspective of a psychiatrist—he told me that his training background in psychiatry has given him a "pragmatic and symptom-based lens" through which to view IH. A major theme of his published work has been on the overlap between IH and depression. "Depression and hypersomnolence are intertwined," he said at a talk for the Hypersomnia Foundation in 2019. "Depression is a common presumptive diagnosis in IH, yet many studies seek to exclude depression in cohorts of people with central nervous system hypersomnia."

While the majority of people experiencing depression report disturbed sleep or insomnia, about a quarter display hypersomnolence, Plante said. Most research on sleep in depression focuses on insomnia, but hypersomnolence has been well documented as a feature of atypical depression. In the context of depression, hypersomnolence has also been associated with impairment in daily life, resistance to treatment, and suicide risk.[9]

Plante noted that in the *DSM-5* diagnostic manual used by psychiatrists, the category that is the equivalent of IH—hypersomnolence disorder—is more flexible than in the International Classification of Sleep Disorders. In particular, it is more driven by subjective symptoms than IH (box 16.1). The MSLT is sometimes used in sleep clinics to differentiate IH from a psychiatric disorder such as depression, but this distinction may not be meaningful, he said.[10]

BOX 16.1: *DSM-5* HYPERSOMNOLENCE DISORDER CRITERIA

Self-reported excessive sleepiness (hypersomnolence) despite main sleep period lasting at least 7 hours, with at least one of the following symptoms:

- Recurrent naps or lapses into sleep per day
- A prolonged main sleep period of more than 9 hours that is unrefreshing
- Difficulty being fully awake after abrupt awakening
- Occurs at least 3x per week
- Accompanied by significant distress/impairment in cognitive, social, or occupational functioning
- Not better explained by another sleep disorder, medical disorder, or external substance

---

In a 2020 paper, Plante has described the use of several measurements to establish objective hypersomnolence, in addition to the MSLT. These included total sleep time, the psychomotor vigilance test, and/or pupillometry. Combining these old and new measurements, his team has been able to demonstrate "objective hypersomnolence" in twice as many patients at the Wisconsin clinic than with the MSLT alone.[11]

This demonstrates how research teams studying IH and hypersomnolence have been gathering groups of patients with characteristics that are not the same, because of lack of agreement about whom to include and which criteria to use. One group of investigators may hold that only people who fit ICSD-3 criteria or who can be verified to sleep for eleven hours per night should be given the IH label, while others, including Plante, have applied a more flexible framework.

As a consequence of a looser approach to diagnosis, the people in the Wisconsin study were not uniformly the extreme long sleepers studied by some other researchers. Those who went through a MSLT had an average sleep latency of 9.8 minutes—more than eight minutes, the ICSD-3 cutoff. Only a few displayed total sleep times of over eleven hours. Keep that in mind as we examine the results, which are still helpful in understanding the hypersomnolent brain.

SLOW DOWN FOR SLOW WAVES

Plante's study used high-density EEG to look at the patterns of electrical activity the brain generates during sleep. People with hypersomnolence displayed a localized deficiency in the slow brain waves that occur during the deepest part of non-REM sleep.[12] Slow waves are considered a marker of restorative sleep, so this finding could explain why people with hypersomnolence are not getting the same benefits as healthy people from sleep and thus feel the need to sleep for longer time periods.

Slow wave sleep occurs between 10 to 25 percent of total sleep time in young adults and declines with age. The slow waves appear in short bouts and show up more at the beginning of the night, presumably when the brain is catching up on what it needs. They also occur at a higher density after someone has gone without sufficient sleep, so they are part of the brain's homeostatic response to staying awake for a long time.

Slow waves on an EEG recording represent neurons within the brain firing in synchronous oscillations, with active and quiet phases coming about once per second. Slow wave oscillations appear to extend across several regions of the brain, not residing in any one part. The oscillations link the thalamus, in the middle of the brain, and the cortex, closer to the surface, but EEGs mainly detect what is happening in the cortex. The influential sleep researchers Giulio Tononi and Chiara Cirelli, also from the University of Wisconsin, have proposed that slow wave sleep allows a nightly recalibration of the brain's synapses, facilitating the strengthening of memory for events that occurred during the previous day.

The deficiency in slow wave activity seen in people with hypersomnolence was not apparent in most previous studies of sleep architecture in IH, which had shown that people with IH don't display more or less REM sleep than healthy controls.[13] A few previous studies did show similar results,[14] but they generally took information from the standard number of EEGs used in a polysomnogram.

High-density EEGs put a considerably larger number of electrodes (256, compared to six) on someone's scalp, providing more fine-grained spatial information.

What the additional resolution revealed is that people with hypersomnolence still experience slow wave sleep, but parts of their brains are not participating as much in the synchronized activity. Reductions in slow wave activity appeared on both sides of the brain and were stronger on the left side. The regions of the brain where the differences were the greatest included the somatosensory cortex, where the brain processes the sense of touch, and the supramarginal gyrus. When someone is awake, the supramarginal gyrus has important functions in language processing and interpreting others' emotions, but it appears to have a role in sleep quality as well.[15]

One major point of the Wisconsin study is that people with hypersomnolence displayed the same characteristics in their high-density EEGs whether they reported symptoms of depression or not. It bolsters Plante's point about the overlap of psychiatric hypersomnolence with IH and suggests a "shared brain abnormality" in persons with hypersomnolence disorder, independent of the presence or absence of depression.

The localization of the deficiency in slow wave activity around the parietal lobe was distinct from what Wisconsin researchers had previously observed in a subset of people with obstructive sleep apnea who did *not* display high levels of subjective sleepiness.[16] Also, the limited anatomical extent of the difference in slow wave activity may distinguish hypersomnolence disorder from general aging and neurodegenerative disorders, in which slow wave sleep weakens across a larger extent of the brain. On average, participants in the Wisconsin study were in their twenties. More research is necessary on how slow wave activity in IH varies with age and changes over time.

The Wisconsin study does not answer the question of *why* people with hypersomnolence have a partial reduction in slow wave activity. However, it does point to potential ways to address the deficiency. One possible option would be devices that can enhance slow wave sleep through acoustic stimulation.[17] These devices harness the observation, dating to the earliest days of EEG studies in the 1930s, that sleepers' brain waves respond to noises in the room—the effect is sometimes called a K-complex. The response to sound may represent the brain's attempt to get back on track when external stimuli are knocking at the gates of the thalamus.

Recent research from Wisconsin and elsewhere shows that acoustic stimulation can enhance memory for vocabulary words learned the previous day, if the device's sounds are timed to nudge the brain's oscillations at the right moment.

These devices, produced by the electronics giant Philips and also the French company Dreem, have now made it to the consumer market. These devices are attractive for IH treatment because they are nonpharmaceutical interventions, in contrast to stimulants or other medications. However, slow wave–promoting devices' long-term effects remain untested, and their efficacy in IH has not been examined. Plante had originally planned to include such devices in his study; he told me his team was unable to for practical reasons of scope.[18]

Enhancing slow wave sleep through pharmaceutical means is possible, but it is unclear how to do it cleanly, as the history of GHB/Xyrem illustrates. As another example, the compound gaboxadol was studied as a sleep aid by Merck and Lundbeck and was reported to increase slow wave sleep and reduce sleepiness in healthy sleep-deprived adults.[19] However, the companies stopped research on the drug in 2007 because of side effects including dizziness, headaches, hallucinations, and tachycardia.[20]

BARELY AWAKE

A second set of studies on the biology of IH comes from a research group led by Thien Thanh Dang-Vu in Montreal. Dang-Vu said his interest in hypersomnia had been piqued during his training. He had published several studies of brain imaging and cognition in other sleep disorders such as sleepwalking or REM sleep behavior disorder. "I began to think about—where are the gaps in knowledge?" Dang-Vu said. "We don't have a clear idea of what's happening in NT2 and IH, and filling those gaps could be important both for the field and for patients."

His lab's studies have shown that people with IH display a pattern of reduced cerebral blood flow while awake, compared with controls. The pattern resembles what researchers have observed in healthy young people who are in the middle of slow wave sleep.[21] For people with IH, it appears that part of the brain is still in a sleep mode when the person is awake. Partly, this study tells us what we already know: for people with IH, sleep is encroaching on them even when they are awake. Transient automatic behavior, or "microsleeps," has not been explored in experiments like this but may be consistent with this finding.

The region of the brain most strongly affected by reduced blood flow was the medial prefrontal cortex, which is involved in activities such as decision making, regulating emotions, and memory retrieval. The reduced blood flow appears as a

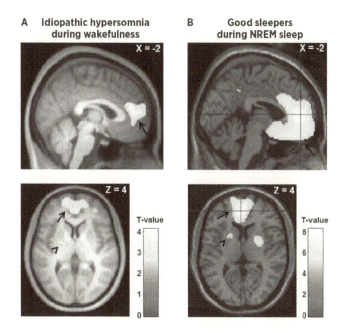

FIGURE 16.1. Brain imaging when people with IH are awake reveals patterns similar to when healthy people are asleep.

Source: Soufiane Boucetta et al., "Altered Regional Cerebral Blood Flow in Idiopathic Hypersomnia," *Sleep* 40 (2017): zsx140.

splotch above and behind the eyes (figure 16.1). Overall, the areas that showed reduced blood flow were part of the default mode network, or DMN, a dispersed set of brain regions responsible for internal awareness and consciousness.

The default mode network consists of regions that are active when someone is not doing anything in particular, especially something that requires focused attention. When someone engages in a specific task, such as reading or solving puzzles, activity in the DMN decreases and others perk up. During sleep, the various parts of the DMN interact with one another less; they appear to disconnect from one another. A way to describe the DMN is as a network responsible for daydreaming, or like the machinery that maintains the idle on a car: the level that the engine returns to when the gas pedal is not pressed.

To be confident about their results, Dang-Vu and his colleagues had to be certain that their study subjects were not actually sleeping, which is a strong possibility when people with IH are put into a scanner for an extended period of time.

The type of imaging they used was SPECT (single photon emission computed tomography), with a radioactive probe that is quickly metabolized. The critical time window lasts just a few minutes, while the probe is taken up by brain cells. Technicians monitored participants to check that they weren't asleep during this time.

By current criteria, this was an orthodox IH group. Study participants all fell asleep in an average of eight minutes or less on their MSLTs, and all reported habitual sleep durations of eleven hours or more. If someone reported a higher sleepiness score or fell asleep faster on the MSLT, the pattern of decreased blood flow tended to be stronger.

The pattern of reduced blood flow in IH didn't resemble that seen in healthy people who were asked to stay up all night, and it also didn't look what has been observed in narcolepsy type 1.[22] Although the specific neurological injury in narcolepsy type 1 is to the hypothalamus, the consequences of that injury and downstream functional changes can be seen elsewhere in the brain. People with narcolepsy type 1 tend to show reduced blood flow at other nodes of the brain in the frontal and temporal lobes; this is not the same as with IH.[23]

The differences suggest that the differences in blood flow between IHers and healthy controls are more connected to the trait of hypersomnolence, rather than the acute state of sleepiness. More imaging research on IHers in different states, awake and asleep, as well as comparisons with other groups, such as people with sleep apnea, could flesh this out.

As with the Wisconsin study, the patterns of reduced blood flow in IH do not say much about a cause. They do resemble the effects of benzodiazepines, although the reductions in blood flow occurring with benzodiazepines extend more broadly across the brain.[24] The findings might fit Rye's somnogen theory, but they do not point specifically to hyperactive GABA inhibitory circuits.

In another paper published in 2019, Dang-Vu's team used MRI (magnetic resonance imaging) to examine differences in brain structure and volume in people with IH.[25] They also monitored functional connectivity; this is a way to infer whether separate brain regions are interacting, by observing whether their activities rise and fall together. Functional connectivity within parts of the DMN, such as the medial prefrontal cortex, was less in people with IH than in healthy controls. But in contrast to what the researchers expected, some of the same regions that showed decreased blood flow measurements in IH displayed *increased* cortical thickness and gray matter volume, in comparison to controls. According to Dang-Vu, this may be a result of the chronically sleepy brain compensating for impaired function.

IMAGING SLEEP DRUNKENNESS

Many IHers have the symptom of severe sleep inertia in common: difficulty waking up and impaired alertness lasting minutes to hours after waking. A survey of the Hypersomnia Foundation registry found that a majority of respondents with IH endorsed difficulty waking and the need for multiple alarms (79 percent and 69 percent). Severe sleep inertia can also exist independently of IH or excessive daytime sleepiness. The more evocative term "sleep drunkenness" describes an extension of severe sleep inertia. Sleep drunkenness is defined as confusion or clumsiness upon awakening, with impairments of speech, motor control, or cognition. "I was walking into walls, and I felt like I was in a bubble," one IHer said at a support group meeting. When experienced clinicians have examined groups of people with IH, they have found that 21 percent to 55 percent of them have sleep drunkenness.[26] Yet to qualify as sleep drunkenness, duration and severity of symptoms have not been standardized.

Sleep inertia and sleep drunkenness can be some of the most disabling aspects of IH because they contribute directly to absences at school or work. The IHer has to depend on someone else to rouse them out of bed and cajole or force them to take their medicine. There may be some overlap with autonomic nervous dysfunction, such that dizziness upon standing induces someone to lie down, facilitating falling asleep again.

Several medications have been reported to alleviate sleep inertia, including antidepressants, nicotine patches, flumazenil, and gamma-hydroxybutyrate. Taken at bedtime, delayed-release methylphenidate or delayed-release bupropion resolved severe sleep inertia for nineteen out of twenty-two patients, according to the Minnesota sleep specialist Carlos Schenck.[27] While some of these medications, such as methylphenidate and bupropion, may be inexpensive and familiar to clinicians, controlled studies focusing on sleep inertia and sleep drunkenness in IH have not been performed until recently.

Through experiments with healthy volunteers, researchers have learned a lot about sleep inertia, which tends to be greater when someone is awakened out of slow wave sleep or when their body temperature is at its lowest, in the middle of the night.[28] Since it is more intense after an interruption of recovery sleep, sleep inertia appears to correlate with slow wave density.[29] Sleep inertia may result from the brain waking up piece by piece. Upon waking, blood flow increases first around the brainstem and midbrain, with the anterior cortex, a major element of the default mode network, coming online later.[30]

An unresolved question is whether sleep drunkenness, as experienced by people with IH, has the same physiological and neurological characteristics as sleep inertia in healthy people. Bedřich Roth's work pointed more toward sleep drunk-enness being distinct in the setting of IH. He thought sleep drunkenness was more dependent on the patient's "pathological disposition" rather than on how much time they had been sleeping or their previous stage of sleep.[31] In one paper, Roth and his colleagues tried with several healthy control subjects to provoke sleep drunkenness, but they managed to do so only once, by having that person stay up all night beforehand. Through neurological examinations, he observed that sleep drunkenness among IHers included impaired motor coordination, dimin-ished reflexes, and vestibular disturbance. He suggested that the cerebellum may be late to reengage with the rest of the brain upon waking.

When Lynn Marie Trotti was beginning a brain imaging study at Emory, she said: "We want to find out if sleep drunkenness in IH is the same as what hap-pens to healthy people with sleep inertia and is more pronounced, or whether it's something different." In designing her study, Trotti wanted to catch people with IH in the fuzzy state just after they wake up. Participants were asked to take a nap in a designated room near the MRI scanner and upon waking were wheeled into the scanner. To gauge functional impairment, they were given a test of work-ing memory called "N-back," in which they are supposed to recognize repeated numbers when the numbers are presented one at a time. Afterward, they could relax and were not given any test. Brain imaging requires the subject to stay still, so simultaneous tests of coordination or balance were not possible.

The study called for participants to go off whatever wake-promoting medica-tion they were on. One participant, a member of the Atlanta hypersomnia sup-port group, reported that this requirement presented some challenges. "I was NOT supposed to fall asleep inside the scanner, which really worried me," this person told me by email. "The only reason I was able to stay awake for that long was because I had to do something, and then I made up songs to the noises that the machine was making. It was extremely difficult, but I did manage to stay awake."

The first paper to be completed from Trotti's study reported findings simi-lar to Dang-Vu's.[32] This paper used a different type of imaging, PET (positron emission tomography), which measures metabolic activity by looking at how much radioactive glucose brain cells take up. Again, the patterns of metabolic activity were different between IH and narcolepsy type 1. Trotti and her col-leagues were able to see several clusters in parts of the default mode network where there was more metabolic activity in IH than in controls. This sug-gested that IHers' brains may be working harder to stay awake—echoing a previous study from France. In an earlier PET imaging study of IH, Montpellier researchers had detected only *hyper*metabolism during the awake state. The authors interpreted this as indicating that IH patients' brains were working harder to stay awake.[33]

Trotti and her colleagues were concerned enough about participants dozing off during imaging to monitor them via EEG, revealing that falling asleep in the scanner was common—even among controls. It suggests that imaging may be capturing a mixture of states when people are either asleep or trying strenuously to stay awake, rather than placidly daydreaming. In future studies, EEG during imaging to control for actual sleep could become standard practice.

THE WRIST IS NOT THE BRAIN

We've been discussing what neurobiology IHers may have in common. One factor that separates them may be the sheer amount of time they spend sleeping, since the long sleep form of IH may be distinct in its pathophysiology. Over the last decade, American, European, and Japanese sleep neurologists have begun to agree that a measure of habitual sleep time or overall sleep need should be used to diagnose IH, rather than relying on the MSLT. Several research groups have found that how quickly someone enters sleep on the MSLT does not match up well with increased sleep need or decreased alertness.

It's difficult to argue that someone is getting insufficient sleep when they report sleeping eleven hours a day or more. But then the question becomes how a doctor or other clinician establishes whether someone really does sleep that much. In practice, listening to the patient is important, but simply asking them what they remember is not that reliable. People often make errors in estimating actual shut-eye time, both in recall and in sleep diaries.

A currently recommended clinical practice is to have a patient wear a wrist actigraphy device for a week or two, particularly before an MSLT. The devices are similar or the same as those that tell people how many steps they've taken and are convenient to take home. The devices don't detect sleep but can tell whether someone is motionless for enough time that they are probably asleep.

Plenty of studies compare actigraphy to the gold standard of polysomnography in the sleep lab—and some findings have been discouraging. In a study from Trotti and colleagues, actigraphy devices did not appear to perform consistently enough to help make decisions for individuals.[34] In one study, 102 patients, most of whom were undergoing evaluation for sleep apnea, were asked to wear actigraphy devices while they were undergoing an overnight sleep exam. This allowed investigators to compare their measurement of sleep time against standard EEG.

The Emory study evaluated the performance of two devices: the Philips Actiwatch, which is FDA cleared for clinical and research use, and the Jawbone UP3, a consumer product. On average, the Actiwatch overestimated sleep time by more than twenty-five minutes, and the Jawbone did so by almost one hour. On the individual level, the authors concluded that "a clinician attempting to apply the Actiwatch result would not know whether it was overestimating sleep time by more than 2 hours or underestimating it by almost 1.5 hours."

Plante's group has performed similar studies on consumer sleep devices' shortcomings.[35] A Finnish study was more optimistic, finding close agreement between actigraphy and polysomnography when performed at home in a group of 281 patients being evaluated for sleep disorder diagnosis.[36] The authors concluded: "Actigraphy is far from being perfect, but there is no better method to assess sleep time across several nights."

It's an academic concern for now, but in the future, precise diagnosis may depend on accurately measuring how many hours someone sleeps per week *at home*. In the United States, insurance reimbursement for actigraphy with sleep disorder diagnoses has been inconsistent, and many insurers' policies say that the practice is not reliable enough to be reimbursed.[37]

Better technology could help. People with IH and related disorders could benefit from knowing whether medication changes or low-key interventions make a verifiable difference in their sleep quality. An overnight sleep test is expensive, costing thousands of dollars—and it only takes a one-night snapshot of someone's sleep. The tension between accuracy in the sleep lab versus convenience and lower cost at home exists in other aspects of sleep medicine. In the last decade, sleep apnea diagnosis has been shifting to the home, in response to pressure from insurance.[38] The end result is more convenient, but both patients and physicians get less information out of the encounter.

While actigraphy devices that only track motion may not be accurate enough to measure sleep for diagnostic purposes, multisensor trackers, monitoring heart rate and sometimes temperature and skin conductance, may approach the level of accuracy needed.[39] Portable EEG headsets, earbuds, or patches that adhere to the body may be able to record sleep phases and measure sleep duration at home with sufficient accuracy. Such devices represent the future of the field, whether clinicians want to deal with the flood of consumer technology or not. Apple and other consumer electronics companies clearly see the potential for a broad market and are likely to continue to produce more convenient gadgets in the future.

MARATHON SLEEP TESTS

In contrast to home sleep studies, a few European sleep centers have devised diagnostic procedures in which patients are asked to sleep in the laboratory for as long as they can. The idea is to allow the patient to fully saturate their need for sleep and gauge just how excessive it is. An advantage here: the patient is hooked up to EEG leads, and the amount of time they spend sleeping can be verified directly.

The most arduous of these is the protocol used in Montpellier, which calls for patients to sleep as long as they can over thirty-two hours (two nights and the day between).[40] The conditions of enforced bed rest with dim light and no screens or reading material sound punitive; they are designed to be as boring as possible. The protocols in Paris and Bologna are less stringent and allow patients to move around or read during the day.

As Dauvilliers and his colleagues note, many healthy adolescents or young adults will be able to sleep for eleven hours in the first twenty-four, because they habitually get less sleep than needed during the week and then catch up on the weekend. The second night is necessary to separate IHers from everyone else. The ideal cutoff for separating IH from controls was nineteen hours of sleep over the thirty-two-hour period. These standards may need to be adjusted for age, body mass index, or other factors once enough information is available.

Dauvilliers has advocated use of a strict laboratory protocol, such as the one his group uses, to gauge how much sleep someone needs for IH diagnosis. And recently, that approach has yielded some rewards. His group has shown that long sleep IHers show an association between sleep inertia and slower reaction time, as measured by the venerable psychomotor vigilance test.[41]

Here, we see a clash between the needs of clinical practice and of research. In clinical practice, marathon sleep tests would be inconvenient and costly, especially in the United States. For research studies, it is essential to have homogeneous and well-defined groups of patients. Otherwise long-sought biomarkers for IH, like the GABA-enhancing somnogen, will continue to slip away.

CHAPTER 17

THE FDA OPENS A DOOR

Sometimes, our sciences create kinds of people that in a certain sense did not exist before.

—Ian Hacking, 2006

In March 2021, the Hypersomnia Foundation and Jazz Pharmaceuticals announced the launch of "I Have IH," a campaign aimed at increasing awareness of IH among both health care professionals and the general public. On Facebook, the Hypersomnia Foundation posted photos showing that the campaign's content was visible on a large video screen in New York City's Times Square. Reacting to the post, Diana Kimmel expressed satisfaction. This was what she and other people with IH wanted. "Fellow IHers, we have been heard and seen and it feels good," she wrote. The public display and the awareness campaign showed that IH's time had come. They were the culmination of events that Anna Sumner's interactions with Parker, Rye, and Jenkins had nudged forward several years before.

Let's acknowledge what Anna's round of publicity did not accomplish. Despite the Emory researchers' hopes, it has not brought about the identification of another endogenous sleep-promoting substance. The sleepy stuff still lurked in patients' CSF samples, waiting for discovery. And despite Anna's and others' experiences, most sleep neurologists have not embraced flumazenil as an alternative to conventional stimulants. When IHers clamored for it, the drug did become available to those able to find a willing prescriber.

Instead, the more far-reaching downstream effects of Anna's speaking up have been to encourage IH patients' efforts to organize themselves. Anna may have stood apart from the hypersomnia community, but as "a legend to everyone in the room," she inspired others. Her example accelerated the community's coalescence in the United States online and in person.

Others around the world have played critical roles in IH's emergence; this book tells just one part of the story. Many sleep neurologists had grown frustrated with the MSLT's categories and had been dealing with difficult-to-treat IH patients for years. Starting with the reference point of Bedřich Roth and his colleagues in Prague, researchers in France have built up our current knowledge of what IH is, while investigators in North America have achieved advances with EEG markers and brain imaging. On the advocacy side, Rye and others close to him had direct roles forming the Hypersomnia Foundation, but other groups such as Hypersomnolence Australia started independently.

While they disagreed with him about his research, Rye's critics agreed with him that MSLT-based distinctions between narcolepsy type 2 and IH were flimsy. In terms of pathophysiology, narcolepsy type 1 was well defined, based on a deficiency in hypocretin and the loss of a specific group of neurons. In contrast, narcolepsy type 2 and what distinguished it from IH were not well defined. There was a growing consensus among sleep medicine specialists that IH deserved to be on an equal footing with narcolepsy type 2. In fact, a coming realignment may join those groups together.

REDRAWING THE LINES

Sleep specialists have been signaling that the section of the current International Classification of Sleep Disorders dealing with narcolepsy and idiopathic hypersomnia will be revised. This will affect how members of the hypersomnia and narcolepsy community deal with the world and how they see themselves.

The revision was certainly needed. The evidence was clear that variability in MSLT results, along with the effects of antidepressants, can obscure the ability to distinguish between narcolepsy type 2 and IH.[1] Rye and Trotti's 2013 publication on repeated MSLTs giving discordant results helped open up the discussion; others have since confirmed their findings.[2] Since current distinctions between narcolepsy type 2 and IH are not meaningful, it makes more sense to divide people up by different criteria.

Accordingly, two recent papers from sleep neurologists have called for merging narcolepsy type 2 and IH without long sleep; IH with long sleep should be carved off into a separate category. Looking ahead to ICSD-4, both proposals would narrow how the term "idiopathic hypersomnia" is applied: to people with an excessive need for sleep, rather than excessive daytime sleepiness by itself.

Where the proposals differed was in what the merged category would be called. The first proposal came from a transatlantic group including Trotti, Arnulf, and the Harvard sleep neurologist Kiran Maski. They published a paper that grew out of discussions at the 2018 International Symposium on Narcolepsy, concluding: "It is not justifiable to exclude patients having IH, especially those without a long sleep time, from pharmacological treatments currently approved or being tested for NT2."[3]

These authors proposed to name the new merged category "narcolepsy spectrum disorder."[4] The term, meant to acknowledge the heterogeneity in the category, had already been used to describe relatives of people with narcolepsy type 1. In this group, it also referred to the appearance of symptoms such as sleep paralysis and hypnogogic hallucinations at a rate intermediate between narcolepsy type 1 and IH.

The second paper, from European experts, proposed to name the merged category "idiopathic excessive sleepiness."[5] It also contained a system of diagnostic criteria for narcolepsy and other disorders, labeling them "definite" or "probable," along with possible subcategories. The European proposal was more rigorous

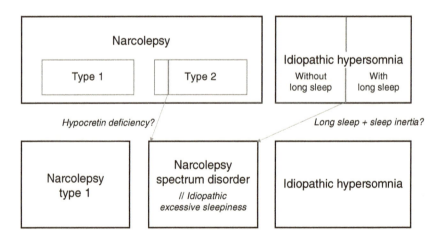

FIGURE 17.1. Proposals for the fourth edition of the International Classification of Sleep Disorders.

Source: Adapted from Rolf Fronczek et al., "To Split or to Lump? Classifying the Central Disorders of Hypersomnolence," *Sleep* 43 (2020).

about applying what has been learned about narcolepsy since 1999. It would shift the definition of narcolepsy toward hypocretin deficiency and away from a collection of symptoms connected to REM sleep. Are narcolepsy and hypersomnia two parts of a spectrum of excessive sleepiness? Or does the term "narcolepsy" only apply to hypocretin deficiency?

The European proposal also raised questions about what would happen when some diagnoses were redefined as *not* narcolepsy, especially if hypocretin CSF measurements remained rare. After all, the FDA approved newer drugs such as pitolisant and solriamfetol for something *called* narcolepsy. Maski, together with Emmanuel Mignot and Tom Scammell, emphasized access in a commentary on the European proposal: "A new nomenclature may force clinicians to change diagnostic labels for established patients because suggested diagnostic tests are inaccessible, thus compromising care and drug reimbursement."[6] At the same time, recent approvals have left people with hypersomnolence resulting from *other* psychiatric or medical disorders (besides narcolepsy or IH) on the sidelines.

The physicians who wrote the European proposal did not dismiss the issue of access to treatment. One of the lead authors, Gert Jan Lammers, has used the example of a patient who falls between gaps in current diagnostic categories.[7] His hypothetical patient habitually sleeps 10.5 hours per day, has difficulty waking in the morning, experiences sleep drunkenness, and reports regularly feeling drowsy or falling asleep during the day. During an MSLT, he displays an average sleep latency of 9.3 minutes, which is above the current cutoff for narcolepsy type 2 or IH. But his overall sleep duration per twenty-four hours is under the ICSD-3 boundary of eleven hours. According to Lammers, this patient has a serious problem, but no diagnosis can be made when applying current criteria. At the same time, sleep deprivation as a possible cause for complaints can be overlooked, he said.

Previous revisions of the ICSD touched on similar issues. Since the first ICSD, IH was split into two, then fused back together. The idea of merging narcolepsy type 2 and IH without long sleep had come up before.[8] This time, the science has advanced far enough to bring about a shift whose implications go beyond drug reimbursement.

If either proposal for reclassifying IH and narcolepsy type 2 is approved, thousands of people diagnosed with narcolepsy will be told that what they live with every day is now named something else. In the United States, people with narcolepsy type 2 diagnoses outnumber those with narcolepsy type 1. They currently have explanations for what they experience, and many have made career and life choices based on their understanding of their disorders. Some have found a community where they belong. Redrawing boundaries and renaming diagnoses might disrupt those supportive associations—although new ones can form.

The proposed name change for narcolepsy type 2 carries some similarities to the redefinition of autism spectrum disorder in the *DSM-5* (*Diagnostic and Statistical Manual of Mental Disorders*) a decade ago, along with current discussions around renaming schizophrenia.[9] For narcolepsy and IH, issues may be sharper around medication access, since no medications are approved for the core symptoms of autism.[10]

One prominent voice in the narcolepsy community, Julie Flygare, CEO of Project Sleep, called for more discussion before renaming and reclassifying. Flygare said there should be a broader process of consultation among patient groups and other stakeholders considering both the science and questions of identity and medication access. "We are not the only disease community to face an issue like this one, and I think sometimes we forget the wealth of information and advice we could get by looking out beyond the sleep field horizon to leverage expertise from the broader healthcare landscape," Flygare said in an email. "My biggest opinion is that there are lots of opinions to be considered."

Hypersomnia Foundation board members Diane Powell and Betsy Ashcraft told me they were following the reclassification discussion and had not made a decision about it but were concerned about insurance coverage and future FDA approvals. "The question is really, what gets us to more research and more treatments approved?" Ashcraft said. "I'm not going to be fighting over what we call it."

When he was revising his book on narcolepsy and hypersomnia in the 1970s, Bedřich Roth seemed to anticipate future controversy over how to delineate the two conditions. He said a decision should be made "jointly by a group of specialists who have worked in this field and discussed the issue thoroughly."[11] In this context, it may be helpful to rename IH—possibly, a narrower redefinition—after Roth. The condition could be called *Roth hypersomnia syndrome*, honoring Roth as "the true father of sleep medicine" and removing the discouraging word *idiopathic*. It could also serve as a reminder of IH's background in postwar Czechoslovakia, an environment where pharmaceutical companies had minimal influence.

DEFINING THE GROUP OF PEOPLE WITH IH

However the nomenclature gets updated, the self-organization of the hypersomnia community means that people with IH and related disorders will have a voice in matters they care about. In the past, people with IH appeared in research papers as controls: sleepy people who didn't have narcolepsy or sleep apnea. Now, more recent research is focused on them, and non-IHers have someone to ask about the experience of having IH. For related sleep disorders, the Circadian Sleep Disorders Network and the Kleine Levin Syndrome Foundation have played a similar role. Together, their representatives can have seats at the table when dealing with pharmaceutical companies, insurers, the FDA, the National Institutes of Health, and the American Academy of Sleep Medicine.

If or when narcolepsy type 2 and IH are redefined, some open questions for research are: How many people have these disorders? What co-morbidities, such as depression or cardiovascular disease, do they have? What is their life expectancy? In population studies, researchers have observed increased mortality for people who report extra short or extra-long habitual sleep duration—the so-called U-shaped curve.[12] Does that increased mortality apply to people with IH? What are the risk factors, such as viral infections (including COVID-19), anesthesia, or environmental exposures?

The classic way of determining the prevalence of a disease is to examine everyone, without selecting people who are already in the doctor's office. The subjective symptom of excessive daytime sleepiness is common among the general population, and a substantial fraction of people may appear to have narcolepsy or IH if they go through an MSLT. A rigorous study would require more than asking people about their sleep habits. It might involve administering the MSLT or other more invasive procedures such as lumbar puncture and eliminating common causes of sleepiness such as sleep apnea or hypothyroidism. This may be impractical on a large scale.

So far, a workaround has been to examine health insurance databases, catching the people who have already gone through the process of diagnosis under current categories. This type of analysis has its limitations, because insurance coverage has provided a disincentive for the IH diagnosis. Some people may be uncounted because they lack access to health care or because their condition is obscured by other diagnoses. Even so, in the United States, the number of people with IH diagnoses has been rising in recent years (2013–2016), reaching one in ten thousand people.[13] IH's prevalence is about a quarter of narcolepsy's, but that could change if narcolepsy type 2 and IH without long sleep are merged.

ALL THAT JAZZ

The study on the prevalence of IH did not materialize out of thin air. Jazz Pharmaceuticals was releasing a flurry of research on IH at the same time as it was mounting its awareness campaign, in anticipation of an expected FDA approval for the company's drug Xywav—the first ever for the indication of IH.

The results of a trial of Xywav and IH, conducted by investigators in Europe and the United States, were presented at neurology and sleep research conferences that spring. Xywav showed strong effects on daytime sleepiness as well as sleep inertia, an outcome measured for the first time in any industry-sponsored clinical trial.[14] Lead author Yves Dauvilliers and his colleagues had developed a composite measure of IH symptoms, the Idiopathic Hypersomnia Severity Scale, which included questions about sleep inertia. The IHSS was similar to the questionnaire Tom Roth and colleagues created for Balance Therapeutics' IH trial.

In the Xywav clinical trial, most of the participants had the form of IH without long sleep. The short-term efficacy for both groups was about the same. The study was set up so that all participants received the drug first and were able to grow accustomed to it over a few months. Then half the group was put on placebo for two weeks, to see how much their symptoms worsened. Because Xywav has acute sedative effects, a truly blind placebo would be difficult to implement.

The "randomized discontinuation" or "maintenance of effect" study design was previously used to test medications for restless leg syndrome, as well as antidepressants and antipsychotics. The format weeds out patients who can't tolerate the tested medication, but some experts have questioned the format, arguing that classic randomized controlled trials are more valid.[15]

In the Xywav study, more than half of the participants had been taking stimulant medications and were allowed to continue using those drugs. Common adverse events included nausea, headache, dizziness, anxiety, and vomiting, which can be experienced with Xyrem as well. Some people who reported sleep inertia or long sleep time were advised to try a once-per-night regimen, a more flexible arrangement than the standard two doses per night for narcolepsy. More followup was needed to determine whether Xywav's benefits are stable over time.

The FDA was willing to approve Xywav for IH based on subjective outcome measures only. With the wake-promoting medication pitolisant, the FDA had softened its previous requirement for objective outcome measures because there was relevant backup data. However, for Xywav and IH, all the outcomes for establishing success were subjective, nonphysiological measures of sleepiness, such as the Epworth Sleepiness Scale and the Idiopathic Hypersomnia Severity Scale. This is not to say that the Idiopathic Hypersomnia Severity Scale is faulty or unreliable—but it is not an objective, physiological measure of sleepiness the field has previously relied on. Sleep specialists and IHers are still figuring out the best ways to measure the symptoms of IH, both objectively and subjectively, and how to gauge success. What is more important: a boost in the number of hours awake during the day or the quality of wakefulness during that time?

ONE APPROVED, MORE TO COME?

In August 2021, the FDA announced its approval of Xywav for the indication of IH. The approval was celebrated by the Hypersomnia Foundation, as well as by many people with IH, because it sent a signal of legitimacy and recognition for IH to the medical community. It was expected to encourage other pharmaceutical companies, such as Takeda and Harmony Biosciences, to follow.

Takeda had already initiated a pilot study for people with IH in 2020, testing one of its hypocretin receptor agonists. A Takeda spokesperson declined to discuss the company's decision making, but they did acknowledge the Hypersomnia Foundation as a partner. The company's representatives have appeared at foundation events to recruit study participants. In 2021, the company announced plans to move a related drug into clinical trials for narcolepsy type 2 and IH together, after narcolepsy type 1. Later that year, Takeda halted its clinical research program studying hypocretin receptor agonists because of liver toxicity concerns.[16]

Harmony's narcolepsy drug pitolisant became available in the United States in 2020. At the end of 2021, Harmony announced plans for a study of pitolisant for people with IH, after beginning studies on pitolisant for people with myotonic dystrophy and Prader-Willi syndrome.[17] Jazz appeared to have more expansive aims for its drug solriamfetol, having previously revealed discussions with regulators regarding excessive daytime sleepiness in the context of depression, potentially a large market that could overlap with IH.[18] Jazz later sold the rights to solriamfetol to another company.

The recent approval of medications such as Aduhelm for Alzheimer's disease has led some critics to charge that the FDA's rigorous standards for clinical trial efficacy have slipped.[19] With Xywav and IH, the FDA's relaxation of its previous preference for "objective" outcome measures in sleep disorders could be seen as part of this pattern. However, it is difficult to see how the agency could have proceeded otherwise, when experts in the sleep medicine field are still figuring out appropriate outcome measures for IH. An FDA spokesperson confirmed that the agency has sought advice on sleep disorder outcomes through informal meetings with advisors such as Trotti and Mignot, but the agency declined to answer questions about how outcome measures for Xywav and IH were set.[20]

A SUBJECTIVE SUCCESS

Despite the possibility of greater acceptance of IH, the prospect of Xywav as the only FDA-approved medication for idiopathic hypersomnia raises concerns. Before 2021, Xywav's predecessor Xyrem had been the expensive and intimidating option, only tried after other medications were unsatisfactory. In Europe, Xywav and IH may face more obstacles because of preferences for objective outcome measures. And while the establishment of Xywav as "on label" for IH may facilitate IH patients' access and insurance coverage in the United States, it was unclear how the change might affect coverage for other medications, which would continue to be off label.

For Xywav and IH, there was no advisory committee meeting, which might have allowed discussion of safety issues. For example, Jazz's data indicates that half of all newly diagnosed IH patients in the United States had been previously diagnosed with sleep apnea.[21] It is not surprising that IHers are often diagnosed with other conditions first, but in 2012, the FDA had warned against Xyrem's use in people with sleep apnea because of the risk of respiratory depression.[22] Some clinicians prescribe Xyrem for people with dual diagnoses of narcolepsy and sleep apnea if they are successfully treated through CPAP, but the safety of this practice is debatable.[23] Lynn Marie Trotti signaled some ambivalence about Xywav in her comments to the *New York Times*, saying: "If you gave me a list of medicines and said, 'Which one do you want approved for idiopathic hypersomnia?' I don't know that I would have picked Xywav."[24]

That same year, Trotti and others had finished an update of the American Academy of Sleep Medicine's practice guidelines for narcolepsy and IH. The update had recommended modafinil as a first-line treatment for IH, based on clinical trials from Europe and Japan. Still, Trotti said an FDA-approved drug for IH was a positive change: "For some patients, it is going to be the right medication, and they should be able to access it."

Despite fears expressed before it was approved for narcolepsy, the availability of Xyrem has not contributed to widespread diversion or abuse. Xyrem did not become "the next OxyContin," partly because it was so expensive. Xyrem and Xywav represent a more restrictive model for the controlled distribution of abuse-prone drugs. Some 40 percent of newly approved drugs include REMS (Risk Evaluation and Mitigation Strategies) programs such as Xyrem/Xywav's, although the FDA has mostly let pharmaceutical companies such as Jazz grade and police themselves on their implementation. A 2013 inspector general's report for the Department of Health and Human Services concluded: "FDA lacks comprehensive data to determine whether risk evaluation and mitigation strategies improve drug safety."

Recognition of IH and a better public understanding of it continue as works in progress. Whether someone's family member or friend recognizes IH as something real won't change overnight. Awareness will take time to build, through media campaigns and more targeted outreach to medical schools and professional groups such as pharmacists and school nurses.

Jazz's sponsorship of a joint awareness campaign with the Hypersomnia Foundation meant that the company was able to play a role in defining how IH was perceived by the wider public. To its credit, Jazz avoided disease mongering Cephalon-style by not blending together IH with everyday sleepiness or fatigue. (In contrast, marketing for Sunosi was broader.) On a Jazz-sponsored website designed for health care professionals, one of the videos showed a man in his thirties with severe sleep inertia, whose partner struggles to have him wake up at around 3 pm.[25] After almost twenty minutes, the man is sitting up but can barely speak. This is not an experience most people have.

Through insurance database research, Jazz has sketched a picture of how people with IH pass through the health care system. One of the company's abstracts revealed that only a minority of new IH patients in the United States were diagnosed after an overnight sleep study and MSLT. It was more common for the diagnosing physician to document long sleep time in an unspecified way.[26] Balance Therapeutics–associated investigators had encountered a similar phenomenon, finding that a majority of patients with a preliminary diagnosis of IH failed study screening "due to inability to meet inclusion criteria necessary for the IH diagnosis."[27]

DO NARCOLEPSY OR IH HAVE A SILVER LINING?

Anna Sumner Pieschel's story, detailed at the beginning of this book, still stands as part of an inspirational, hopeful strand of communication about IH and other sleep disorders. Her progress through family and professional milestones—baby, marriage, law partnership—after her recovery demonstrated that IH did not curtail her life. In a Georgia Public Broadcasting video, as her young son was seen running around her backyard, she told an interviewer: "If you told me six years ago that this would be my life, I wouldn't have thought it was possible. . . . Today I have everything I never thought I would have."

In a similarly optimistic tone, beginning in 2018, other people with IH have been part of a scholarship program sponsored by the patient advocacy group Project Sleep, in which young people with narcolepsy and IH receive $1,000 scholarships. Smiling students explain that they have idiopathic hypersomnia—and they are managing it well enough to attend college. According to their online biographies, they were able to get good grades in high school and participate in sports, band, and church mission trips. "We can do it, despite how debilitating IH can be," they are saying.

In serving their communities, patient advocacy groups like Project Sleep and the Hypersomnia Foundation have to balance how they display examples of people who are successfully navigating challenges such as college or career and, on the other hand, provide resources for those who are encountering obstacles or may need to apply for disability from Social Security or private insurance. The foundation has held informational sessions on the process of applying for disability, in an effort to destigmatize the process. "Our role is to be an empathetic resource and also to be realistic about what people need," Diane Powell told me. "We recognize that IH can make ordinary things difficult to navigate. I don't know how many other places have a guide on how to take a nap in your car."

In the narcolepsy community, the positive, optimistic mode has extended to the point where a few people with narcolepsy say that it provides them with an advantage. One proponent of this view has been the Harvard geneticist George Church, who has narcolepsy type 1. Church, an innovator in the field of DNA sequencing, is also known for nodding off during conferences and discussion panels. He told me that while narcolepsy affects him every day, cataplexy for him is "infrequent and mild—usually laughter-induced." In interviews, Church has said that many ideas and solutions to scientific problems have come to him while he was dreaming or almost asleep. However, it took him until he was in his fifties to realize that narcolepsy is "a feature not a bug," as he told the science writer Sharon Begley.[28] To be sure, Church's personal choices may not be practical for others. He did not take medication and managed his sleepiness partly by not eating during daylight hours.

Church has suggested that people with narcolepsy could be seen as part of the neurodiversity movement, which holds that brains different from the typical are not necessarily in need of medication or correction. Neurodiversity is usually defined as the idea that neurological differences such as autism, ADHD, and dyslexia are the result of natural variation in human brain development, which may be advantageous in specific situations.[29] In particular, some autistic people have championed neurodiversity, saying that they don't need to be "fixed."

Parallel to this line of thinking are studies of creativity among people with narcolepsy. Arnulf and colleagues in France found that people with narcolepsy type 1 and 2 displayed higher creativity scores than a demographically matched control group.[30] The study included questionnaires assessing personality and formal tests of creativity, such as drawing and making up stories. Arnulf's paper attributes higher creativity in people with narcolepsy to lucid dreaming and "life-long privileged access to rapid eye movement sleep and dreams." "One may wonder whether the frequent sleep bouts that subjects with narcolepsy perform in passive conditions offer them the opportunity to mind wander and to incubate complex problems," the authors write.

Within the hypersomnia community, I haven't heard much enthusiasm for this idea. Diana Kimmel and others told me that the neurodiversity/creativity tune did not resonate with them, while a few people said they had come up with creative ideas while sleepy. This brings us to more pessimistic modes of communication. When sleep and brain fog consume a larger and larger proportion of the hours and days in someone's life, there is no bright side. In a video posted online by the Hypersomnia Foundation in 2020, board member Michelle Emrich argued forcefully that IH severely limits her life without enhancing any corner of it.[31] Emrich had to give up her medical practice in 2012 after an IH diagnosis the year before, and she decided not to have children because she was too sick to care for them.

At the suggestion of her therapist, Emrich said she "set my baseline minimum for a successful day at feeding myself. That's it. No showering, no getting dressed, no washing the dishes, no paying the bills, no calling a friend. I can't predictably do any of these things." "It is a testament both to an excellent therapist and my immense willpower that I have learned to cope with IH well enough to want to remain alive," she said in the video. "My entire world revolves around sleep need. . . . Because of my sleep schedule and severe brain fog, I only have a few hours in which to live my life."

Talk therapy can have a role in the management of IH, even if many IHers want to draw a distinction between sleepiness and mood. Investigators in Illinois have developed an adaptation of cognitive behavioral therapy for hypersomnia, and discussing coping strategies with a therapist who understands IH can mitigate depressive symptoms.[32]

Left untreated, people with IH may be impaired in their ability to drive safely, support their families around the house, or sustain enjoyable activities such as watching a movie. The standard for successful treatment must go beyond medication to include peer support, coping strategies, and accommodations.

FIGURE 17.2. Cartoon depicting daily life with idiopathic hypersomnia.

Source: Milo Hiisikolo.

SEX AND RACE DISPARITIES IN IH DIAGNOSES

According to insurance databases, IH appears to be slightly more common in women than men: 11.1 versus 9.5 diagnoses per 100,000 people in 2016. Yet the IH CoRDS registry is almost 85 percent female, reflecting a combination of factors. For cultural reasons, men may be less likely to volunteer personal information for a chronic illness registry. In sleep apnea studies, men were less likely to report symptoms of sleepiness or fatigue than women, despite similar levels of physiological sleepiness.[33]

It is possible that biological differences may be driving higher underlying rates of IH in women, but we can only speculate about mechanisms. Obstructive sleep apnea is more common in men and generally should be absent for an IH diagnosis. Clinicians may be more willing to consider IH for a slender young woman than for an overweight man. Narcolepsy diagnoses, both with and without cataplexy, are also more common among women than men in the United States.[34] More women than men have autoimmune diseases, probably because of endocrine influences on the immune system.[35]

Variations in the prevalence of narcolepsy and IH by race in the United States are currently unknown. One recent effort in addressing this issue was a 2015 analysis of the Stanford Center for Narcolepsy's patient database.[36] This found that African Americans with narcolepsy type 1 were more likely to have low hypocretin CSF levels but at the same time were less likely to display cataplexy. This suggests that hypocretin testing may be more important for accurate narcolepsy diagnosis in African Americans.

For comparison, obstructive sleep apnea is thought to be more prevalent among African Americans and Latinos than among whites, partly because of differences in rates of obesity.[37] Among autoimmune diseases, systemic lupus erythematosus is more likely to affect African Americans and Latinos than whites, while multiple sclerosis appears more among whites.

So far, the population captured by the Hypersomnia Foundation's CoRDS registry appears to be disproportionately white, with more than 90 percent of participants identified as white as of 2020.[38] In a statement on its website, the Hypersomnia Foundation attributed the disparity primarily to racial minorities experiencing barriers to diagnosis of sleep disorders—a reasonable assumption that could be investigated in more detail. Discrimination and inclusion are issues that rare disease advocacy groups have been paying more attention to, and it comes down to making more effort to listen to people and meet them where they are. At a 2021 diversity-, equity-, and inclusion-oriented workshop held by the foundation, a few participants recalled carefully choosing their clothes, so that physicians would listen to what they said about their hypersomnia symptoms and take them seriously.

An essay by James Stevens, a young African American man who was diagnosed with IH in high school in New Jersey, indicates that lack of social support made his experience dealing with IH worse.[39] One of his teachers suspected he was smoking too much marijuana. The vice-principal at his high school told him, "Look, I know it's hard to get up in the morning, but we're all tired." His mother, who had grown up "poor and angry" in Atlantic City, called him lazy and eventually kicked him out of the house in an attempt at tough love.

James did find a neurologist and was diagnosed with IH after a sleep study. He tried a long list of medications, which didn't help. Adderall gave him headaches that felt like "beads of sand whishing around my brain." He lost several jobs because of oversleeping. In many ways, James's narrative is archetypical for IH, except that he experienced remission, regaining a typical sleep schedule in 2017. "The lack of support I needed at such a vulnerable time was closer to killing me than hypersomnia ever had," he concluded.

His essay is another reminder of young people who may be struggling and require more understanding and recognition. The hypersomnia community can both learn from and expand to include others' stories.

A CHALLENGE OF RISK INTERPRETATION

In 2019, the narcolepsy and hypersomnia communities were jolted by news about modafinil. For several medications commonly prescribed for sleep disorders, solid information is scarce about the risks of exposure during pregnancy, such as an elevated probability of miscarriage or birth defects. This encapsulates the uncertainty that many with narcolepsy and hypersomnia have to deal with if they are planning for pregnancy.

Isabelle Arnulf has given an example of a woman with narcolepsy who was then taking modafinil and the antidepressant venlafaxine and had contacted her for advice: "My neurologist said I should stop taking my medication immediately. I use a computer for work, and without the medication, I would be unable to do my job. I am also worried I would be depressed if I have to go off my meds and stay at home during the pregnancy." Previously, expert opinion held that modafinil would pose less risk, in comparison to conventional stimulants. For several years, Arnulf had recommended that her patients switch to modafinil from other stimulants before pregnancy—and other specialists followed her lead.[40]

The Food and Drug Administration has been requiring drug manufacturers to establish pregnancy registries to gather data on newly approved medications' safety. However, participation in registries is voluntary, and with medications for neurological disorders, the number of mothers involved is often low enough to make statistical analyses difficult.

In 2019, an interim report from the Provigil/Nuvigil Pregnancy Registry revealed a rate of 15 percent for major malformations, in comparison to a rate of 3 percent in the general population. A rate of cardiac anomalies of 5 percent was also higher than expected for infants exposed to modafinil during pregnancy. The number of modafinil-exposed pregnancies on which the interim report was based was small (78),[41] but it took almost a decade since the registry's establishment to say anything at all. Around the same time, a Danish study also reported an increased risk of malformations with modafinil, compared to either methylphenidate or nonexposed pregnancies.[42] This shows that expert opinion sometimes has to catch up to real-world data; earlier unpublished safety data had been reassuring. Health officials in France and Canada promptly recommended that pregnant women and those trying to become pregnant not take modafinil. Arnulf has agreed with this assessment, and Trotti told me that her recommendations to patients have also changed as a result.

An analysis of the full registry report, with a larger number of prospective pregnancies, was published in 2021. It confirmed the increased risk, although the mechanism was unclear because several types of adverse outcomes were reported.[43] As of 2022, the U.S. FDA has not updated modafinil's label.

A surprising counterpoint came in the International Pregnancy Safety Study consortium, which included investigators from Harvard and several Nordic countries. This reported that neither methylphenidate nor amphetamines were associated with an increased risk of malformations. Compared with the registry report for modafinil, this study included a much larger number of pregnancies exposed to the drugs in question: several thousand.[44] This study does not mean that conventional stimulants are risk-free during pregnancy, since other long-range outcomes were not tracked. More such studies are needed on pregnancy outcomes with medications for neurological disorders.

THE FUTURE OF PTZ, FLUMAZENIL, AND CLARITHROMYCIN

One of the patients featured in Jazz's 2021 awareness campaign was Meghan Mallare (from chapter 10). As the campaign started, sleep continued to take up more than one hundred hours of Meghan's week. She made no testimonial for Xywav because she had not tried it yet. Out of many medications she did try, one of the few that made her feel truly awake was the experimental drug PTZ, which she was able to access as part of Balance Therapeutics' clinical trial. "I woke up in the morning on my own and felt awake for the first time. It was such an incredible feeling," she said.

At that time, Balance Therapeutics had been dissolved. It was one more disappointment for proponents of the GABA hypothesis of IH. Jenkins and Rye's search for the GABA-enhancing peptide they detected in Anna and other patients remained inconclusive, and they were not able to announce their findings. Jenkins had been contributing to efforts to automate old-school patch clamp experiments, an advance that sped up his research.[45] In 2019, Jenkins acquired sophisticated equipment that would allow him to use smaller volumes of precious CSF and process many more samples per day. But after the end of the federal grant he and Rye were awarded, he had to transition to a smaller laboratory space. By mid-2022, Jenkins had left Emory for a position in Connecticut.

After the challenge from Dauvilliers, Rye and Jenkins's findings did not get much respect from some experts in sleep medicine. Their critics seemed to assume: *if they were right, they should have found the somnogen by now.* Emmanuel Mignot was more generous when asked for his opinion of their work.[46] He said that excessive GABA signaling as an explanation for hypersomnia was a reasonable hypothesis. However, one attempt to replicate Jenkins's laboratory results was unsuccessful, and the placebo effect might explain flumazenil's apparent clinical benefits, he said. Mignot left the door open: "Flumazenil is a relatively safe medication, why not try it? . . . I think it could help some people."

Other sleep researchers have told me they thought a GABA-based mechanism for hypersomnia was appealing or had prescribed some of their patients clarithromycin when more conventional medications were unsatisfactory.

A possibility to repurpose flumazenil is still open. Researchers at the University of Michigan have been testing whether the drug could be used for Parkinson's disease.[47] Starting in 2018, they have been investigating intravenous flumazenil's effects on motor symptoms of Parkinson's, such as postural instability and gait impairment. The results have not yet been reported.

A FEW COMPARISONS

The number of people in the hypersomnia community who use flumazenil may diminish. Some report that its efficacy has faded over time. However, I predict that some will stay with it, despite the expense and inconvenience. One woman told me that before trying flumazenil, she had progressed from working as a full-time lawyer and active mother to only having a few hours a day when she wasn't sleeping or resting. Stimulants had helped a bit, but nothing else had come close to restoring "normal" for her. Before, it was almost impossible to get out of bed until noon or later. She said: "The benefit of flumazenil is that I feel normal. I do not feel anything unusual or different: I am just awake like a normal person."

Despite anecdotal evidence for flumazenil's efficacy, the drug has not been tested for IH in the same rigorous clinical trial format as modafinil or other medications. Determining who is likely to respond positively to it remains difficult; CSF patch clamp assays do not predict response. Someone has to try it to find out if it works for them. "These medicines work—but who is the target popula-tion?" David Rye once asked me. He compared successfully treating hypersom-nia to harpooning a swordfish—with Anna as the swordfish. "Now we're trawling the whole North Atlantic."

Without a corporate or university sponsor, a trial of flumazenil for idiopathic hypersomnia is not currently on the horizon. The drug is available for off-label prescriptions through a few compounding pharmacies in the United States, an intermediate state between FDA approval and complete scarcity.

Members of the hypersomnia community embraced Rye's "sleepy juice" the-ory even though the biochemical evidence for it was incomplete, because his expla-nation for their sleepiness was appealing and he voiced the frustration that many of them felt. A pertinent comparison for Rye might be to Martin Scharf, who pio-neered the use of GHB as a treatment for narcolepsy in the United States in the 1980s.[48] Scharf may have tripped over the FDA's documentation requirements, but he helped demonstrate that GHB for narcolepsy was clinically feasible, pav-ing the way for Xyrem and Xywav.

The community of IHers and others who find benefit from flumazenil, along with the physicians who prescribe it, might evolve into something like the Low-Dose Naltrexone Trust, whose annual conference is sponsored by compounding pharmacies.[49] Naltrexone is FDA approved for the treatment of alcohol and opioid use disorders and has also been prescribed at lower doses for the off-label treatment of inflammation and pain. A clear mechanistic explanation for its anti-inflammatory properties is lacking, and low-dose naltrexone's activity has been described as "paradoxical."[50]

Clarithromycin's future is also uncertain. IHers' experimentation with it might remind some observers of the subculture of "Lyme-literate" doctors willing to treat Lyme disease patients with antibiotics over long periods.[51] However, enthu-siasm for clarithromycin was never that great—it literally left a bad taste in IHers' mouths. Cautious clinicians such as Trotti, as much as they wanted to give their patients more options, did not want to lead them astray. Trotti's work on clarithromycin continued with a clinical study, funded by the National Institute of Neurological Disorders and Stroke. She and her team planned to measure clar-ithromycin's effects on GABA signaling, as well as probing inflammation and changes in the intestinal microbiome as possible explanations for how the anti-biotic may act against excessive daytime sleepiness.

Rye and Jenkins may still find the hidden treasure they were looking for, but it is equally likely that other researchers will transform our understanding of IH and its pathophysiology. IH could splinter into more than one category as a result. The rewards of studying IH will include a deeper knowledge of why we sleep and how it restores us. People with IH can advise others and contribute to research on why sleep is consuming their lives. The next person to write a chapter in the story of IH may be someone who lives with it every day.

POSTSCRIPT ON COVID-19

Most of the events in this book took place before COVID-19, a global challenge for public health whose effects continue to unfold. At the beginning of the pandemic, I wondered whether the world might see an eerie repeat of encephalitis lethargica and the variety of neurological injuries that resulted. Some sleep researchers had proposed that COVID-19 might offer an opportunity to understand better how narcolepsy develops, in a way analogous to what occurred with the 2009 H1N1 influenza pandemic.[52] While life altering for those involved, the surge in European narcolepsy diagnoses did help investigators clarify the autoimmune mechanism behind the disease.

Instead, a new category of postviral chronic illness has emerged: postacute sequelae of COVID-19, or "long COVID," which may eventually imprint itself upon millions of people around the world.[53] Patient groups have pushed officials at the National Institutes of Health and elsewhere to fund research on long COVID. The similarities between long COVID and ME/CFS (myalgic encephalomyelitis/chronic fatigue syndrome) could end up propelling advances in the understanding of ME/CFS.

Where does that leave IH and narcolepsy type 2? It depends whether we view neurological sleep disorders such as IH and narcolepsy in a silo or as part of a broader continuum of chronic illnesses, such as autoimmune disorders.[54] Long COVID can have neurological components, but it varies from individual to individual and extends far beyond a disturbance of sleep. It can include brain fog and postural orthostatic tachycardia, symptoms that people with IH experience as well.

Narcolepsy and IH are studied by a relatively small group of investigators around the world. Yet IH may have elements in common with disorders that are not usually considered sleep disorders, such as myotonic dystrophy and hepatic encephalopathy. Mapping the common elements may yield insights both in patient care and in research. Similarly, mutual benefits could come from exchanges with other fields, such as psychiatry, rheumatology, and addiction medicine.

ACKNOWLEDGMENTS

This section highlights my thanks to more than one hundred people who made this book possible. The writing of this book was generously supported by a 2019 grant from the Alfred P. Sloan Foundation. I am grateful to Chris Gunter and Andrew Miller for early encouragement, to Tom Scammell for allowing me to attend the seventh International Symposium on Narcolepsy, and to Holly Korschun and Vince Dollard at Emory's Division of Communications and Public Affairs for tolerating my work on this book. Special thanks to Alexander Bieri and Svenja Egli at the Roche archive in Basel and National Institutes of Health and Food & Drug Administration staff responsible for Freedom of Information Act requests. Additional thanks to Jordan Tomes and Blanka Owensova for Czech translations and to Tara Coyt, Chuck Clark, and Glenn Hutchinson for English critique. In the text, IH and narcolepsy patients are identified by their first name; researchers and clinicians by last name.

THE STARTING POINT

Anna Sumner Pieschel, James and Ward Sumner, and James Sumner III.

MEMBERS OF THE ATLANTA HYPERSOMNIA SUPPORT GROUP, NARCOLEPSY ATLANTA, AND THE BROADER HYPERSOMNIA COMMUNITY

Diana Kimmel, Saraiah Naps, Julie Flygare, Thomas Fast, Diane Powell, Betsy Ashcraft, Evelyn Honig, Keith Harper, Michelle Chadwick, Matt Baker, Dean Jordheim, Danielle Hulshizer, Kristin Loomis, Michelle Emrich, Michael Sparace, Linda Johnson, David Kellogg, Meghan Mallare, Amy Desmarais, Romy Baudois, Stacy Erickson Edwards, Cate Murray, Catherine Page-Rye, Rebecca King, James Stevens, Alex Haagaard.

SOURCES

David Rye, Andrew Jenkins, Kathy Parker, Lynn Marie Trotti, Prabhjyot Saini, Bob Baker, Karel Šonka, Soňa Nevšímalová, Jan Roth, Anniki Rothova, Evžen Růžička, Donald Bliwise, Amanda Freeman, Glenda Keating, Gillian Hue, Paul Garcia, Mike Owens, Olivia Moody, Carol Clark, Allan Rechtschaffen, Bruce Wainer, Allan Levey, John Roback, Cliff Saper, Clete Kushida, Mahlon DeLong, Mary Carskadon, Tom Roth, Jim Krueger, Jamie Zeitzer, Phyllis Zee, Emmanuel Mignot, Richard Bogan, Amita Sehgal, Robert Thomas, George O'Neil, Erin Kelty, Luis de Lecea, Jerry Siegel, Tom Kilduff, Gary Aston-Jones, Mehdi Tafti, Giuseppe Plazzi, Hanns Möhler, Alessandro Guidotti, Phil Skolnick, Steven Paul, Roger Butterworth, Giuseppe Scollo-Lavizzari, William Ondo, Thien Thanh Dang Vu, David Plante, Jesse Cook, and Morgan Lam (to whom I owe a beer). Informal discussions with Isabelle Arnulf, Yves Dauvilliers, Todd Swick, Roger Broughton, Peter Young, Anne Heidbreder, Markku Partinen, Guy Leschziner, Ramin Khatami, Birgitte Kornum, Geert Mayer, Makoto Honda, and Kiran Maski.

BOOK BUSINESS

Fact-checking services: Kelly Hills and Cadence Bambenek.
Representation: Jessica Papin at Dystel, Goderich & Bourret.
Editor: Miranda Martin at Columbia University Press.

THE BOTTOM LINE

Edith Eastman for love and support.

NOTES

INTRODUCTION

1. Thomas Fast, "Snooze TV Episode 1: Parents with Children with Idiopathic Hypersomnia," Hypersomnia Foundation, YouTube video posted January 9, 2015, https://www.youtube.com/watch?v=xgbuCiZ9rME.

2. Colin L. Talley, "The Emergence of Multiple Sclerosis, 1870–1950: A Puzzle of Historical Epidemiology," *Perspectives in Biology and Medicine* 48 (2005): 383–95.

3. Joanna Kempner, *Not Tonight: Migraine and the Politics of Gender and Health* (Chicago: University of Chicago Press, 2014).

4. Richard Hargreaves and Jes Olesen, "Calcitonin Gene-Related Peptide Modulators—the History and Renaissance of a New Migraine Drug Class," *Headache* 59 (2019): 951–70.

1. ANNA SLEEPS A LOT, AND WE DON'T KNOW WHY

1. Anna Sumner, "Waking Sleeping Beauty: An Antidote for Hypersomnia," Emory University YouTube video, posted November 21, 2012, https://www.youtube.com/watch?v=V9gnvWtta4M.

2. David B. Rye et al., "Modulation of Vigilance in the Primary Hypersomnias by Endogenous Enhancement of GABA-A Receptors," *Science Translational Medicine* 4 (2012): 161ra151.

3. Timothy I. Morgenthaler et al., "Practice Parameters for the Treatment of Narcolepsy and Other Hypersomnias of Central Origin," *Sleep* 30 (2007): 1705.

4. Elisabeth I. Penninga et al, "Adverse Events Associated with Flumazenil Treatment for the Management of Suspected Benzodiazepine Intoxication," *Basic & Clinical Pharmacology & Toxicology* 118 (2016): 37–44.

5. Virginia Hughes, "Re-Awakenings," *Last Word on Nothing* (blog), November 22, 2012, https://www.lastwordonnothing.com/2012/11/22/re-awakenings/.

6. Bedřich Roth, "Sleep Drunkenness and Sleep Paralysis," *Ceskoslovenská neurologie* 19 (1956): 48–58; translation by Jordan Tomes.

7. Michel Billiard, "Development of Sleep Medicine in Europe," in *Sleep Medicine*, ed. Sudhansu Chokroverty (New York: Springer, 2015), 113–24.

8. Bedřich Roth, "Narcolepsy and Hypersomnia: Review and Classification of 642 Personally Observed Cases," *Schweizer Archiv für Neurologie Neurochirurgie und Psychiatrie* 119 (1976): 31–41.
9. Soňa Nevšímalová, "Idiopathic Hypersomnia," in *Sleep Medicine*, ed. Chokroverty, 223–28.
10. Michel Billiard, "Idiopathic Hypersomnia," in *Sleepiness: Causes, Consequences, and Treatment*, ed. Michael J. Thorpy (Cambridge: Cambridge University Press, 2011), 130.
11. Maurice M. Ohayon, Yves Dauvilliers, and Charles F. Reynolds III, "Operational Definitions and Algorithms for Excessive Sleepiness in the General Population," *Archives of General Psychiatry* 69 (2012): 71–79.
12. David E. McCarty, "Resolution of Hypersomnia Following Identification and Treatment of Vitamin D Deficiency," *Journal of Clinical Sleep Medicine* 6 (2010): 605–6; Imran Khawaja et al., "Vitamin B12 Deficiency: A Rare Cause of Excessive Daytime Sleepiness," *Journal of Clinical Sleep Medicine* 15 (2019): 1365–67.
13. Saraiah Naps, interview, email and phone, April 2020.
14. Jimmy Kimmel, "What It Feels Like to Have Narcolepsy," *Esquire*, August 2003.
15. Russell Rosenberg and Ann Y. Kim, "The AWAKEN survey: Knowledge of Narcolepsy Among Physicians and the General Population," *Postgraduate Medicine* 126 (2014): 78–86.
16. Michelle Emrich, "A Doctor's Once Agile Brain Broken by IH," Hypersomnia Foundation, YouTube video, posted November 23, 2020, https://www.youtube.com/watch?v=Ww_6LfS4dFc.
17. Caroline Maness et al., "Systemic Exertion Intolerance Disease/Chronic Fatigue Syndrome Is Common in Sleep Centre Patients with Hypersomnolence: A Retrospective Pilot Study," *Journal of Sleep Research* 28 (2019): e12689.
18. Todd Swick, interview, September 2018.
19. Yoko Komada et al., "Difference in the Characteristics of Subjective and Objective Sleepiness Between Narcolepsy and Essential Hypersomnia," *Psychiatry and Clinical Neuroscience* 59 (2005): 194–99; Patrice Bourgin, Jamie M. Zeitzer, and Emmanuel Mignot, "CSF Hypocretin-1 Assessment in Sleep and Neurological Disorders," *Lancet Neurology* 7 (2008): 649–62.
20. Carlos H. Schenck et al., "English Translations of the First Clinical Reports on Narcolepsy and Cataplexy by Westphal and Gélineau in the Late 19th Century," *Journal of Clinical Sleep Medicine* 3 (2007): 301–11.
21. Daniela Latorre et al., "T Cells in Patients with Narcolepsy Target Self-Antigens of Hypocretin Neurons," *Nature* 562 (2018): 63–68.
22. Soňa Nevšímalová-Bruhova and Bedřich Roth, "Heredofamilial Aspects of Narcolepsy and Hypersomnia," *Schweizer Archiv für Neurologie, Neurochirurgie und Psychiatrie* 110 (1972): 45–54.
23. Gary S. Richardson et al., "Excessive Daytime Sleepiness in Man: Multiple Sleep Latency Measurement in Narcoleptic and Control Subjects," *Electroencephalography and Clinical Neurophysiology* 45 (1978): 621–27.
24. Chad Ruoff et al., "The MSLT Is Repeatable in Narcolepsy Type 1 but Not Narcolepsy Type 2: A Retrospective Patient Study," *Journal of Clinical Sleep Medicine* 14 (2018): 65–74.
25. Karel Šonka, Marek Šusta, and Michel Billiard, "Narcolepsy with and Without Cataplexy, Idiopathic Hypersomnia with and Without Long Sleep Time: A Cluster Analysis," *Sleep Medicine* 16 (2015): 225–31.
26. Elisa Evangelista et al., "Alternative Diagnostic Criteria for Idiopathic Hypersomnia: A 32-Hour Protocol," *Annals of Neurology* 83 (2018): 235–47; Fabio Pizza et al., "Daytime Continuous Polysomnography Predicts MSLT Results in Hypersomnias of Central Origin," *Journal of Sleep Research* 22 (2013): 32–40.
27. Cyrille Vernet and Isabelle Arnulf, "Idiopathic Hypersomnia with and Without Long Sleep Time: A Controlled Series of 75 Patients," *Sleep* 32 (2009): 753–59.
28. Michel Billiard and Karel Šonka, "Idiopathic Hypersomnia," *Sleep Medicine Reviews* 29 (2016): 23–33.

29. Yves Dauvilliers et al., "Normal Cerebrospinal Fluid Histamine and Tele-Methylhistamine Levels in Hypersomnia Conditions," *Sleep* 35 (2012): 1359–66; Antonio Baruzzi et al., "Cerebrospinal Fluid Homovanillic Acid and 5-Hydroxyindoleacetic Acid in Hypersomnia with Periodic Apneas or Idiopathic Hypersomnia: Preliminary Results," *Sleep* 3 (1980): 247–49; Kim F. Faull et al. "Cerebrospinal Fluid Monoamine Metabolites in Narcolepsy and Hypersomnia," *Annals of Neurology* 13 (1983): 258–63.

30. Yves Dauvilliers et al., "Absence of γ-Aminobutyric Acid—a Receptor Potentiation in Central Hypersomnolence Disorders," *Annals of Neurology* 80 (2016): 259–68.

31. Olivia A. Moody et al. "Rigor, Reproducibility, and *in Vitro* CSF Assays: The Devil in the Details," *Annals of Neurology* 81 (2017): 904–7.

2. THE DOCTORS AND GABA

1. Kathy P. Parker, "Sleep and Dream Patterns in Patients Experiencing Dialysis," PhD diss., Georgia State University, 1990.

2. Will Astor, "Nurse and Researcher Becomes a Leader," *Rochester Business Journal*, September 18, 2009.

3. Kathy P. Parker et al., "Lowering Dialysate Temperature Improves Sleep and Alters Nocturnal Skin Temperature in Patients on Chronic Hemodialysis," *Journal of Sleep Research* 16 (2007): 42–50.

4. Dellarie L. Shilling and Donna Hodnicki, "APRN Prescribing in Georgia: An Evolving Environment," *Journal of the American Association of Nurse Practitioners* 27 (2015): 300–7.

5. Possible for PhDs until 2007. Kingman P. Strohl, "Sleep Medicine Training Across the Spectrum," *Chest* 139 (2011): 1221–31.

6. David B. Rye and Lynn M. Trotti, "Restless Legs Syndrome and Periodic Leg Movements of Sleep," *Neurologic Clinics* 30 (2012): 1137–66.

7. Gregory Berns, *Satisfaction: The Science of Finding True Fulfillment* (New York: Henry Holt, 2005).

8. David B. Rye, "The Genetics and Pathogenesis of Restless Legs Syndrome: Implications for the Clinician," *Medscape*, 2008, https://www.medscape.org/viewarticle/572098.

9. Restless Legs Syndrome Foundation blog, "Name Change Talking Points," March 2013, http://rlsfoundation.blogspot.com/2013/03/name-change-talking-points.html.

10. Steven Woloshin and Lisa M. Schwartz, "Giving Legs to Restless Legs: A Case Study of How the Media Helps Make People Sick," *PLoS Medicine* 3 (2006): e170.

11. Rob Stein, "Marketing the Illness and the Cure? Drug Ads May Sell People on the Idea That They Are Sick," *Washington Post*, May 30, 2006.

12. Hreinn Stefansson et al., "A Genetic Risk Factor for Periodic Limb Movements in Sleep," *New England Journal of Medicine* 357 (2007): 639–47.

13. Karen Barrow, "Patient Voices: Restless Leg Syndrome," *New York Times*, 2008.

14. Claudio Bassetti and Michael Aldrich, "Idiopathic Hypersomnia: A Series of 42 Patients," *Brain* 120 (1997): 1423–35.

15. Bethany Brookshire, "Dopamine Is ___," *Slate*, July 3, 2013, https://slate.com/technology/2013/07/what-is-dopamine-love-lust-sex-addiction-gambling-motivation-reward.html.

16. Josh Berke, "What Does Dopamine Mean?," *Nature Neuroscience* 21 (2018): 787–93.

17. Andrew Jenkins, Nicholas P. Franks, and William R. Lieb, "Effects of Temperature and Volatile Anesthetics on GABA-A receptors," *Anesthesiology* 90 (1999): 484–91.

18. Meagan Ward, "Mind the Gap: Dr. Andrew Jenkins," *Central Sulcus* (blog), September 2006, https://thecentralsulcus.wordpress.com/2013/08/16/mind-the-gap-dr-andrew-jenkins/.

19. O. P. Hamill et al., "Improved Patch-Clamp Techniques for High-Resolution Current Recording from Cells and Cell-Free Membrane Patches," *Pflugers Archiv.* 391 (1981): 85–100.

20. From a 2007 Parker/Rye/Jenkins letter to Roche.

21. A concentration of 5 micromolar flumazenil was used in patch clamp experiments. Rye later presented data showing that Anna's plasma level averaged around 60 nanomolar.

22. Brian J. Ruscito and Neil L. Harrison, "Hemoglobin Metabolites Mimic Benzodiazepines and Are Possible Mediators of Hepatic Encephalopathy," *Blood* 102 (2003): 1525–28.

23. Elio Lugaresi et al., "Suspected Covert Lorazepam Administration Misdiagnosed as Recurrent Endozepine Stupor," *Brain* 121 (1998): 2201. For more on idiopathic recurring stupor, see Quinn Eastman, "The Downfall of 'Idiopathic Recurring Stupor,'" https://quinneastman.medium.com /the-downfall-of-idiopathic-recurring-stupor-3ad8c2330ea4.

24. William G. Ondo and Y. S. Silay, "Intravenous Flumazenil for Parkinson's Disease: A Single Dose, Double Blind, Placebo Controlled, Cross-Over Trial," *Movement Disorders* 21 (2006): 1614–7; William G. Ondo and Christine Hunter, "Flumazenil, a GABA Antagonist, May Improve Features of Parkinson's Disease," *Movement Disorders* 18 (2003): 683–5.

25. Marco L. A. Sivilotti, "Flumazenil, Naloxone and the 'Coma Cocktail,'" *British Journal of Clinical Pharmacology* 81 (2016): 428–36.

26. Donna L. Seger, "Flumazenil—Treatment or Toxin," *Journal of Toxicology: Clinical Toxicology* 42 (2004): 209–16.

27. H. J. O'Connor, "Status Epilepticus Following Administration of Flumazenil After Endoscopy," *Endoscopy* 23 (1991): 53.

28. Barbara Gordon, *I'm Dancing as Fast as I Can* (New York: Moyer Bell, 2011).

3. THE ANTIDOTE

1. William H. Spivey, "Flumazenil and Seizures: Analysis of 43 Cases," *Clinical Therapeutics* 14 (1992): 292–305.

2. Following an incomplete dilation and extraction, the head of the fetus emerged after the woman had been sent home. Anna and the supervising partner at her firm won an award for their handling of the case. June D. Bell, "Grisly Tape Backfired in Malpractice Case," *National Law Journal*, April 17, 2006.

3. Mathias Basner and David F. Dinges, "Maximizing Sensitivity of the Psychomotor Vigilance Test to Sleep Loss," *Sleep* 34 (2011): 581–91.

4. Danielle Moron et al., "The Psychomotor Vigilance Task as a Diagnostic Tool for Hypersomnolence," *Sleep* 42S (2019): A247–48.

5. Carol Clark, "'Sleeping Beauty' Case Awakens Hope for Cure," *Emory Magazine*, 2008, www .emory.edu/EMORY_MAGAZINE/2008/winter/sleeping.html.

6. F. X. Breheny, "Reversal of Midazolam Sedation with Flumazenil," *Critical Care Medicine* 20 (1992): 736–39.

7. Lewis Kraufkopf, "He Has Roche's Rx; CEO of Drug Maker Knows His Prescriptions," *Bergen County Record*, July 3, 2003.

8. United States Patent and Trademark Office website, "Patent Terms Extended Under 35 USC § 156," https://www.uspto.gov/patents/laws/patent-term-extension/patent-terms-extended-under-35 -usc-156.

9. Philip J. Hilts, "More Eligible for AIDS Drug," *New York Times*, June 1, 1990.

10. FDA Website, IND Applications for Clinical Treatment (Expanded Access): Overview, https:// www.fda.gov/drugs/investigational-new-drug-ind-application/ind-applications-clinical -treatment-expanded-access-overview.

11. General Accounting Office, "Investigational New Drugs: FDA Has Taken Steps to Improve the Expanded Access Program but Should Further Clarify How Adverse Events Data Are Used," GAO-17-564, July 11, 2017, https://www.gao.gov/products/gao-17-564.

12. Tim K. Mackey and Virginia J. Schoenfeld, "Going 'Social' to Access Experimental and Potentially Life-Saving Treatment: An Assessment of the Policy and Online Patient Advocacy Environment for Expanded Access," *BMC Medicine* 14 (2016): 17.

13. Roche newsletter describing May 2008 visit.

14. Yoko Komada et al., "Difference in the Characteristics of Subjective and Objective Sleepiness Between Narcolepsy and Essential Hypersomnia," *Psychiatry and Clinical Neuroscience* 59 (2005): 194–99.

15. United States Patent & Trademark Office, "Use of Gabba [*sic*] Receptor Antagonists for the Treatment of Excessive Sleepiness and Disorders Associated with Excessive Sleepiness," US2011/0028418 A1, February 3, 2011.

16. David B. Rye, 2009 National Institute of Neurological Disorders and Stroke grant application, 3R01NS00551015-03S1, obtained through a Freedom of Information Act request.

17. Valerie is not her actual name; she asked for her identity to be concealed.

4. RYE VERSUS MSLT

1. Allan I. Levey et al., "Monoclonal Antibodies to Choline Acetyltransferase: Production, Specificity, and Immunohistochemistry," *Journal of Neuroscience* 3 (1983): 1–9.

2. Francis Crick, "Function of the Thalamic Reticular Complex: The Searchlight Hypothesis," *Proceedings of the National Academy of Sciences* 81 (1984): 4586–90.

3. Wainer had achieved early recognition for vaccinating monkeys against opiates, an antiaddiction strategy still discussed today. Kathryn F. Bonese et al., "Changes in Heroin Self-Administration by a Rhesus Monkey After Morphine Immunisation," *Nature* 252 (1974): 708–10.

4. David B. Rye et al., "Pedunculopontine Tegmental Nucleus of the Rat," *Journal of Comparative Neurology* 259 (1987): 483–528.

5. Interview with John Roback.

6. Jon Van, "Diagnosticians Go to Head of Matter," *Chicago Tribune*, January 14, 1979; Ronald Kotulak, "Patient Chats with Doctors During His Brain Surgery," *Chicago Tribune*, July 16, 1981.

7. Clete A. Kushida et al., "Cortical Asymmetry of REM Sleep EEG Following Unilateral Pontine Hemorrhage," *Neurology* 41 (1991): 598–601.

8. J. William Langston and Jon Palfreman, *The Case of the Frozen Addicts* (Amsterdam: IOS, 2014).

9. Paul C. Bucy, "A Biographical Memoir of Percival Bailey," *Biographical Memoirs* 58 (National Academies Press, 1989), https://www.nap.edu/read/1645/chapter/3.

10. David B. Rye et al., "Presentation of Narcolepsy After 40," *Neurology* 50 (1998): 459–65.

11. William J. Adie, "Idiopathic Narcolepsy: A Disease *Sui Generis*, with Remarks on the Mechanism of Sleep," *Brain* 49 (1926): 257–306.

12. Robert E. Yoss and David D. Daly, "Criteria for the Diagnosis of the Narcoleptic Syndrome," *Proceedings of the Staff Meetings of the Mayo Clinic* 32 (1957): 320–28.

13. Vogel's paper goes into detail about his patient's dreams, fantasies, and paintings but does not appear to describe cataplexy. Gerald Vogel, "Studies in Psychophysiology of Dreams III: The Dream of Narcolepsy," *Archives of General Psychiatry* 3 (1960): 421–28.

14. Allan Rechtschaffen et al., "Nocturnal Sleep of Narcoleptics," *Electroencephalography and Clinical Neurophysiology* 15 (1963): 599–609. Five out of nine people with narcolepsy studied in this paper displayed cataplexy.

15. William Dement, Allan Rechtschaffen, and George Gulevich, "The Nature of the Narcoleptic Sleep Attack," *Neurology* 16 (1966): 18–33.

16. David Raynal, "Polygraphic Aspects of Narcolepsy," *Proceedings of the First International Symposium on Narcolepsy* (1975): 671–84.

17. Mary A. Carskadon and William C. Dement, "Effects of Total Sleep Loss on Sleep Tendency," *Perceptual and Motor Skills* 48 (1979): 495–506.

18. Merrill M. Mitler et al., "REM Sleep Episodes During the Multiple Sleep Latency Test in Narcoleptic Patients," *Electroencephalography and Clinical Neurophysiology* 46 (1979): 479–81.

19. Timothy A. Roehrs et al., "Daytime Sleepiness and Antihistamines," *Sleep* 7 (1984): 137–41.

20. Robert E. Yoss, "The Pupillogram and Narcolepsy: A Method to Measure Decreased Levels of Wakefulness," *Neurology* 19 (1969): 921–28.

21. Roger Broughton et al., "Excessive Daytime Sleepiness and the Pathophysiology of Narcolepsy-Cataplexy: A Laboratory Perspective," *Sleep* 9 (1986): 205–15.

22. American Academy of Sleep Medicine, *International Classification of Sleep Disorders: Diagnostic and Coding Manual*, 2nd ed. (Westchester, IL: AASM, 2005).

23. Theodore L. Baker et al., "Comparative Polysomnographic Study of Narcolepsy and Idiopathic Central Nervous System Hypersomnia," *Sleep* 9 (1986): 232–42.

24. Gert J. Lammers and J. G. van Dijk, "The Multiple Sleep Latency Test: A Paradoxical Test?," *Clinical Neurology and Neurosurgery* 94 (1992): S108–10.

25. Kristyna M. Hartse, Thomas Roth, and Frank J. Zorick, "Daytime Sleepiness and Daytime Wakefulness: The Effect of Instruction," *Sleep* 5 (1982): S107–18.

26. David B. Rye et al., "Daytime Sleepiness in Parkinson's Disease," *Journal of Sleep Research* 9 (2000): 63–69.

27. Kathy P. Parker et al., "Daytime Sleepiness in Stable Hemodialysis Patients," *American Journal of Kidney Diseases* 41 (2003): 394–402.

28. David B. Rye, Bhupesh Dihenia, and Donald L. Bliwise, "Reversal of Atypical Depression, Sleepiness and REM Sleep Propensity in Narcolepsy with Buproprion," *Depression and Anxiety* 7 (1998): 92–95.

29. Adam Moscovitch, Markku Partinen, and Christian Guilleminault, "The Positive Diagnosis of Narcolepsy and Narcolepsy's Borderland," *Neurology* 43 (1993): 55–60; Michael S. Aldrich, "The Clinical Spectrum of Narcolepsy and Idiopathic Hypersomnia," *Neurology* 46 (1996): 393–401.

30. Emmanuel Mignot et al., "Correlates of Sleep-Onset REM Periods During the Multiple Sleep Latency Test in Community Adults," *Brain* 129 (2006): 1609–23.

31. Phone interview with Roth, 2019; Meeta Singh, Christopher L. Drake, and Thomas Roth, "The Prevalence of Multiple Sleep-Onset REM Periods in a Population-Based Sample," *Sleep* 29 (2006): 890–95.

32. David J. Kupfer, "REM Latency: A Psychobiologic Marker for Primary Depressive Disease," *Biological Psychiatry* 11 (1979): 159–74.

33. Jerome M. Siegel, "REM Sleep: A Biological and Psychological Paradox," *Sleep Medicine Reviews* 15 (2011): 138–42.

34. Merrill Wise, "An Update on the MSLT and MWT Practice Parameters Released Last Year," *Sleep Review*, May 7, 2006, https://sleepreviewmag.com/sleep-diagnostics/in-lab-tests/mslt-mwt-studies/a-new-approach/.

35. Michael R. Littner et al., "Practice Parameters for Clinical Use of the Multiple Sleep Latency Test and the Maintenance of Wakefulness Test," *Sleep* 28 (2005): 113–21.

36. Margaret S. Blattner et al., "Quantification of Late REM Periods in Patients with Prolonged Sleep Duration," *Sleep* 43S (2020): A288.

37. Lois E. Krahn et al., "Recommended Protocols for the Multiple Sleep Latency Test and Maintenance of Wakefulness Test in Adults: Guidance from the American Academy of Sleep Medicine," *Journal of Clinical Sleep Medicine* 17 (2021): 2489–98.

38. Alex Haagaard, "My Invisible Illness Story," *Medium*, September 28, 2016, https://bullshit.ist/my -invisible-illness-story-57563706fa5f.

39. Terri E. Weaver and Ronald R. Grunstein, "Adherence to Continuous Positive Airway Pressure Therapy," *Proceedings of the American Thoracic Society* 5 (2008): 173–78.

40. Chitra Lal et al., "Excessive Daytime Sleepiness in Obstructive Sleep Apnea. Mechanisms and Clinical Management," *Annals of the American Thoracic Society* 18 (2021): 757–68.

41. David A. Schulman et al., "A Sleep Medicine Curriculum for Pulmonary and Pulmonary/Critical Care Fellowship Programs," *Chest* 155 (2019): 554–56.

42. Emerson Wickwire, "Are Ethics a Taboo Subject in Sleep Medicine?," *Sleep Review*, September 4, 2014, https://sleepreviewmag.com/uncategorized/ethics-taboo-sleep-medicine/.

43. Stephen Talsness et al., "Thinking Outside the 'Sleep Logs,'" *Sleep* 42S (2019): A413–14.

44. Paul E. Peppard et al., "Increased Prevalence of Sleep-Disordered Breathing in Adults," *American Journal of Epidemiology* 177 (2013): 1006–14.

45. Raphael Heinzer, Helena Marti-Soler, and Jose Haba-Rubio, "Prevalence of Sleep Apnoea Syndrome in the Middle to Old Age General Population," *Lancet Respiratory Medicine* 4 (2016): e5–e6.

46. Raphael Heinzer et al., "Prevalence of Sleep-Disordered Breathing in the General Population: The HypnoLaus Study," *Lancet Respiratory Medicine* 3 (2015): 310–18.

47. Melissa C. Lipford et al., "Correlation of the Epworth Sleepiness Scale and Sleep-Disordered Breathing in Men and Women," *Journal of Clinical Sleep Medicine* 15 (2019): 33–38.

5. BEHIND THE CURTAIN

1. Charles L. Dana, *On Morbid Drowsiness and Somnolence: A Contribution to the Pathology of Sleep* (New York: G. P. Putnam & Sons, 1884).

2. Nathaniel Kleitman, *Sleep and Wakefulness* (Chicago: University of Chicago Press, 1939).

3. Simon Shorvon, "Fashion and Cult in Neuroscience—the Case of Hysteria," *Brain* 130 (2007): 3342–48.

4. John A. Goodfellow and Hugh J. Willison, "Guillain–Barré Syndrome: A Century of Progress," *Nature Reviews Neurology* 12 (2016): 723–31.

5. Michel Billiard, *Sleep Medicine: A Comprehensive Guide* (New York: Springer, 2015), 121.

6. William Mahoney, *The History of the Czech Republic and Slovakia* (Westport, CT: Greenwood, 2011).

7. Maximilien Vessier, *La Pitie-Salpetriere* (Paris: Assistance Hopitaux Publique de Paris, 1999).

8. From Roth's promotion materials, part of Kamil Henner's papers at the Archives of the Czech Academy of Sciences, trans. Jordan Tomes. See also John Connelly, *Captive University: Sovietization of East German, Czech, and Polish Higher Education, 1945–1956* (Chapel Hill: University of North Carolina Press, 2000).

9. Interview with Charles University's neurology department chair Evžen Růžička.

10. Bedřich Roth, "Sleep Drunkenness and Sleep Paralysis," *Československá neurologie* 19 (1956): 48; translation by Jordan Tomes.

11. Soňa Nevšímalová-Bruhova, "On the Problem of Heredity in Hypersomnia, Narcolepsy and Dissociated Sleep Disturbances," *Acta Universitatis Carolinae Medica* 18:109–80 (1973).

12. Soňa Nevšímalová-Bruhova and Bedřich Roth, "Heredofamilial Aspects of Narcolepsy and Hypersomnia," *Schweizer Archiv für Neurologie, Neurochirurgie und Psychiatrie* 110 (1972): 45–54.

13. Bedřich Roth and Soňa Nevšímalová, "Depression in Narcolepsy and Hypersomnia," *Schweizer Archiv für Neurologies, Neurochirurgie und Psychiatrie* 116 (1975): 291–300.

14. I relied on a German translation of the first edition of Roth's book. Bedřich Roth, *Narkolepsie und Hypersomnie* (Berlin: VEB Verlag Volk und Gesundheit, 1962), 162, 208.

15. The only paper from Dement cited in Roth's 1950s book: William C. Dement, "The Occurrence of Low Voltage, Fast, Electroencephalogram Patterns During Behavioral Sleep in the Cat," *Electroencephalography and Clinical Neurophysiology* 10 (1958): 291–96.

16. Roth, *Narkolepsie und Hypersomnie*, 361.

17. Norman Goldstein and Mary E. Giffin, "Psychogenic Hypersomnia," *American Journal of Psychiatry* 115 (1959): 922–28.

18. Cyril Höschl, Petr Winkler, and Ondrej Peec, "The State of Psychiatry in the Czech Republic," *International Review of Psychiatry* 24 (2012): 278–85.

19. Henri Gastaut and M. Fischer-Williams, "Electroencephalographic Study of Syncope," *Lancet* 273 (1957): 1018–25.

20. Henri Gastaut and Bedřich Roth, "Electroencephalographic Manifestations in 150 Cases of Narcolepsy with & Without Cataplexy," *Revue Neurologique* 97:388–93 (1957).

21. Bedřich Roth, "The Clinical and Theoretical Importance of EEG Rhythms Corresponding to States of Lowered Vigilance," *Electroencephalography and Clinical Neurophysiology* 13 (1961): 395–99.

22. Bedřich Roth and Eliska Klimkova-Deutschova, "Effect of the Chronic Action of Industrial Poisons on the Electroencephalogram of Man," *Ceskoslovenská neurologie* 27 (1964): 40–47.

23. Nada Rothova and Bedřich Roth, "A Case of Focal Reflex Epilepsy Caused by Tactile Stimulation in a Child," *Ceskoslovenská neurologie* 26 (1963): 33–35.

24. Soňa Nevšímalová, "Idiopathic Hypersomnia," in *Sleep Medicine*, ed. Sudhansu Chokroverty (New York: Springer, 2015), 223–28.

25. Bedřich Roth, Soňa Nevšímalová, and Allan Rechtschaffen, "Hypersomnia with 'Sleep Drunkenness,'" *Archives of General Psychiatry* 26 (1972): 456–62.

26. Allan Rechtschaffen and Anthony Kales, *A Manual of Standardized Terminology, Techniques, and Scoring System for Sleep Stages of Human Subjects* (National Institute of Neurological Diseases and Blindness, 1968).

27. Max Hirshkowitz, "Lessons from the Forefathers of Sleep," *Sleep Review*, November 2007, http://www.sleepreviewmag.com/2007/11/lessons-from-the-forefathers-of-sleep/.

28. Vladimir V. Kusin, *From Dubček to Charter 77: A Study of "Normalization" in Czechoslovakia, 1968–1978* (New York: St. Martin's, 1978).

29. Interviews with Karel Šonka and Evžen Růžička.

30. Bedřich Roth, *Narcolepsy and Hypersomnia* (Basel: Karger, 1980).

31. Roth, *Narcolepsy and Hypersomnia*, 222.

32. Sleep Disorders Classification Committee, "Diagnostic Classification of Sleep and Arousal Disorders," *Sleep* 2 (1979): 5–137.

33. Christian Guilleminault, R. Phillips, and William C. Dement, "A Syndrome of Hypersomnia with Automatic Behavior," *Electroencephalography and Clinical Neurophysiology* 38: 403–13 (1975).

34. Theodore L. Baker et al., "Comparative Polysomnographic Study of Narcolepsy and Idiopathic Central Nervous System Hypersomnia," *Sleep* 9 (1986): 232–42.

35. Bedřich Roth et al., "A Study of the Occurrence of JLA DR2 in 124 Narcoleptics: Clinical Aspects," *Schweizer Archiv für Neurologie Neurochirurgie und Psychiatrie* 139 (1985): 41–51.

36. Bedřich Roth et al., "An Alternative to the MSLT for Determining Sleepiness in Narcolepsy and Hypersomnia: Polygraphic Score of Sleepiness," *Sleep* 9 (1986): 243–45.

37. Roger Broughton and Michelle Chadwick, "Bedřich Roth: Pioneer in Sleep Medicine," *Sleep Medicine* 76 (2020): 160–69.

6. THE ESSENCE OF SLEEPINESS

1. Paul J. Marangos et al., "Demonstration of an Endogenous, Competitive Inhibitor(s) of 3H Diazepam Binding in Bovine Brain," *Life Sciences* 5 (1978): 1893–1900; Erminio Costa et al., "On a Brain Polypeptide Functioning as a Putative Effector for the Recognition Sites of Benzodiazepine and Beta-carboline Derivatives," *Neuropharmacology* 22 (1983): 1481–92.

2. David B. Rye, National Institute of Neurological Disorders and Stroke application, 3R01NS055015-03S1.

3. Beryl L. Benderly, "Shovel-Ready Science," *Science*, March 6, 2009.

4. Alexis Fedorchak, *Public Library of Science Blogs*, December 2013, https://blogs.plos.org/thestudentblog/2013/12/13/an-ode-to-patch-clamping/.

5. Areles Molleman, *Patch Clamping: An Introductory Guide to Patch Clamp Electrophysiology* (Wiley, 2002).

6. Chris Bladen, "#LabHacks: 14 Sharp Tips for Patch Clamping," *NeuroWire*, https://www.scientifica.uk.com/neurowire/14-sharp-tips-for-patch-clampers.

7. Andrew Jenkins, "IH & GABA," Hypersomnia Foundation 2018 conference, YouTube video posted February 1, 2020, https://www.youtube.com/watch?v=4iEVmRXHOc4.

8. Eric L. Bittman et al., "Animal Care Practices in Experiments on Biological Rhythms and Sleep," *Journal of the American Association for Laboratory Animal Science* 52 (2013): 437–43.

9. Kisou Kubota, "Kuniomi Ishimori and the First Discovery of Sleep-Inducing Substances in the Brain," *Neuroscience Research* 6 (1989): 497–518.

10. Kenton Kroker, *The Sleep of Others* (Toronto: University of Toronto Press, 2007).

11. John R. Pappenheimer, Thomas B. Miller, and Clark A. Goodrich, "Sleep-Promoting Effects of Cerebrospinal Fluid from Sleep-Deprived Goats," *Proceedings of the National Academy of Sciences* 58 (1967): 513–17.

12. Vladimir Fencl, G. Koski, and John R. Pappenheimer, "Factors in Cerebrospinal Fluid from Goats That Affect Sleep and Activity in Rats," *Journal of Physiology* 216 (1971): 565–89.

13. John R. Pappenheimer, "The Sleep Factor," *Scientific American* 235 (1976): 24–29.

14. Bryan Marquard, "John R. Pappenheimer, 92; Taught Physiology at Harvard," *Boston Globe*, December 27, 2007.

15. James M. Krueger, J. Bacsik, and J. Garcia-Arraras, "Sleep-Promoting Material from Human Urine and Its Relation to Factor S from Brain," *American Journal of Physiology* 238 (1980): E116–23.

16. October 2019 interview.

17. B. M. Bergmann et al., "Are Physiological Effects of Sleep Deprivation in the Rat Mediated by Bacterial Invasion?," *Sleep* 19:554–62 (1996).

18. James M. Krueger et al., "Sleep: A Physiological Role for IL-1 Beta and TNF-Alpha," *Annals of the New York Academy of Sciences* 29 (1998): 148–59.

19. Michel Jouvet, "Sleep and Serotonin: An Unfinished Story," *Neuropsychopharmacology* 21 (1999): S24–27.

20. Grigorios Oikonomou et al., "The Serotonergic Raphe Promote Sleep in Zebrafish and Mice," *Neuron* 103 (2019): 686–701.

21. "Scientists Are Extracting the Sleep Potion," *New York Times*, April 14, 1974.

22. Abba J. Kastin et al., "DSIP—More Than a Sleep Peptide?," *Trends in Neuroscience* 3 (1980): 163–65.

23. Dietrich Schneider-Helmert, M. Graf, and G. A. Schoenenberger, "Synthetic Delta-Sleep-Inducing Peptide Improves Sleep in Insomniacs," *Lancet* 317 (1981): 1256; U.S. Patent & Trademark Office, "Nonapeptide for Treating Addictive Drug Withdrawal Conditions," 4444758, April 24, 1984, https://patents.google.com/patent/US4444758A/en.

24. Gerhilde Reheis-Melcher, "Delta Sleep-Inducing Peptide," *Arznei Telegramm* 5, no. 47 (1990).

25. V. I. Odin et al., "Diabetes Mellitus in Elderly: Geroprotective and Antidiabetic Properties of Delta-Sleep Induced Peptide," *Advances in Gerontology* 15 (2004): 101–14.

26. Vladimir M. Kovalzon and Tatyana V. Strekalova, "Delta Sleep-Inducing Peptide (DSIP): A Still Unresolved Riddle," *Journal of Neurochemistry* 97 (2006): 303–9.

27. Wilhelm Feldberg and S. L. Sherwood, "Injections of Drugs Into the Lateral Ventricle of the Cat," *Journal of Physiology* 123:148–67 (1954).

28. Matthew Walker, *Why We Sleep* (New York: Scribner, 2017).

29. Tarja Porkka-Heiskanen et al., "Adenosine: A Mediator of the Sleep-Inducing Effects of Prolonged Wakefulness," *Science* 276 (1997): 1265–68.

30. "New Clues to Why We Snooze," *Science*, May 22, 1997, https://www.science.org/content/article/new-clues-why-we-snooze.

31. Robert E. Strecker et al., "Another Chapter in the Adenosine story," *Sleep* 29 (2006): 426–28.

32. Jamie M. Zeitzer et al., "Extracellular Adenosine in the Human Brain During Sleep and Sleep Deprivation: An In Vivo Microdialysis Study," *Sleep* 29:455–61 (2006).

33. J. V. Retey et al., "A Functional Genetic Variation of Adenosine Deaminase Affects the Duration and Intensity of Deep Sleep in Humans," *Proceedings of the National Academy of Sciences* 102 (2005): 15676–81.

34. Valerie Bachmann et al., "Functional ADA Polymorphism Increases Sleep Depth and Reduces Vigilant Attention in Humans," *Cerebral Cortex* 22 (2012): 962–70.

35. Theresa E. Bjorness et al., "Control and Function of the Homeostatic Sleep Response by Adenosine A1 Receptors," *Journal of Neuroscience* 29 (2009): 1267–76.

36. Ehsan Shokri-Kojori et al., "β-Amyloid Accumulation in the Human Brain After One Night of Sleep Deprivation," *Proceedings of the National Academy of Sciences* 115:4483–88 (2018).

37. David Elmenhorst et al., "Recovery Sleep After Extended Wakefulness Restores Elevated A1 Adenosine Receptor Availability in the Human Brain," *Proceedings of the National Academy of Sciences* 114 (2017): 4243–48.

38. Hirofumi Toda et al., "A Sleep-Inducing Gene, *nemuri*, Links Sleep and Immune Function in Drosophila," *Science* 363 (2019): 509–15.

39. Daniel A. Lee et al., "Genetic and Neuronal Regulation of Sleep by Neuropeptide VF," *eLife* 6 (2017): e25727.

40. Hiromasa Funato et al., "Forward Genetic Analysis of Sleep in Randomly Mutagenized Mice," *Nature* 539 (2016): 378–83.

41. Zhiqiang Wang et al., "Quantitative Phosphoproteomic Analysis of the Molecular Substrates of Sleep Need," *Nature* 558 (2018): 435–39.

42. Takato Honda et al., "A Single Phosphorylation Site of SIK3 Regulates Daily Sleep Amounts and Sleep Need in Mice," *Proceedings of the National Academy of Sciences* 115 (2018): 10458–63.

43. Zachary Zamore and Sigrid C. Veasey, "Neural Consequences of Chronic Sleep Disruption," *Trends in Neurosciences* 45 (2022): 678–91.

44. David Zada et al., "Parp1 Promotes Sleep, Which Enhances DNA Repair in Neurons," *Molecular Cell* 81 (2021): 4979–93.

7. MY FAVORITE MISTAKE

1. Andria Simmons, "Keeper of Loud Dogs Gets His Day in Court," *Atlanta Journal Constitution*, August 29, 2008.
2. Joel M. Geiderman, "Central Nervous System Disturbances Following Clarithromycin Ingestion," *Clinical Infectious Diseases* 29 (1999): 464–65.
3. Paul S. Garcia and Andrew Jenkins, "Inhibition of the GABA(A) Receptor by a Macrolide but Not by a Lincosamide Antibiotic," American Society of Anesthesiologists 2009 meeting, A1385, http://www.asaabstracts.com/strands/asaabstracts/abstract.htm?year=2009&index=10&absnum=477.
4. Lynn M. Trotti et al., "Improvement in Daytime Sleepiness with Clarithromycin in Patients with GABA-Related Hypersomnia: Clinical Experience," *Journal of Psychopharmacology* 28 (2014): 697–702.
5. Christian M. Jespersen et al., "Randomised Placebo Controlled Multicentre Trial to Assess Short Term Clarithromycin for Patients with Stable Coronary Heart Disease," *BMJ* 332 (2006): 22–27.
6. Cory Acuff, November 20, 2008, Office of Technology Transfer "Breakfast Club" presentation.
7. The raw numbers were posted on https://www.clinicaltrials.gov/ct2/show/NCT01183312 in 2017.
8. Lynn M. Trotti, Lorne A. Becker, and Cochrane Movement Disorders Group, "Iron for the Treatment of Restless Legs Syndrome," *Cochrane Database of Systematic Reviews* (2019), CD007834; Lynn M. Trotti et al., "Medications for Daytime Sleepiness in Individuals with Idiopathic Hypersomnia," *Cochrane Database of Systematic Reviews* (2021), CD012714.
9. Catherine A. Hosmer, *Taking the Pulse of the U.S. Health Care System* (iUniverse, 2007).
10. October 2018 Emory Radiology grand rounds.
11. Interviews with Anna's parents and Saraiah Naps.
12. Gabe Gutierrez, "Disorder Causes Lawyer to Sleep up to 18 Hours a Day," *Today/MSNBC*, November 25, 2012.
13. Melinda Beck, "Scientists Try to Unravel the Riddle of Too Much Sleep," *Wall Street Journal*, December 10, 2012.
14. Interview with Betsy Ashcraft.
15. Gene J. Koprowski, " 'Sleeping Beauty' Gene Proves Beastly for Sufferers," *Fox News*, November 21, 2012.
16. Greg Miller, "Putting Themselves to Sleep," *Science*, November 21, 2012.
17. Aisha Tyler, "Sleeping Beauty Syndrome Makes It Hard to Wake Up," WCSC, https://www.live5news.com/story/29629513/sleeping-beauty-syndrome-makes-it-hard-to-wake-up/.
18. Johannes A. Romijn, "Pituitary Diseases and Sleep Disorders," *Current Opinion in Endocrinology, Diabetes and Obesity* 23 (2016): 345–51.

8. THE ATLANTA SLEEPERS CLUB

1. Living with Hypersomnia website, archived in 2014: http://web.archive.org/web/20140607062515/http://www.livingwithhypersomnia.com/fr/podcast-001-hypersomnia-conference/.
2. Lynn M. Trotti, Beth A. Staab, and David B Rye, "Test-Retest Reliability of the Multiple Sleep Latency Test in Narcolepsy Without Cataplexy and Idiopathic Hypersomnia," *Journal of Clinical Sleep Medicine* 9 (2013): 789–95.
3. David B. Rye, "What's in a Name? Understanding the Origins of the Terminologies for the Family of Hypersomnias," YouTube, Hypersomnia Foundation video posted February 25, 2015,

https://youtu.be/qUB1NquAxCI. Rye enthuses over Gowers's term "somnosis" but does not discuss Bedřich Roth extensively.

4. Virginia Hughes, "Wake No More," *Matter*, January 26, 2015, https://medium.com/matter/wake-no-more-8bbd49528b9.

5. Jackie Sturt et al., "Neurolinguistic Programming: A Systematic Review of the Effects on Health Outcomes," *British Journal of General Practice* 62 (2012): e757–e764.

6. Archived at http://web.archive.org/web/20120217090029/http://www.confidentfuture.com.au/.

7. Lloyd Johnson, "NLP—The Secrets to Learning Neuro-Linguistic Programming Online," learn-NLPonline YouTube video posted May 27, 2012, https://www.youtube.com/watch?v=9DwP6 2Xrt_k.

8. Christian Guilleminault and Susanna Mondini, "Mononucleosis and Chronic Daytime Sleepiness. A Long-Term Follow-Up Study," *Archives of Internal Medicine* 146 (1986): 1333–35.

9. Erin Kelty et al., "Use of Subcutaneous Flumazenil Preparations for the Treatment of Idiopathic Hypersomnia: A Case Report," *Journal of Psychopharmacology* 28 (2014): 703–6.

10. National Institutes of Health website archived in 2013: http://web.archive.org/web/20130128095235 /http://rarediseases.info.nih.gov/RareDiseaseList.aspx?StartsWith=I. An entry for idiopathic hypersomnolence was added by the end of 2013.

11. Mayo Clinic website accessed via http://web.archive.org for 2013–2016.

12. Talk About Sleep, archived in 2010, http://web.archive.org/web/20100204051157/http://www .talkaboutsleep.com/idiopathic-hypersomnia/.

13. Analysis of comments to FDA Docket 2013-N-0815, obtained by Freedom of Information Act.

14. Twitter/email conversations with Chadwick. Also Eva Lewicki, "Aussie Mum's Medical Mystery," *That's Life!*, May 9, 2020, https://www.thatslife.com.au/aussie-mums-medical-mystery-help-i-cant -stop-sleeping.

15. Gary Hulse et al., "Withdrawal and Psychological Sequelae, and Patient Satisfaction Associated with Subcutaneous Flumazenil Infusion for the Management of Benzodiazepine Withdrawal," *Journal of Psychopharmacology* 27 (2013): 222–27.

16. Sujata Gupta, "Blocking the High: One Man's Quixotic Quest to Cure Addiction," *Mosaic Science*, March 17, 2015, https://mosaicscience.com/story/blocking-the-high/.

17. Stefano Tamburin et al., "Low Risk of Seizures with Slow Flumazenil Infusion and Routine Anti-convulsant Prophylaxis for High-Dose Benzodiazepine Dependence," *Journal of Psychopharmacology* 31 (2017): 1369–73.

18. Gary Hulse et al., "Novel Indications for Benzodiazepine Antagonist Flumazenil in GABA Mediated Pathological Conditions of the Central Nervous System," *Current Pharmaceutical Design* 21 (2015): 3325–42.

19. Hughes, "Wake No More," last paragraph.

20. Samantha Bresnahan, "Living with Hypersomnia: The Woman Who Slept, but Got No Rest," CNN, April 22, 2015, https://www.cnn.com/2015/04/22/health/hypersomnia-woman-always -tired/index.html.

21. Lynn M. Trotti et al., "Clarithromycin in γ-Aminobutyric Acid–Related Hypersomnolence: A Randomized, Crossover Trial," *Annals of Neurology* 78 (2015): 454–65.

22. Vincent LaBarbera et al., "Central Disorders of Hypersomnolence, Restless Legs Syndrome and Surgery with General Anesthesia: Patient Perceptions," *Frontiers in Human Neuroscience* 12 (2018): 99.

23. Emails from Kristin Loomis, executive director of the Human Herpesvirus-6 Foundation.

24. ProPublica Nonprofit Explorer, Hypersomnia Foundation, https://projects.propublica.org /nonprofits/organizations/464162735. Also Hypersomnia Foundation Inc. (14019717) documents, downloaded from the Georgia Secretary of State website.

25. Hypersomnia Foundation announcement, "Passing the Torch," January 16, 2017, https://www.hypersomniafoundation.org/passing-the-torch/.

26. Page-Rye stepped down from the board in late 2019.

27. Susannah L. Rose, "Patient Advocacy Organizations: Institutional Conflicts of Interest, Trust, and Trustworthiness," *Journal of Law, Medicine & Ethics* 41 (2013): 680–87.

28. Christine Miserandino, "The Spoon Theory," *But You Don't Look Sick*, 2003, https://butyoudontlooksick.com/articles/written-by-christine/the-spoon-theory/.

9. THE STORY OF FLUMAZENIL

1. Willy Haefely and Walter Hunkeler, "The Story of Flumazenil," *European Journal of Anaesthesiology* 2S (1988): 3–13.

2. Andrea Tone, *The Age of Anxiety* (New York: Basic Books, 2008).

3. "Valium Patent Expires, but Generic Equivalent Still Months Away," Associated Press, March 2, 1985; "Stay Calm," *Economist*, September 28, 1991.

4. U.S. Senate, Subcommittee on Health and Scientific Research, "Use and Misuse of Benzodiazepines," 96th Congress, September 10, 1979 (Washington, DC: U.S. Gov't Printing Office, 1980).

5. Hans C. Peyer, "Roche, a Company History 1896–1996" (Editiones Roche, 1996).

6. Hanns Möhler and Mose Da Prada, "The Challenge of Neuropharmacology: A Tribute to the Memory of Willy Haefely" (Editiones Roche, 1994), 9.

7. Willy Haefely et al., "Possible Involvement of GABA in the Central Actions of Benzodiazepines," *Advances in Biochemistry and Psychopharmacology* 14 (1975): 131–51.

8. Hanns Möhler and Toshikazu Okada, "Benzodiazepine Receptor: Demonstration in the Central Nervous System," *Science* 198 (1977): 849–51.

9. Walter Hunkeler et al., "Selective Antagonists of Benzodiazepines," *Nature* 290 (1981): 514–16.

10. Hanns Möhler and J. G. Richards, "Agonist and Antagonist Benzodiazepine Receptor Interaction *in vitro*," *Nature* 294 (1981): 763–65.

11. "Ein Benzodiazepin, das die Wirkung von Benzodiazepinen aufhebt," *Roche* 14 (1982): 11–17. Original quote: "Ich bin über die Versuchsituation voll orientiert, fühle mich wie vom Wecker aus dem Schlaf gerissen und möchte aufstehen."

12. Austin Darragh et al., "Reversal of Benzodiazepine-Induced Sedation by Intravenous Ro 15-1788," *Lancet* 318 (1981): 1042.

13. Austin Darragh et al., "Absence of Central Effects in Man of the Benzodiazepine Antagonist Ro 15-1788," *Psychopharmacology* 80 (1983): 192–95.

14. Leslie Iversen, "Anti-Anxiety Receptors in the Brain?," *Nature* 266 (1977): 678.

15. Solomon Snyder , *Brainstorming: The Science and Politics of Opiate Research* (Cambridge, MA: Harvard University Press, 1989).

16. Walter E. Müller, *The Benzodiazepine Receptor* (Cambridge: Cambridge University Press, 1987), 78.

17. Willy Haefely, "Alleviation of Anxiety: The Benzodiazepine Saga," in *Discoveries in Pharmacology: Psycho-and Neuro-pharmacology*, ed. M. J. Parnham and J. Bruinvels (Amsterdam: Elsevier, 1983), 293.

18. Jeff Goldberg, *Anatomy of a Scientific Discovery* (New York: Bantam, 1988), 141–59.

19. Giuseppe Scollo-Lavizzari, "The Clinical Anti-Convulsant Effects of Flumazenil, a Benzodiazepine Antagonist," *European Journal of Anesthesiology* 2S (1998): 129–38.

20. Anna Higgitt, Malcolm Lader, and Peter Fonagy, "The Effects of the Benzodiazepine Antagonist Ro 15-1788 on Psychophysiological Performance and Subjective Measures in Normal Subjects," *Psychopharmacology* 89 (1986): 395–403.

21. Möhler and Da Prada, "The Challenge of Neuropharmacology," 125–29.

22. Peretz Lavie, "Ro 15-1788 Decreases Hypnotic Effects of Sleep Deprivation," *Life Sciences* 41 (1987): 227–33; Peretz Lavie, "Intrinsic Effects of the Benzodiazepine Receptor Antagonist Ro 15-1788 in Sleepy and Alert Subjects," *International Journal of Neuroscience* 46 (1989): 131–37.

23. H. M. Emrich, P. Sonderegger, and N. Mai, "Action of the Benzodiazepine Antagonist Ro 15-1788 in Humans After Sleep Withdrawal, *Neuroscience Letters* 47 (1984): 369–73; Axel Steiger et al., "Flumazenil Exerts Intrinsic Activity on Sleep EEG and Nocturnal Hormone Secretion in Normal Controls," *Psychopharmacology* 113 (1994): 334–38; Erich Seifritz et al., "Effects of Flumazenil on Recovery Sleep and Hormonal Secretion After Sleep Deprivation in Male Controls," *Psychopharmacology* 120 (1995): 449–56; Ulrich Hemmeter et al., "Effect of Flumazenil Augmentation on Microsleep and Mood in Depressed Patients During Partial Sleep Deprivation," *Journal of Psychiatric Research* 41 (2007): 876–84.

24. David J. Nutt et al., "Flumazenil Provocation of Panic Attacks. Evidence for Altered Benzodiazepine Receptor Sensitivity in Panic Disorder," *Archives of General Psychiatry* 47 (1990): 917–25.

25. Thaddeus J. Marczynski, J. Artwohl, and B. Marczynska, "Chronic Administration of Flumazenil Increases Life Span and Protects Rats from Age-Related Loss of Cognitive Functions," *Neurobiology of Aging* 15 (1994): 69–84.

26. Jamie Talan, "Help for AIDS, Alcoholism, Memory?," *Newsday*, March 30, 1993.

27. John Travis, "Biologists Visit New Orleans (Under an Assumed Name)," *Science* 260 (1993): 162–63.

28. Nick Neave et al., "Dose-dependent Effects of Flumazenil on Cognition, Mood, and Cardio-respiratory Physiology in Healthy Volunteers," *British Dental Journal* 189 (2000): 668–74.

29. Zoya Farzampour, Richard J. Reimer, and John Huguenard, "Endozepines," *Advances in Pharmacology* 72 (2015): 147–64.

30. Erminio Costa, *An Early Attempt to Foster Neuroscience Globalization* (Good Life, 2003); Robert Kanigel, *Apprentice to Genius: The Making of a Scientific Dynasty* (Baltimore, MD: Johns Hopkins University Press, 1993).

31. Alan P. Kozikowski, "On the Path from Chemistry to Neuroscience: Early Explorations in Chemical Medicine Under the Mentorship of Dr. Erminio Costa, a Neuroscientist with a Big Brain and a Bigger Heart," *Pharmacological Research* 64 (2011): 327–29.

32. Pamela Taulbee, "Solving the Mystery of Anxiety," *Science News* 124 (1983): 45.

33. Erminio Costa et al., "Pharmacology of Neurosteroid Biosynthesis," *Annals of the New York Academy of Sciences* 746 (1994): 223–42.

34. Maria D. Majewska et al., "Steroid Hormone Metabolites Are Barbiturate-Like Modulators of the GABA Receptor," *Science* 232 (1986): 1004–7.

35. Graziano Pinna, "Allopregnanolone, the Neuromodulator Turned Therapeutic Agent: Thank You, Next?," *Frontiers in Endocrinology* 11 (2020).

36. Johanna Berthier et al., "Menstruation-Related Hypersomnia: Three Adolescent Cases Responding to Treatment with the Oral Contraceptive Pill," *Journal of Neuroscience and Neuropsychology* 3 (2019): 102.

37. Shaotong Zhu et al., "Structure of a Human Synaptic GABA-A Receptor," *Nature* 559 (2018): 67–72.

38. Michelle D. Brot et al., "The Anxiolytic-Like Effects of the Neurosteroid Allopregnanolone: Interactions with GABA-A Receptors," *European Journal of Pharmacology* 325 (1997): 1–7.

39. Ciaran O'Boyle, Ronan Lambe, and Austin Darragh, "Central Effects in Man of the Novel Schistosomicidal Benzodiazepine Meclonazepam," *European Journal of Clinical Pharmacology* 29 (1985) 105–8.

40. Roche Products Ltd., Hypnovel and Anexate product literature, c. 1987, Roche archives, HAR PD.3.1 DMC-110342.

41. Warren E. Leary, "F.D.A. Assailed on Approval of Drug," *New York Times*, May 6, 1988.

42. Rainer Dorow et al., "Clinical Perspectives of Beta-Carbolines from First Studies in Humans," *Brain Research Bulletin* 19 (1987): 319–26; Peter R. Bieck, "Human Pharmacology of CGS 8216, a Benzodiazepine Antagonist," *Clinical Neuropharmacology* 7S (1984): S365.

43. Rainer Dorow et al., "Severe Anxiety Induced by FG 7142, a Beta-Carboline Ligand for Benzodiazepine Receptors," *Lancet* 322 (1983): 98–99.

44. Wallace B. Mendelson et al., "A Benzodiazepine Receptor Antagonist Decreases Sleep and Reverses the Hypnotic Actions of Flurazepam," *Science* 219 (1983): 414–16.

45. Philip J. Hilts, "Chemical Discoveries Open 'Inner Space,'" *Washington Post*, September 6, 1982.

46. Rachel Nave, Paula Herer, and Peretz Lavie, "The Intrinsic Effects of Sarmazenil on Sleep Propensity and Performance Level of Sleep-Deprived Subjects," *Psychopharmacology* 115 (1994): 366–70.

47. Minutes of the Central Nervous System Task Force meeting, January 1989, report from Peter Schoch, pp. 14–16, Roche archives, HAR FE.0.4-105726a.

48. Minutes of the Pharmaceutical Research Steering Committee, January 28, 1987, p. 7, Roche archives, HAR FE.0.4-105782.

49. E. Pietro Bonetti et al., "Ro 15-4513: Partial Inverse Agonism at the BZR and Interaction with Ethanol," *Pharmacology, Biochemistry, and Behavior* 31 (1988): 733–49.

50. Gina Kolata, "New Drug Counters Alcohol Intoxication," *Science* 234: 1198–99 (1986).

51. Robert Langreth, "High Anxiety," *Forbes,* November 27, 2000.

52. Daniel F. Schafer and E. Anthony Jones, "Hepatic Encephalopathy and the γ-Aminobutyric-Acid Neurotransmitter System," *Lancet* 319 (1982): 18–20.

53. Mario Baraldi et al., "Supersensitivity of Benzodiazepine Receptors in Hepatic Encephalopathy Due to Fulminant Hepatic Failure in the Rat: Reversal by a Benzodiazepine Antagonist," *Clinical Science* 67 (1984): 167–75.

54. Giuseppe Scollo-Lavizzari, "First Clinical Investigation of the Benzodiazepine Antagonist Ro 15-1788 in Comatose Patients," *European Neurology* 22 (1983): 7–11.

55. Giuseppe Scollo-Lavizzari et al., "Reversal of Hepatic Coma by Benzodiazepine Antagonist Ro 15-1788," *Lancet* 1324/1325 (1985).

56. Giuseppe Scollo-Lavizzari and H. Mathis, "Benzodiazepine Antagonist (Ro 15-1788) in Ethanol Intoxication: A Pilot Study," *European Neurology* 24 (1983): 352–54; European Patent Office, EP 0-398-171-A2, May 11, 1990, https://patents.google.com/patent/EP0398171A2/en.

57. Denis A. Burke et al., "Reversal of Hepatic Coma with Flumazenil with Improvement in Visual Evoked Potentials," *Lancet* 332 (1988): 505–6.

58. Peter Ferenci et al., "Successful Long-term Treatment of Portal-Systemic Encephalopathy by the Benzodiazepine Antagonist Flumazenil," *Gastroenterology* 96 (1989): 240–43.

59. Anthony S. Basile et al., "Elevated Brain Concentrations of 1,4-Benzodiazepines in Fulminant Hepatic Failure," *New England Journal of Medicine* 325 (1991): 473–78.

60. Jamie Talan, "Study of Brain Yields Hope of Treating Hepatic Comas," *Newsday*, October 24, 1990.

61. Cihan Yurdaydin et al., "Gut Bacteria Provide Precursors of Benzodiazepine Receptor Ligands in a Rat Model of Hepatic Encephalopathy," *Brain Research* 679 (1995): 42–48.

62. Gilles Pomier-Layrargues et al., "Flumazenil in Cirrhotic Patients in Hepatic Coma: A Randomized Double-Blind Placebo-Controlled Crossover Trial," *Hepatology* 19 (1994) 32–37.

63. Pascal Perney et al., "Plasma and CSF Benzodiazepine Receptor Ligand Concentrations in Cirrhotic Patients with Hepatic Encephalopathy," *Metabolic Brain Diseases* 13 (1998): 201–10.

64. Giuseppe Barbaro et al., "Flumazenil for Hepatic Encephalopathy Grade III and IVa in Patients with Cirrhosis," *Hepatology* 28 (1998): 374–78.

65.   Samir Ahboucha et al., "Increased Levels of Pregnenolone and Its Neuroactive Metabolite Allopregnanolone in Autopsied Brain Tissue from Cirrhotic Patients Who Died in Hepatic Coma," *Neurochemistry International* 49 (2006): 372–78.

66.   Stacia Brown, letter posted on Hypersomnia Research site, archived 2014, http://web.archive.org /web/20140802210733/http://hypersomniaresearch.org/media/Hypersomnia-Research-Project .pdf.

67.   Catherine A. Christian et al., "Endogenous Positive Allosteric Modulation of GABA-A Receptors by Diazepam Binding Inhibitor," *Neuron* 7 (2013): 1063–74.

68.   Others have proposed DBI as a "hunger factor" that stimulates appetite and is overabundant in humans and animals with obesity. José Manuel Bravo-San Pedro et al., "Acyl-CoA-Binding Protein (ACBP): The Elusive 'Hunger Factor' Linking Autophagy to Food Intake," *Cell Stress* 3 (2019): 312–18.

69.   Olivia Moody, "Pharmacologically Targeting the GABA-A Receptors in Neurological Disease," PhD diss., Emory University, 2017, chap. 4, https://etd.library.emory.edu/concern/etds/0c483j38t.

70.   Test subjects watch a light that flickers at a changing frequency and report when the light appears to become continuous. Natalia D. Mankowska et al., "Critical Flicker Fusion Frequency: A Narrative Review," *Medicina* 57 (2021): 1096.

10. WEIRD DRUGS

1.   Hypersomnia Research webpage, archived 2014, http://web.archive.org/web/20140802210733 /http://hypersomniaresearch.org/.

2.   David B. Rye, National Institute for Neurological Disorders and Stroke application, R01NS089719-01A1, 2015.

3.   Lynn M. Trotti, National Institute for Neurological Disorders and Stroke application, K23NS083748-01A1, 2014.

4.   Lynn M. Trotti et al, "Flumazenil for the Treatment of Idiopathic Hypersomnia: Clinical Experience with 153 Patients," *Journal of Clinical Sleep Medicine* 12 (2016): 1389–94.

5.   Bruce Jancin, "Flumazenil Effective in Refractory Hypersomnolence," *Chest Physician/MDEdge News,* July 12, 2016.

6.   Yves Dauvilliers et al., "Absence of γ-Aminobutyric Acid-a Receptor Potentiation in Central Hypersomnolence Disorders," *Annals of Neurology* 80 (2016): 250–68.

7.   Yves Dauvilliers et al., "Normal Cerebrospinal Fluid Histamine and Tele-methylhistamine Levels in Hypersomnia Conditions," *Sleep* 35 (2012): 1359–66.

8.   Olivia A. Moody et al., "Rigor, Reproducibility and in vitro CSF Assays: The Devil in the Details," *Annals of Neurology* 81 (2017): 904–7.

9.   Hanieh Toosi, Esther del Cid-Pellitero, and Barbara E. Jones, "GABA Receptors on Orexin and Melanin-Concentrating Hormone Neurons Are Differentially Homeostatically Regulated Following Sleep Deprivation," *eNeuro* 3 (2016).

10.   Richard W. Olsen and Jing Liang, "Role of GABAA Receptors in Alcohol Use Disorders Suggested by Chronic Intermittent Ethanol (CIE) Rodent Model," *Molecular Brain* 10 (2017): 45.

11.   Amy Desmarais, "Rising Voices of Narcolepsy," Project Sleep, YouTube video posted September 11, 2020, https://www.youtube.com/watch?v=9nxgc2VSP7Y.

12.   Hui Shen et al. "A Stress Steroid Triggers Anxiety Via Increased Expression of $\alpha 4\beta\delta$ GABA-A Receptors in Methamphetamine Dependence," *Neuroscience* 254 (2013).

13.   Aarti Kuver and Sheryl S. Smith, "Flumazenil Decreases Surface Expression of $\alpha 4\beta 2\delta$ GABAA Receptors by Increasing the Rate of Receptor Internalization," *Brain Research Bulletin* 120 (2016): 131–43.

14. Francesca Biggio et al., "Flumazenil Selectively Prevents the Increase in Alpha(4)-Subunit Gene Expression and an Associated Change in GABA(A) Receptor Function Induced by Ethanol Withdrawal," *Journal of Neurochemistry* 102 (2007): 657–66.

15. Shari Roan, "Addiction Treatment, Novel but Unproved," *Los Angeles Times*, October 9, 2006.

16. Walter Ling et al., "Double-blind Placebo-Controlled Evaluation of the PROMETA™ Protocol for Methamphetamine Dependence," *Addiction* 107 (2012): 361–69.

17. Marco Faccini et al., "Slow Subcutaneous Infusion of Flumazenil for the Treatment of Long-Term, High-Dose Benzodiazepine Users: A Review of 214 Cases," *Journal of Psychopharmacology* 30, no. 10 (2016).

18. Fabio Lugoboni and Roberto Leone, "What Is Stopping Us from Using Flumazenil?," *Addiction* 107 (2012): 1369.

19. Gordon Research Conferences, "The Science of Sleep: Emerging Themes and Paradigm Shifts," March 16–21, 2014.

20. Ren-Qi Huang et al., "Pentylenetetrazole-Induced Inhibition of Recombinant γ-Aminobutyric Acid Type A (GABAA) Receptors: Mechanism and Site of Action," *Journal of Pharmacology and Experimental Therapeutics* 298 (2001): 986–95.

21. Hermann Ruef, "Über Klinische Erfahrungen Mit Cardiazol," *Klinische Wochenschrift* 4 (1925): 1680–1.

22. M. Herbert Barker and Samuel A. Levine, "Cardiazol," *Archives of Internal Medicine* 42 (1928): 14–22.

23. Niall McCrae, "'A Violent Thunderstorm': Cardiazol Treatment in British Mental Hospitals," *History of Psychiatry* 17 (2006): 67–90.

24. Jesse Ballenger, *Treating Dementia—Do We Have a Pill for It?* (Baltimore, MD: Johns Hopkins University Press, 2009).

25. Emily Underwood, "Can Down Syndrome Be Treated?," *Science* 343 (2014): 964–67. Roche took a descendant of flumazenil into Down syndrome clinical trials, but its drug was ultimately unsuccessful. Joerg F. Hipp et al., "Basmisanil, a Highly Selective GABAA-α5 Negative Allosteric Modulator: Preclinical Pharmacology and Demonstration of Functional Target Engagement in Man," *Scientific Reports* 11 (2021): 7700.

26. April 2020 interview with Garner. Doses used were typically 100 times lower than seizure-inducing doses.

27. Fabian Fernandez et al., "Pharmacotherapy for Cognitive Impairment in a Mouse Model of Down Syndrome," *Nature Neuroscience* 10 (2007): 411–13.

28. Norman F. Ruby et al., "Circadian Locomotor Rhythms Are Normal in Ts65Dn Down Syndrome Mice and Unaffected by Pentylenetetrazole," *Journal of Biological Rhythms* 25 (2010): 63–66.

29. Damien Colas et al., "Short-term Treatment with Flumazenil Restores Long-Term Object Memory in a Mouse Model of Down Syndrome," *Neurobiology of Learning and Memory* 140 (2017): 11–16.

30. Australian New Zealand Clinical Trials Registry, ACTRN12612000652875, https://www.anzctr.org.au/Trial/Registration/TrialReview.aspx?id=362609.

31. Maria S. Trois et al., "Obstructive Sleep Apnea in Adults with Down Syndrome," *Journal of Clinical Sleep Medicine* 5 (2009): 317–23.

32. Damien Colas et al., "Sleep and EEG Features in Genetic Models of Down Syndrome," *Neurobiology of Disease* 30 (2008): 1–7.

33. David B. Rye et al, "An Open-Label Study of the Efficacy, Safety and Tolerability of Oral BTD-001 in Adults with Idiopathic Hypersomnia or Narcolepsy Type 2," *Sleep Medicine* 40S (2017): e285–86.

34. U.S. National Library of Medicine, "A Study of Safety and Efficacy of BTD-001 in Treatment of Patients with Idiopathic Hypersomnia (IH) or Narcolepsy Type 2," https://clinicaltrials.gov/ct2/show/NCT02512588; "A Study of Oral BTD-001 in Adults With Idiopathic Hypersomnia (ARISE2)," https://clinicaltrials.gov/ct2/show/NCT03542851.

35. Confirmed via multiple sources, but this person did not agree to discuss PTZ sourcing on the record.

36. Steven R. Gerbsman, "Date Certain M&A of Balance Therapeutics, Inc.," Gerbsman Partners blog, August 11, 2020, https://blog.gerbsmanpartners.com/2020/08/11/gerbsman-partners-date-certain -ma-of-balance-therapeutics-inc/.

11. THE HEART OF THE BRAIN

1. Ulrich Voderholzer et al., "A 19-H Spontaneous Sleep Period in Idiopathic Central Nervous System Hypersomnia," *Journal of Sleep Research* 7 (1998): 101–3.

2. Yves A. Dauvilliers and Luc Laberge, "Myotonic Dystrophy Type 1, Daytime Sleepiness and REM Sleep Dysregulation," *Sleep Medicine Reviews* 16 (2012): 539–45.

3. Constantin von Economo, "Encephalitis Lethargica," *Wiener Klinsche Wochenschrift* 30 (1917): 581–85.

4. Paul B. Foley, *Encephalitis Lethargica: The Mind and Brain Virus* (New York: Springer, 2018).

5. L. L. Anderson, Joel A. Vilensky, and R. C. Duvoisin, "Neuropathology of Acute Phase Encephalitis Lethargica: A Review of Cases from the Epidemic Period," *Neuropathology and Applied Neurobiology* 35 (2009): 462–72.

6. John A. Sours, "Narcolepsy and Other Disturbances in the Sleep-Waking Rhythm: A Study of 115 Cases with Review of the Literature," *Journal of Nervous and Mental Disease* 137 (1963): 525–42.

7. Luman Daniels, "Narcolepsy," *Medicine* 13 (1934): 1–134.

8. Eleanore Carey, "I Recover from Sleeping Sickness," *American Mercury* 32 (1934): 165–69.

9. Western Samoa was hit by both encephalitis lethargica and flu between 1918 and 1922, but American Samoa was not affected by either. R. T. Ravenholt and William H. Foege, "1918 Influenza, Encephalitis Lethargica, Parkinsonism," *Lancet* 320 (1982): 860–64.

10. Sherman McCall et al., "Influenza RNA Not Detected in Archival Brain Tissues from Acute Encephalitis Lethargica Cases or in Postencephalitic Parkinson Cases," *Journal of Neuropathology and Experimental Neurology* 60 (2001): 696–704; K. C. Lo et al., "Lack of Detection of Influenza Genes in Archived Formalin-Fixed, Paraffin Wax-Embedded Brain Samples of Encephalitis Lethargica Patients from 1916 to 1920," *Virchows Archiv* 442 (2003): 591–96.

11. Robert R. Dourmashkin et al., "Evidence for an Enterovirus as the Cause of Encephalitis Lethargica," *BMC Infectious Diseases* 12 (2012): 136.

12. Russell C. Dale et al., "Encephalitis Lethargica Syndrome: 20 New Cases and Evidence of Basal Ganglia Autoimmunity," *Brain* 127 (2004): 21–33.

13. Clifford B. Saper, Thomas E. Scammell, and Jun Lu, "Hypothalamic Regulation of Sleep and Circadian Rhythms," *Nature* 437 (2005): 1257–63.

14. Clifford B. Saper and Bradford B. Lowell, "The Hypothalamus," *Current Biology* 24 (2014): PR1111–16.

15. Bedřich Roth, *Narcolepsy and Hypersomnia* (Basel: Karger, 1980), 224.

16. Ayelet Snow et al., "Severe Hypersomnolence After Pituitary/Hypothalamic Surgery in Adolescents: Clinical Characteristics and Potential Mechanisms," *Pediatrics* 110 (2002): e74.

17. Gerald M. Rosen et al., "Sleep in Children with Neoplasms of the Central Nervous System: Case Review of 14 Children," *Pediatrics* 112 (2003): e46–54.

18. Mitchell G. Miglis et al., "Frequency and Severity of Autonomic Symptoms in Idiopathic Hypersomnia," *Journal of Clinical Sleep Medicine* 16 (2020): 749–56.

19. Lynn M. Trotti and Donald L. Bliwise, "Brain MRI Findings in Patients with Idiopathic Hypersomnia," *Clinical Neurology and Neurosurgery* 157 (2017): 19–21.

20. Walle J. Nauta, "Hypothalamic Regulation of Sleep in Rats," *Journal of Neurophysiology* 9 (1946): 285–316.

21. In addition to GABA, hypothalamic preoptic neurons produce neuropeptides such as cholecystokinin, which can induce sleep in the context of satiety. Shinjae Chung et al., "Identification of Preoptic Sleep Neurons Using Retrograde Labelling and Gene Profiling," *Nature* 545 (2017): 477–81.

22. Jonathan E. Sherin et al., "Activation of Ventrolateral Preoptic Neurons During Sleep," *Science* 271 (1996): 216–19.

23. Mohammed Aftab Alam et al., "Neuronal Activity in the Preoptic Hypothalamus During Sleep Deprivation and Recovery Sleep," *Journal of Neurophysiology* 111 (2014): 287–99.

24. Ronald Szymusiak et al., "Sleep-Waking Discharge Patterns of Ventrolateral Preoptic/Anterior Hypothalamic Neurons in Rats," *Brain Research* 803 (1998): 178–88.

25. Andrew S. P. Lim et al., "Sleep Is Related to Neuron Numbers in the Ventrolateral Preoptic/Intermediate Nucleus in Older Adults with and Without Alzheimer's Disease," *Brain* 137 (2014): 2847–61.

26. Kirsi-Marja Zitting et al., "Young Adults Are More Vulnerable to Chronic Sleep Deficiency and Recurrent Circadian Disruption Than Older Adults," *Scientific Reports* 8 (2018): 11052.

27. Daniel Kroeger et al., "Galanin Neurons in the Ventrolateral Preoptic Area Promote Sleep and Heat Loss in Mice," *Nature Communications* 9 (2018): 4129.

28. Edward C. Harding, Nicholas P. Franks, and William Wisden, "The Temperature Dependence of Sleep," *Frontiers in Neuroscience*, April 24, 2019.

29. Soňa Nevšímalová et al., "A Contribution to Pathophysiology of Idiopathic Hypersomnia," *Clinical Neurophysiology* 53S (2000): 366–70.

30. Daniel Aeschbach et al., "A Longer Biological Night in Long Sleepers Than in Short Sleepers," *Journal of Clinical Endocrinology and Metabolism* 88 (2003): 26–30.

31. Robert J. Thomas and Matt T. Bianchi, "A Circadian Mechanism for Idiopathic Hypersomnia—a Long Biological Night," *Sleep Medicine* 74 (2020): 31–32.

32. Friedrich K. Stephan and Irving Zucker, "Circadian Rhythms in Drinking Behavior and Locomotor Activity of Rats Are Eliminated by Hypothalamic Lesions," *Proceedings of the National Academy of Sciences* 69 (1972): 1583–86.

33. Dale M. Edgar, William C. Dement, and C. A. Fuller, "Effect of SCN Lesions on Sleep in Squirrel Monkeys: Evidence for Opponent Processes in Sleep-Wake Regulation," *Journal of Neuroscience* 13 (1993): 1065–79.

34. David P. Breen, "Hypothalamic Volume Loss Is Associated with Reduced Melatonin Output in Parkinson's Disease," *Movement Disorders* 31 (2016): 1062–66.

35. Aleksandar Videnovic et al., "Circadian Melatonin Rhythm and Excessive Daytime Sleepiness in Parkinson's Disease," *JAMA Neurology* 71 (2014): 463–69.

36. K. L. Toh et al., "An hPer2 Phosphorylation Site Mutation in Familial Advanced Sleep Phase Syndrome," *Science* 291 (2001): 1040–43; Alina Patke et al., "Mutation of the Human Circadian Clock Gene *CRY1* in Familial Delayed Sleep Phase Disorder," *Cell* 169 (2017): 203–15.

37. Julian Lippert et al., "Altered Dynamics in the Circadian Oscillation of Clock Genes in Dermal Fibroblasts of Patients Suffering from Idiopathic Hypersomnia," *PLoS One* 9 (2014): e85255.

38. Linus Materna et al., "Idiopathic Hypersomnia Patients Revealed Longer Circadian Period Length in Peripheral Skin Fibroblasts," *Frontiers in Neurology* 9 (2018): 424.

39. Lucia Pagani et al., "Serum Factors in Older Individuals Change Cellular Clock Properties," *Proceedings of the National Academy of Sciences* 108 (2011); 7218–23.

40. Brendan M. Gabriel et al., "Disrupted Circadian Oscillations in Type 2 Diabetes Are Linked to Altered Rhythmic Mitochondrial Metabolism in Skeletal Muscle," *Science Advances* 7 (2021);

William H. Walker et al., "Circadian Rhythm Disruption and Mental Health," *Translational Psychiatry* 10 (2020): 28.

41. Aditya Ambati et al., "Proteomic Biomarkers of Circadian Time," *Sleep* 42S (2019): A17–18; Dasha Cogswell et al., "Identification of a Preliminary Plasma Metabolome-Based Biomarker for Circadian Phase in Humans," *Journal of Biological Rhythms* 36 (2021): 369–83.

42. Soňa Janáčková et al., "Idiopathic Hypersomnia: A Report of Three Adolescent-Onset Cases in a Two Generation Family," *Journal of Child Neurology* 26 (2011): 522–25.

43. Michel Billiard and Yves Dauvilliers, "Idiopathic Hypersomnia," *Sleep Medicine Reviews* 5 (2001): 351–60.

44. In 2002, Emmanuel Mignot's group described an African American family including seven members with narcolepsy and two with idiopathic hypersomnia. Some DAN family members displayed cataplexy despite having normal hypocretin CSF levels. One with IH, and another with narcolepsy, also did not have the usual HLA risk factor. A search for responsible genes was unproductive, Mignot said. Emmanuel Mignot et al., "The Role of Cerebrospinal Fluid Hypocretin Measurement in the Diagnosis of Narcolepsy and Other Hypersomnias," *Archives of Neurology* 59 (2002): 1553–62.

45. Kris A. Wetterstrand, "DNA Sequencing Costs: Data from the NHGRI Genome Sequencing Program," https://www.genome.gov/about-genomics/fact-sheets/DNA-Sequencing-Costs-Data.

46. Gregory Costain et al., "Clinical Application of Targeted Next-Generation Sequencing Panels and Whole Exome Sequencing in Childhood Epilepsy," *Neuroscience* 418 (2019): 291–310; Siddharth Srivastava et al., "Meta-analysis and Multidisciplinary Consensus Statement: Exome Sequencing Is a First-Tier Clinical Diagnostic Test for Individuals with Neurodevelopmental Disorders," *Genetics in Medicine* 21 (2019): 2413–21.

47. Taku Miyagawa et al., "A Variant at 9q34.11 Is Associated with HLA-DQB1*06:02 Negative Essential Hypersomnia," *Journal of Human Genetics* 63 (2018): 1259–67.

48. People with two copies of the risk allele made up 14 percent of the essential hypersomnia cases and 6 percent of the controls (odds ratio 2.63).

49. Taku Miyagawa et al., "A Rare Genetic Variant in the Cleavage Site of Prepro-orexin Is Associated with Idiopathic Hypersomnia," *NPJ Genomic Medicine* 7 (2022): 29.

50. Soňa Nevšímalová et al., "Idiopathic Hypersomnia: A Homogeneous or Heterogeneous Disease?," *Sleep Medicine* 80 (2021): 86–91.

51. Charles Nunn and David Samson, "Sleep in a Comparative Context: Investigating How Human Sleep Differs from Sleep in Other Primates," *American Journal of Biological Anthropology* 166 (2018): 601–12.

52. Heming Wang et al., "Genome-wide Association Analysis of Self-Reported Daytime Sleepiness Identifies 42 Loci That Suggest Biological Subtypes," *Nature Communications* 10 (2019): 3503.

53. Hassan S. Dashti et al., "Genome-wide Association Study Identifies Genetic Loci for Self-Reported Habitual Sleep Duration Supported by Accelerometer-Derived Estimates," *Nature Communications* 10: 1100 (2019).

54. Yukihide Momozawa and Keijiro Mizukami, "Unique Roles of Rare Variants in the Genetics of Complex Diseases in Humans," *Journal of Human Genetics* 66 (2021): 11–23.

55. Chris R. Jones et al., "Familial Advanced Sleep-Phase Syndrome: A Short-Period Circadian Rhythm Variant in Humans," *Nature Medicine* 5 (1999): 1062–65.

56. Kong L. Toh et al., "An hPer2 Phosphorylation Site Mutation in Familial Advanced Sleep Phase Syndrome," *Science* 291 (2001): 1040–43.

57. Ying Xu et al., "Functional Consequences of a CKIdelta Mutation Causing Familial Advanced Sleep Phase Syndrome," *Nature* 434 (2005): 640–44; Arisa Hirano et al., "A Cryptochrome 2 Mutation Yields Advanced Sleep Phase in Humans," *eLife* 5 (2016): e16695.

58. Luoying Zhang et al., "A PERIOD3 Variant Causes a Circadian Phenotype and Is Associated with a Seasonal Mood Trait," *Proceedings of the National Academy of Sciences* 113 (2016): E1536–44; K. C. Brennan et al., "Casein Kinase Iδ Mutations in Familial Migraine and Advanced Sleep Phase," *Science Translational Medicine* 5 (2013): 183ra56–11.

59. Lijuan Xing et al., "Mutant Neuropeptide S Receptor Reduces Sleep Duration with Preserved Memory Consolidation," *Science Translational Medicine* 11 (2019): eaax2014.

60. Ying He et al., "The Transcriptional Repressor DEC2 Regulates Sleep Length in Mammals," *Science* 325 (2009): 866–70.

61. Guangsen Shi et al., "A Rare Mutation of β1-Adrenergic Receptor Affects Sleep/Wake Behaviors," *Neuron* 103 (2019): 1044–55.

62. Daniel Aeschbach et al., "Evidence from the Waking Electroencephalogram That Short Sleepers Live Under Higher Homeostatic Sleep Pressure Than Long Sleepers," *Neuroscience* 102 (2001): 493–502.

63. Karen Weintraub, "Why Do Some People Need Less Sleep? It's in Their DNA," *Scientific American*, October 16, 2019.

64. Taku Miyagawa et al., "A Missense Variant in *PER2* Is Associated with Delayed Sleep-Wake Phase Disorder in a Japanese Population," *Journal of Human Genetics* 64 (2019): 1219–25; Anja Schirmacher et al., "Sequence Variants in Circadian Rhythmic Genes in a Cohort of Patients Suffering from Hypersomnia of Central Origin," *Biological Rhythm Research* 42 (2011): 407–16.

65. Courtney E. Casale and Namni Goel, "Genetic Markers of Differential Vulnerability to Sleep Loss in Adults," *Genes* 12 (2021): 1317.

12. IMMOBILIZED BY HAPPINESS

1. Kiran Maski et al., "Listening to the Patient Voice in Narcolepsy: Diagnostic Delay, Disease Burden, and Treatment Efficacy," *Journal of Clinical Sleep Medicine* 13 (2017): 419–25.

2. Ginger Carls et al., "Burden of Disease in Pediatric Narcolepsy: A Claims-Based Analysis of Health Care Utilization, Costs, and Comorbidities," *Sleep Medicine* 66 (2020): 110–18.

3. Maski et al., "Listening to the Patient Voice in Narcolepsy."

4. Russell Rosenberg and Ann Y. Kim, "The AWAKEN survey: Knowledge of Narcolepsy Among Physicians and the General Population," *Postgraduate Medicine* 126 (2014): 78–86.

5. Raquel N. Taddei et al., "Diagnostic Delay in Narcolepsy Type 1: Combining the Patients' and the Doctors' Perspectives," *Journal of Sleep Research* 25 (2016): 709–15.

6. Christine Won et al., "The Impact of Gender on Timeliness of Narcolepsy Diagnosis," *Journal of Clinical Sleep Medicine* 10 (2014): 89–95.

7. Farid R. Talih, "Narcolepsy Presenting as Schizophrenia," *Innovations in Clinical Neuroscience* 8 (2011): 30–34.

8. Chris Higgins, "I've Fallen in Love and I Can't Get Up," *This American Life*, June 4, 2010, https://www.thisamericanlife.org/409/held-hostage/act-three-7.

9. Patricia Frerking, "We're Baaaaa-ack! Or Something," *Sleeping Around: Adventures in Narcolepsy* (blog), February 4, 2015, http://sleepingaroundadventuresinnarcolepsy.blogspot.com/2015/02/were-baaaaa-ack-or-something.html.

10. Sebastiaan Overeem et al., "The Clinical Features of Cataplexy," *Sleep Medicine* 12 (2011): 12–18.

11. Fabio Pizza et al., "Clinical and Polysomnographic Course of Childhood Narcolepsy with Cataplexy," *Brain* 136 (2013): 3787–95.

12. Claire C. Wylds-Wright, *Waking Mathilda* (Palace Gate, 2017).

13. Luis de Lecea et al., "The Hypocretins: Hypothalamus-Specific Peptides with Neuroexcitatory Activity," *Proceedings of the National Academy of Sciences* 95 (1998): 322–27; Richard M. Chemelli et al., "Narcolepsy in Orexin Knockout Mice: Molecular Genetics of Sleep Regulation," *Cell* 98 (1999): 437–51.

14. Shi-Bin Li et al., "Hypothalamic Circuitry Underlying Stress-Induced Insomnia and Peripheral Immunosuppression," *Science Advances* 6 (2020): eabc2590.

15. From de Lecea's talk at the 2018 International Symposium on Narcolepsy, Massachusetts.

16. Benjamin Boutrel et al., "Role for Hypocretin in Mediating Stress-Induced Reinstatement of Cocaine-Seeking Behavior," *Proceedings of the National Academy of Sciences* 102 (2005): 19168–73.

17. Thomas C. Thannickal et al., "Opiates Increase the Number of Hypocretin-Producing Cells in Human and Mouse Brain and Reverse Cataplexy in a Mouse Model of Narcolepsy," *Science Translational Medicine* 10 (2018): eaao4953.

18. Morgan H. James et al., "Increased Number and Activity of a Lateral Subpopulation of Hypothalamic Orexin/Hypocretin Neurons Underlies the Expression of an Addicted State in Rats," *Biological Psychiatry* 85 (2019): 925–35.

19. Ashley M. Blouin et al., "Human Hypocretin and Melanin Concentrating Hormone Levels Are Linked to Emotion and Social Interaction," *Nature Communications* 4 (2013): 1547.

20. John Peever and Patrick M. Fuller, "The Biology of REM sleep," *Current Biology* 27 (2017): R1237–48.

21. Anne Vassalli et al., "Electroencephalogram Paroxysmal Theta Characterizes Cataplexy in Mice and Children," *Brain* 136 (2013): 1592–1608.

22. Michael S. Aldrich and Ann E. Rogers, "Exacerbation of Human Cataplexy by Prazosin," *Sleep* 12 (1989): 254–56.

23. Sebastian Overeem, Gert Jan Lammers, and J. Gert van Dijk, "Weak with Laughter," *Lancet* 354 (1999): 838.

24. Stefano Meletti et al., "The Brain Correlates of Laugh and Cataplexy in Childhood Narcolepsy," *Journal of Neuroscience* 35 (2015): 11583–94.

25. Louise Bonnet et al., "The Role of the Amygdala in the Perception of Positive Emotions: An 'Intensity Detector,'" *Frontiers in Behavioral Neuroscience,* July 7, 2015.

26. Ramin Khatami, Steffen Birkmann, and Claudio L. Bassetti, "Amygdala Dysfunction in Narcolepsy-Cataplexy," *Journal of Sleep Research* 16 (2007): 226–29.

27. Carrie E. Mahoney et al., "Cataplexy Triggered by Social Cues: A Role for Oxytocin in the Amygdala," *Sleep* 43S (2020): A2.

28. Erica Seigneur and Luis de Lecea, "Hypocretin (Orexin) Replacement Therapies," *Medicine in Drug Discovery* 8 (2020): 100070.

13. FRUSTRATING AND MOSTLY FRUITLESS

1. Thomas E. Scammell, "The Frustrating and Mostly Fruitless Search for an Autoimmune Cause of Narcolepsy," *Sleep* 29 (2006): 601–2.

2. David D. Daly and Robert E. Yoss, "A Family with Narcolepsy," *Proceedings of the Staff Meetings of the Mayo Clinic* 34 (1959): 313–19. Only three people in this pedigree displayed cataplexy, although others experienced sleep paralysis and hypnogogic hallucinations.

3. Robert E. Yoss and David D. Daly, "Hereditary Aspects of Narcolepsy," *Transactions of the American Neurological Association* 85 (1960): 239–40.

4. Takeo Juji et al., "HLA Antigens in Japanese Patients with Narcolepsy," *Tissue Antigens* 24 (1984): 316–19.

5. Peter V. Markov and Oliver G. Pybus, "Evolution and Diversity of the Human Leukocyte Antigen (HLA)," *Evolution, Medicine and Public Health* 1 (2015).

6. Emmanuel Mignot et al., "DQB1*0602 and DQA1*0102 (DQ1) Are Better Markers Than DR2 for Narcolepsy in Caucasian and Black Americans," *Sleep* 17: S60–67 (1994).

7. Yutaka Honda, "Clinical Features of Narcolepsy: Japanese Experiences," in *HLA in Narcolepsy*, ed. Y. Honda and T. Juji (Berlin: Springer, 1988), 24–57.

8. Emmanuel Mignot, "In Memoriam of Dr Yutaka Honda (1929–2009), a Pioneer in Sleep Medicine and Narcolepsy Research," *Sleep* 32: 1528–29 (2009).

9. Paul Reading, "The Neurological Sleep Clinic—Part 1—The Sleepy Patient," *Advances in Clinical Neuroscience & Rehabilitation* 8 (2008): 6.

10. Emmanuel Mignot et al., "HLA DQB1*0602 is Associated with Cataplexy in 509 Narcoleptic Patients," *Sleep* 20 (1997): 1012–20.

11. Anna Azvolinsky, "In Dogged Pursuit of Sleep," *The Scientist*, February 29, 2016, https://www.the-scientist.com/profile/in-dogged-pursuit-of-sleep-33951.

12. Henry Nicholls, *Sleepyhead* (New York: Basic, 2018), 98–111.

13. Seiji Nishino et al., "Hypocretin (orexin) Deficiency in Human Narcolepsy," *Lancet* 355 (2000): 39–40.

14. Thomas C. Thannickal et al., "Reduced Number of Hypocretin Neurons in Human Narcolepsy," *Neuron* 27 (2000): 469–74.

15. Emmanuel Mignot et al., "The Role of Cerebrospinal Fluid Hypocretin Measurement in the Diagnosis of Narcolepsy and Other Hypersomnias," *Archives of Neurology* 59 (2002): 1553–62.

16. Olivier Andlauer et al., "Predictors of Hypocretin (Orexin) Deficiency in Narcolepsy Without Cataplexy," *Sleep* 35 (2012): 1247–55; Regis Lopez et al., "Temporal Changes in the Cerebrospinal Fluid Level of Hypocretin-1 and Histamine in Narcolepsy," *Sleep* 40 (2017): zsw010.

17. Birgitte R. Kornum, "Narcolepsy Type 1: What Have We Learned from Immunology?," *Sleep* 43 (2020): zsaa055.

18. Adi Aran et al., "Elevated Anti-streptococcal Antibodies in Patients with Recent Narcolepsy Onset," *Sleep* 32 (2009): 979–83.

19. Rebekah H. Borse et al., "Effects of Vaccine Program Against Pandemic Influenza A(H1N1) Virus, United States, 2009–2010," *Emerging Infectious Diseases* 19 (2013): 439–48.

20. Markku Partinen et al., "Increased Incidence and Clinical Picture of Childhood Narcolepsy Following the 2009 H1N1 Pandemic Vaccination Campaign in Finland," *PLOS One* 7 (2012): e33273.

21. Yves Dauvilliers et al., "Post-H1N1 Narcolepsy-Cataplexy," *Sleep* 33 (2010): 1428–30.

22. Louis Jacob et al., "Comparison of Pandemrix and Arepanrix, Two pH1N1 AS03-Adjuvanted Vaccines Differentially Associated with Narcolepsy Development," *Brain Behavior and Immunity* 47 (2015): 44–57.

23. Aditya Ambati et al., "Mass Spectrometric Characterization of Narcolepsy-Associated Pandemic 2009 Influenza Vaccines," *Vaccines* 8 (2020): 630.

24. David Scott and Ann Enander, "Postpandemic Nightmare: A Framing Analysis of Authorities and Narcolepsy Victims in Swedish Press," *Journal of Contingencies & Crisis Management* 25 (2017); Elizabeth Miller et al., "Risk of Narcolepsy in Children and Young People Receiving AS03 Adjuvanted Pandemic A/H1N1 2009 Influenza Vaccine: Retrospective Analysis," *BMJ* 346 (2013): f794.

25. Fang Han et al., "Narcolepsy Onset Is Seasonal and Increased Following the 2009 H1N1 Pandemic in China," *Annals of Neurology* 70: 410–17 (2011).

26. Narong Simakajornboon et al., "Increased Cases of Childhood Narcolepsy After the 2009 H1N1 Pandemics," *Sleep* 40S (2017): 337.

27. Kate Kelland, "How Vaccine Scares Cast Shadows Over Science," *Reuters*, March 21, 2013, https://in.reuters.com/article/vaccines-narcolepsy-science/special-report-how-vaccine-scares-cast-shadows-over-science-idINDEE92K06920130321.

28. Kate Kelland, "Insight: Evidence Grows for Narcolepsy Link to GSK Swine Flu Shot," *Reuters*, January 22, 2013, https://www.reuters.com/article/us-narcolepsy-vaccine-pandemrix/insight-evidence-grows-for-narcolepsy-link-to-gsk-swine-flu-shot-idUSBRE90L07H20130122.

29. Helen Branswell, "A Stubborn Medical Mystery: Was Pandemic Flu Vaccine Tied to an Increase in Narcolepsy Cases?" *STAT*, July 5, 2018.

30. Daniel Weibel et al., "Narcolepsy and Adjuvanted Pandemic Influenza A (H1N1) 2009 Vaccines—Multi-country Assessment," *Vaccine* 36 (2018): 6262–11.

31. Rebekah H. Borse et al., "Effects of Vaccine Program Against Pandemic Influenza A(H1N1) Virus, United States, 2009–2010," *Emerging Infectious Diseases* 19 (2013): 439–48.

32. Pierre Verger, "Prevalence and Correlates of Vaccine Hesitancy Among General Practitioners: A Cross-sectional Telephone Survey in France, April to July 2014," *EuroSurveillance* 21 (2016): 30406.

33. Britta Lundren, "Narrating Narcolepsy—Centering a Side Effect," *Medical Anthropology* 34: 150–65 (2015).

34. Azvolinsky, "In Dogged Pursuit of Sleep."

35. Daniela Latorre et al., "T Cells in Patients with Narcolepsy Target Self-Antigens of Hypocretin Neurons," *Nature* 562 (2018): 63–68.

36. Guo Luo et al., "Autoimmunity to Hypocretin and Molecular Mimicry to Flu in Type 1 Narcolepsy," *Proceedings of the National Academy of Sciences* 115 (2018): E12323–32.

37. Andreas Lutterotti, "Is It Time for Immunotherapy Trials in Narcolepsy?," *Journal of Clinical Sleep Medicine* 13 (2017): 363–64.

38. Thomas E. Scammell et al., "Treatment of Narcolepsy with Natalizumab," *Sleep* 43 (2020): zsaa050.

39. Kelsey Biddle, "Living with Narcolepsy, Running for a Cure," *Invisible Illness* (blog) October 29, 2018, https://medium.com/invisible-illness/living-with-narcolepsy-running-for-a-cure-3242f9040188.

40. Andrew C. Cogswell et al., "Children with Narcolepsy Type 1 Have Increased T-cell Responses to Orexins," *Annals of Clinical and Translational Neurology* 6 (2019): 2566–72.

41. Natasja Wulff Pedersen et al., "CD8+ T cells from Patients with Narcolepsy and Healthy Controls Recognize Hypocretin Neuron-Specific Antigens," *Nature Communications* 10 (2019): 837.

42. Mignot said molecular mimicry is still the most likely explanation for NT1 autoimmunity. Guo Luo et al., "Characterization of T cell Receptors Reactive to HCRT NH2, pHA 273–287, and NP 17–31 in Control and Narcolepsy Patients," *Proceedings of the National Academy of Sciences* 119 (2022): e2205797119.

43. Tobias V. Lanz et al., "Clonally Expanded B cells in Multiple Sclerosis Bind EBV EBNA1 and GlialCAM," *Nature* 603 (2022): 321–27.

44. Masataka Wada et al., "Neuroimaging Correlates of Narcolepsy with Cataplexy: A Systematic Review," *Neuroscience Research* 142 (2019): 16–29.

45. Thomas C. Thannickal et al., "Localized Loss of Hypocretin (Orexin) Cells in Narcolepsy Without Cataplexy," *Sleep* 32 (2009): 993–98.

46. Fabio Pizza et al., "Primary Progressive Narcolepsy Type 1: The Other Side of the Coin," *Neurology* 83 (2014): 2189–90.

47. Sarah W. Black et al., "Partial Ablation of the Orexin Field Induces a Sub-narcoleptic Phenotype in a Conditional Mouse Model of Orexin Neurodegeneration," *Sleep* 41 (2018): zsy116.

48. Leon Rosenthal et al., "2018 Worldwide Survey of Health-Care Providers Caring for Patients with Narcolepsy," *Sleep Medicine* 82 (2021): 23–28.

49. Alyssa Frank, "Orexin-A/Hypocretin-1 for the Diagnosis of Type 1 Narcolepsy," *Mayo Clinic Laboratories*, May 31, 2019, https://news.mayocliniclabs.com/2019/05/31/orexin/.

50. This idea comes from a presentation by Tom Scammell.

51. Zan Wang et al., "Case Report: Dysfunction of the Paraventricular Hypothalamic Nucleus Area Induces Hypersomnia in Patients," *Frontiers in Neuroscience* 16 (2022): 830474.

52. Phillipp Valko et al., "Increase of Histaminergic Tuberomammillary Neurons in Narcolepsy," *Annals of Neurology* 74 (2013): 794–804.

53. Yo Yamada et al., "Chronic Brain Histamine Depletion in Adult Mice Induced Depression-like Behaviors and Impaired Sleep-Wake Cycle," *Neuropharmacology* 175 (2020): 108179.

14. EVERYTHING OFF LABEL

1. Dean Jordheim, Idiopathic Hypersomnia group Facebook thread, February 24, 2014.

2. Lynn Marie Trotti et al., "Disease Symptomatology and Response to Treatment in People with Idiopathic Hypersomnia: Initial Data from the Hypersomnia Foundation Registry," *Sleep Medicine* 75 (2020): 343–49.

3. David Cavalla, *Off-Label Prescribing* (Chichester: Wiley-Blackwell, 2015).

4. Timothy I. Morgenthaler et al., "Practice Parameters for the Treatment of Narcolepsy and Other Hypersomnias of Central Origin," *Sleep* 30 (2007): 1705.

5. "2021 AMA Prior Authorization (PA) Physician Survey," American Medical Association website, February 18, 2022, https://www.ama-assn.org/system/files/prior-authorization-survey.pdf.

6. "Our Struggle for Prescription Drug Coverage—2019 Survey Results," Hypersomnia Foundation website, https://www.hypersomniafoundation.org/our-struggle-for-prescription-drug-coverage-2019-survey-results/.

7. Brian C. Callaghan et al., "Out-of-Pocket Costs Are on the Rise for Commonly Prescribed Neurologic Medications," *Neurology* 92 (2019): e2604–13.

8. United Healthcare, "Clinical Pharmacy Program Guidelines for Provigil, Nuvigil," accessed March 2021, https://www.uhcprovider.com/content/dam/provider/docs/public/commplan/az/pharmacy-clinical-guidelines/AZ-Provigil-Nuvigil-Clinical-Guidelines.pdf.

9. U.S. House of Representatives, Committee on Energy and Commerce, Subcommittee on Health and the Environment, "Health and the Environment Miscellaneous" (testimony for Orphan Drug Bill), 97th Congress, March 8, 1982, 7.367–8, https://catalog.hathitrust.org/Record/100663397.

10. Sarah Jane Tribble and Sydney Lupkin, "Government Investigation Finds Flaws in the FDA's Orphan Drug Program," *Kaiser Health News*, November 30, 2018, https://khn.org/news/government-investigation-finds-flaws-in-the-fdas-orphan-drug-program/.

11. Tufts Center for Study of Drug Development, "Tufts CSDD Impact Report," July/August 2019.

12. Rachel D. Teodorini, Nicola Rycroft, and James H. Smith-Spark, "The Off-Prescription Use of Modafinil," *PLOS One*, February 5, 2020.

13. Lenny Bernstein, "Ambien Should Not Be Handed Out 'Like Candy,' Experts Say of Ronny L. Jackson's Alleged Practices," *Washington Post*, April 25, 2018.

14. Maria J. Hyland, "The Drugs Do Work: My Life on Brain Enhancers," *Guardian*, May 3, 2013.

15. Michel Billiard and Roger Broughton, "Modafinil: Its Discovery, the Early European and North American Experience in the Treatment of Narcolepsy and Idiopathic Hypersomnia, and Its Subsequent Use in Other Medical Conditions," *Sleep Medicine* 49 (2018): 69–72; Helene Bastuji and Michel Jouvet, "Successful Treatment of Idiopathic Hypersomnia and Narcolepsy with Modafinil," *Progress in Neuro-Psychopharmacology & Biological Psychiatry* 12 (1988): 695–700.

16. Helene Bastuji, "Michel Jouvet as a Clinical Neurophysiologist and Neurologist," *Sleep Medicine* 49 (2018): 73–77.

17. Food and Drug Administration, NDA 20–717, Clinical Review Section 2.5, http://web.archive.org/web/20060825161425/http://www.fda.gov/cder/foi/nda/98/020717A_Provigil_medr_P1.pdf.

18. J. S. Lin et al., "Role of Catecholamines in the Modafinil and Amphetamine Induced Wakefulness," *Brain Research* 591 (1992): 19–26.

19. Food and Drug Administration, NDA 20-717, Division Director memo, September 8, 2003: "The sponsor has chosen to assess Provigil's effect on the objective measures MWT or the MSLT. These measures are widely used in the field, but it is worth asking if the sponsor should have, instead, assessed the drug's effect with more direct, or perhaps face valid, measures of sleepiness. For example, one could imagine that, instead of the laboratory-based MSLT or MWT, one could count episodes of falling asleep during the day in patients with narcolepsy, or motor vehicle accidents during the day in patients with OSAHS, or number of work-related accidents while patients with SWSD were working." https://wayback.archive-it.org/7993/20170405140917/https://www.fda.gov/ohrms/dockets/ac/03/briefing/3979B2_02_FDA-Division%20Director%20Memo.htm.

20. U.S. Modafinil in Narcolepsy Multicenter Study Group, "Randomized Trial of Modafinil for the Treatment of Pathological Somnolence in Narcolepsy," *Annals of Neurology* 43 (1998): 88–97; U.S. Modafinil in Narcolepsy Multicenter Study Group, "Randomized Trial of Modafinil as a Treatment for the Excessive Daytime Somnolence of Narcolepsy," *Neurology* 54 (2000): 1166–75. Those with severe cataplexy could not be enrolled.

21. Donald R. Jasinski, "An Evaluation of the Abuse Potential of Modafinil Using Methylphendiate as a Reference," *Journal of Psychopharmacology* 14 (2000): 53–60.

22. Department of Justice, "Schedules of Controlled Substances: Placement of Modafinil Into Schedule IV," *Federal Register* 64 (1999): 4050–52, https://www.deadiversion.usdoj.gov/fed_regs/rules/1999/fr0127.htm.

23. Erica Goode, "New Hope for the Losers in the Battle to Stay Awake," *New York Times*, November 3, 1998.

24. Jonathan P. Wisor et al., "Dopaminergic Role in Stimulant-Induced Wakefulness," *Journal of Neuroscience* 21 (2001): 1787–94.

25. Thomas E. Scammell et al., "Hypothalamic Arousal Regions Are Activated During Modafinil-Induced Wakefulness," *Journal of Neuroscience* 20 (2000): 8620–28.

26. Jonathan Wisor, "Modafinil as a Catecholaminergic Agent: Empirical Evidence and Unanswered Questions," *Frontiers in Neurology* 4 (2013): 139.

27. Food and Drug Administration, NDA 20-717, briefing material for 2003 advisory committee meeting, p. 10, https://wayback.archive-it.org/7993/20170405140858/https://www.fda.gov/ohrms/dockets/ac/03/briefing/3979B2_01_Cephalon-Provigil.htm.

28. Food and Drug Administration, NDA 20-717, Leber approvable memo, p. 12, https://web.archive.org/web/20060825161512/http:/www.fda.gov/cder/foi/nda/98/020717A_Provigil_medr_P4.pdf.

29. Food and Drug Administration, NDA 20-717, January 2002 warning letter from Regulatory Review Officer.

30. Steven R. Brown, "Disease Mongering and Excessive Daytime Sleepiness," *American Family Physician* 80 (2009): 775–78.

31. Aaron Kesselheim et al., "The Prevalence and Cost of Unapproved Uses of Top-Selling Orphan Drugs," *PLOS One* 7 (2012): e31894.

32. Jerome Groopman, "Eyes Wide Open," *New Yorker*, December 3, 2001.

33. Andrew Pollack, "A Biotech Outcast Awakens," *New York Times*, October 20, 2002.

34. Harvard Medical School press release, May 11, 2004, https://www.eurekalert.org/pub_releases/2004-05/hms-thm051104.php.

35. Charles Czeisler et al., "Modafinil for Excessive Sleepiness Associated with Shift-Work Sleep Disorder," *New England Journal of Medicine* 353 (2005): 476–86.

36. Tim Lougheed, "An Assault on Sleep," *Ottawa Citizen*, May 13, 2002.

37. Nancy J. Wesensten et al., "Maintaining Alertness and Performance During Sleep Deprivation: Modafinil Versus Caffeine," *Psychopharmacology* 159 (2002): 238–47.

38. Allan I. Pack, "Should a Pharmaceutical Be Approved for the Broad Indication of Excessive Sleepiness?," *American Journal of Respiratory and Critical Care Medicine* 167 (2003): 109–11.

39. Food and Drug Administration, Peripheral and Central Nervous System Drugs Advisory Committee meeting transcript, September 25, 2003, https://wayback.archive-it.org/7993/20170404075340/https://www.fda.gov/ohrms/dockets/ac/03/transcripts/3979T2.htm.

40. Andrew Pollack, "Regulators Reject a Drug Maker's Plan to Use Its Alertness Pill to Overcome Jet Lag," *New York Times*, March 29, 2010; Russell P. Rosenberg et al., "A Phase 3, Double-Blind, Randomized, Placebo-Controlled Study of Armodafinil for Excessive Sleepiness Associated with Jet Lag Disorder," *Mayo Clinic Proceedings* 85 (July 2010): 630–38.

41. Charles H. Adler et al., "Randomized Trial of Modafinil for Treating Subjective Daytime Sleepiness in Patients with Parkinson's Disease," *Movement Disorders* 18 (2003): 287–93; Kottil W. Rammohan et al., "Efficacy and Safety of Modafinil (Provigil®) for the Treatment of Fatigue in Multiple Sclerosis: A Two Centre Phase 2 Study," *Journal of Neurology Neurosurgery & Psychiatry* 72 (2002): 179–83; Udo A. Zifko et al., "Modafinil in Treatment of Fatigue in Multiple Sclerosis. Results of an Open-Label Study," *Journal of Neurology* 249 (2002): 983–87; Charles DeBattista et al., "Adjunct Modafinil for the Short-Term Treatment of Fatigue and Sleepiness in Patients with Major Depressive Disorder: A Preliminary Double-Blind, Placebo-Controlled Study," *Journal of Clinical Psychiatry* 64 (2003): 1057–64.

42. Department of Justice press release, "Biopharmaceutical Company, Cephalon, to Pay $425 Million & Enter Plea to Resolve Allegations of Off-Label Marketing," September 29, 2008.

43. Food and Drug Administration, NDA 20-717/S-019, briefing document for Psychopharmacologic Drugs Advisory Committee, March 23, 2006, pp. 55–60, https://wayback.archive-it.org/7993/20170405070734/https://www.fda.gov/ohrms/dockets/ac/06/briefing/2006-4212b1-01-cephalon-background.pdf.

44. Jon Hamilton, "FDA Committee Rejects ADHD Use for Modafinil," NPR, March 24, 2006.

45. Lin Cheng, "Current Pharmacogenetic Perspective on Stevens-Johnson Syndrome and Toxic Epidermal Necrolysis," *Frontiers in Pharmacology* 12 (2021): 588063.

46. Food and Drug Administration, *Drug Safety Newsletter* 1 (2007): 5, https://web.archive.org/web/20090119121242/http://www.fda.gov/cder/dsn/2007_fall/2007_fall.pdf.

47. European Medicines Agency, "Assessment Report for Modafinil Containing Medicinal Products," section 1.1, January 27, 2011, https://www.ema.europa.eu/en/documents/referral/modafinil-h-31–1186-article-31-referral-assessment-report_en.pdf.

48. European Medicines Agency, "Assessment Report for Modafinil Containing Medicinal Products," annex I–IV, January 27, 2011, https://www.ema.europa.eu/en/documents/referral/modafinil-article-31-referral-annex-i-ii-iii-iv_en.pdf.

49. Regis Lopez et al., "French Consensus. Management of Patients with Hypersomnia: Which Strategy?," *Revue Neurologique* 173 (2016): 8–18.

50. Julia Chapman et al., "Modafinil/Armodafinil in Obstructive Sleep Apnoea: A Systematic Review and Meta-analysis," *European Respiratory Journal* 47 (2016): 1420–28.

51. Pierre Philip et al., "Modafinil Improves Real Driving Performance in Patients with Hypersomnia: A Randomized Double-Blind Placebo-Controlled Crossover Clinical Trial," *Sleep* 37 (2014): 483–87.

52. Geert Mayer et al., "Modafinil in the Treatment of Idiopathic Hypersomnia Without Long Sleep Time—a Randomized, Double-Blind, Placebo-Controlled Study," *Journal of Sleep Research* 24 (2015): 74–81.

53. Yuichi Inoue, Toshiyuki Tabata, and Naoji Tsukimori, "Efficacy and Safety of Modafinil in Patients with Idiopathic Hypersomnia Without Long Sleep Time: A Multicenter, Randomized, Double-Blind, Placebo-Controlled, Parallel-Group Comparison Study," *Sleep Medicine* 80 (2021): 315–21.

54. Carl A. Roberts et al., "How Effective Are Pharmaceuticals for Cognitive Enhancement in Healthy Adults? A Series of Meta-analyses of Cognitive Performance During Acute Administration of Modafinil, Methylphenidate and D-amphetamine," *European Neuropsychopharmacology* 38 (2020): 40–62.

55. The approval letter for Sunosi says: "Your application for Sunosi was not referred to an FDA advisory committee because the efficacy and safety data were readily interpretable and not controversial, and there was experience with other drugs in this class."

56. Michelle G. Baladi et al., "Characterization of the Neurochemical and Behavioral Effects of Solriamfetol (JZP-110), a Selective Dopamine and Norepinephrine Reuptake Inhibitor," *Journal of Pharmacology and Experimental Therapeutics* 366 (2018): 367–76.

57. Lee Hye-seon, "SK Biopharmaceutical's Sleepiness Drug Initially Aimed to Treat Depression," *Korea Biomedical Review*, August 22, 2019, https://koreabiomed.com/news/articleView.html?idxno =6320.

58. Perri Panula, "Histamine Receptors, Agonists, and Antagonists in Health and Disease," *Handbook of Clinical Neurology* 180 (2021): 377–87.

59. Zoltan Szakacs et al., "Safety and Efficacy of Pitolisant on Cataplexy in Patients with Narcolepsy: A Randomised, Double-Blind, Placebo-Controlled Trial," *Lancet Neurology* 16 (2017): 200–7.

60. Smaranda Leu-Semenscu et al., "Effects of Pitolisant, a Histamine H3 Inverse Agonist, in Drug-Resistant Idiopathic and Symptomatic Hypersomnia: A Chart Review," *Sleep Medicine* 15 (2014): 681–87.

61. Food and Drug Administration, NDA 211150, Pitolisant clinical review, https://www.accessdata .fda.gov/drugsatfda_docs/nda/2019/211150Orig1s000MedR.pdf.

62. U.S. National Library of Medicine, "A Study of Safety and Efficacy of BTD-001 in Treatment of Patients with Idiopathic Hypersomnia (IH) or Narcolepsy Type 2," https://clinicaltrials.gov/ct2 /show/NCT02512588, last update 2019.

63. U.S. National Library of Medicine, "A Study of Oral BTD-001 in Adults with Idiopathic Hypersomnia (ARISE2)," https://clinicaltrials.gov/ct2/show/NCT03542851, last update 2020.

64. Ping Sheng et al., "Efficacy of Modafinil on Fatigue and Excessive Daytime Sleepiness Associated with Neurological Disorders: A Systematic Review and Meta-Analysis," *PLOS One* 8 (2013): e81802.

15. KNOCK YOURSELF OUT

1. U.S. National Library of Medicine, "A Multicenter Study of the Efficacy and Safety of JZP-258 in the Treatment of Idiopathic Hypersomnia (IH) with an Open-label Safety Extension," https:// clinicaltrials.gov/ct2/show/NCT03533114.

2. Gunjan Junnarkar et al., "Development of a Lower–Sodium Oxybate Formulation for the Treatment of Patients with Narcolepsy and Idiopathic Hypersomnia," *Expert Opinion in Drug Discovery* (2021).

3. Jazz Pharmaceuticals, 2020 Annual Report, 5.

4. Julie Vienne et al., "Differential Effects of Sodium Oxybate and Baclofen on EEG, Sleep, Neurobehavioral Performance, and Memory," *Sleep* 35 (2012): 1071–83.

5. Food and Drug Administration, Xyrem label, revised October 2018, https://www.accessdata.fda .gov/drugsatfda_docs/label/2018/021196s030lbl.pdf.

6. Malene Landbo Børresen et al., "Sodium Oxybate (Xyrem) Treatment in Severely Sleep-Deprived Child with Epstein-Barr Virus Encephalitis with Lesion of Sleep-Wake Regulation System: A Case Report," *Sleep Medicine* 62 (2019): 29–31.

7. Lynn M. Trotti, "Conventional (and Not-So-Conventional) Treatments for Idiopathic Hypersomnia," 2018 Hypersomnia Conference, YouTube video posted August 2, 2019, 21–24 minutes, https://www.youtube.com/watch?v=kADh5rCzlRU.

8. Smaranda Leu-Semenescu, Pauline Louis, and Isabelle Arnulf, "Benefits and Risk of Sodium Oxybate in Idiopathic Hypersomnia Versus Narcolepsy Type 1: A Chart Review," *Sleep Medicine* 17 (2016): 38–44.

9. Henri Laborit, "Sodium 4-Hydroxybutyrate," *International Journal of Neuropharmacology* 3 (1964): 433–51.

10. Michael E. Tunstall, "Gamma-OH in Anesthesia for Caesarean Section," *Proceedings of the Royal Society Medicine* 61 (1968): 827–29.

11. Mortimer Mamelak, J. M. Escriu, and O. Stokan, "Sleep-Inducing Effects of Gammahydroxybutyrate," *Lancet* 302 (1973): 328–29.

12. Roger Broughton and Mortimer Mamelak, "Gamma Hydroxy-Butyrate in the Treatment of Narcolepsy: A Preliminary Report," *Proceedings of the First International Symposium on Narcolepsy*, July 1975 (Spectrum Publications).

13. Guilleminault cited his unsatisfactory experiences trying baclofen for people with narcolepsy. Since baclofen chemically resembles GABA, there must be another explanation for GHB's efficacy, he argued.

14. Stanford researchers led by Dement saw no effect of GHB on narcolepsy in dogs. Arthur S. Foutz et al., "Monoaminergic Mechanisms and Experimental Cataplexy," *Annals of Neurology* 10 (1981): 369–76.

15. One of the best sources on GHB's transformation into Xyrem is a paper by a Harvard law student, written under the supervision of a former FDA chief counsel. Ariel Neuman, "GHB's Path to Legitimacy: An Administrative and Legislative History of Xyrem," *Harvard Law School Student Papers* 2004, https://dash.harvard.edu/bitstream/handle/1/9795464/Neuman.html.

16. Martin B. Scharf et al. "The Effects and Effectiveness of γ-Hydroxybutyrate in Patients with Narcolepsy," *Journal of Clinical Psychiatry* 46: 222–25 (1985).

17. Martin Scharf, 2020 interview.

18. Linda Pender, "Better Than a Lullaby," *Cincinnati Magazine*, September 1984.

19. U.S. House of Representatives, Appropriations Committee, Subcommittee on Agriculture, Rural Development, and Related Agencies, "Agriculture, Rural Development, and Related Agencies Appropriations for 1983," 97th Congress, part 7, 1982, p. 96, https://catalog.hathitrust.org/Record /006212564 (American Narcolepsy Association material).

20. Lawrence Scrima et al., "The Effects of Gamma-Hydroxybutyrate on the Sleep of Narcolepsy Patients: A Double-Blind Study," *Sleep* 13 (1990): 479–90; Lawrence Scrima et al., "Efficacy of Gamma-Hydroxybutyrate Versus Placebo in Treating Narcolepsy-Cataplexy: Double-Blind Subjective Measures," *Biological Psychiatry* 26 (1989): 331–43.

21. Jill Barshay, "One Drug, Two Faces," *Minneapolis Star-Tribune*, August 6, 2000.

22. J. Takahara et al., "Stimulatory Effects of Gamma-Hydroxybutyric Acid on Growth Hormone and Prolactin Release in Humans," *Journal of Clinical Endocrinology and Metabolism* 44 (1977): 1014–17.

23. Joseph J. Palamar and Perry N. Halkitis, "A Qualitative Analysis of GHB Use Among Gay Men: Reasons for Use Despite Potential Adverse Outcomes," *International Journal of Drug Policy* 17 (2006): 23–28.

24. J. E. Dyer et al., "Epidemiologic Notes and Reports Multistate Outbreak of Poisonings Associated with Illicit Use of Gamma-Hydroxy Butyrate," *Morbidity and Mortality Weekly Report* 39 (1990): 861–63.

25. Shawn Hubler and Steve Hochman, " 'Designer Drug' Enters Hollywood's Fast Lane," *Los Angeles Times*, November 3 1993.

26. U.S. House of Representatives, Judiciary Committee, Subcommittee on Crime, "Controlled and Uncontrolled Substances Used to Commit Date Rape," 105th Congress, July 30, 1998, p. 125, http://commdocs.house.gov/committees/judiciary/hju62309.000/hju62309_of.htm.

27. Food and Drug Administration, Peripheral and Central Nervous System Drugs Advisory Committee meeting materials, June 6, 2001, https://wayback.archive-it.org/7993/20170404102834/https://www.fda.gov/ohrms/dockets/ac/01/briefing/3754b1.htm; Letter from Division Director, May 9, 2001, https://wayback.archive-it.org/7993/20170405173215/https://www.fda.gov/ohrms/dockets/ac/01/briefing/3754b1_02_section%201.pdf.

28. Gayle Greene, *Insomniac* (Berkeley: University of California Press, 2008).

29. Food and Drug Administration, Peripheral and Central Nervous System Drugs Advisory Committee meeting transcript, June 6, 2001, https://wayback.archive-it.org/7993/20170404102820/https://www.fda.gov/ohrms/dockets/ac/01/transcripts/3754t1.txt.

30. Martin B. Scharf et al., "Effect of Gamma-Hydroxybutyrate on Pain, Fatigue, and the Alpha Sleep Anomaly in Patients with Fibromyalgia," *Journal of Rheumatology* 25 (1998): 1986–90.

31. I. Jon Russell et al., "Sodium Oxybate Relieves Pain and Improves Function in Fibromyalgia Syndrome," *Arthritis & Rheumatism* 60 (2009): 299–309.

32. Jazz annual report from 2007, https://www.sec.gov/Archives/edgar/data/0001232524/000119312508070827/d10k.htm.

33. Jazz annual reports 10K, issued March 2009 and March 2010, for example: https://www.sec.gov/Archives/edgar/data/0001232524/000119312509064139/d10k.htm.

34. Jennifer C. Dooren, "FDA Panel Rejects Jazz Drug as Pain-Disorder Treatment," *Wall Street Journal*, August 21, 2010.

35. Food and Drug Administration, Arthritis & Drug Safety and Risk Management Advisory Committees joint meeting transcript, August 20, 2010, https://wayback.archive-it.org/7993/20170404145710/https://www.fda.gov/downloads/AdvisoryCommittees/CommitteesMeetingMaterials/Drugs/ArthritisAdvisoryCommittee/UCM225445.pdf.

36. Marc I. Rosen et al., "Effects of Gamma-Hydroxybutyric Acid (GHB) in Opioid-Dependent Patients," *Journal of Substance Abuse Treatment* 14 (1997): 149–54.

37. Food and Drug Administration, NDA 21196 Medical Review, part 4, June 15, 2001, https://www.accessdata.fda.gov/drugsatfda_docs/nda/2002/21-196_Xyrem_medr_P4.pdf.

38. Ronald E. Kramer, "The Administrative State and the Death of Peter Gleason, MD: An Off-Label Case Report," *Journal of Clinical Sleep Medicine* 15 (2019).

39. "United States v. Caronia," *Harvard Law Review* 127 (2012): 795–802.

40. Tracy Staton, "10 Big Brands Keep Pumping Out Big Bucks, with a Little Help from Price Hikes," *Fierce Biotech*, May 7, 2014.

41. Daniel M. Hartung et al., "The Cost of Multiple Sclerosis Drugs in the US and the Pharmaceutical Industry," *Neurology* 84 (May 2015): 2185–92.

42. Caroline Pearson, Lindsey Schapiro, and Steven D. Pearson, "The Next Generation of Rare Disease Drug Policy: Ensuring Both Innovation and Affordability," Institute for Clinical and Economic Review white paper, 2022, https://icer.org/wp-content/uploads/2022/04/ICER-White-Paper_The-Next-Generation-of-Rare-Disease-Drug-Policy_040722.pdf.

43. Blue Cross Blue Shield of Illinois, "Xyrem® (sodium oxybate) Prior Authorization Criteria," accessed February 19, 2015, https://web.archive.org/web/20150219031021/https://www.bcbsil.com/provider/pdf/xyrem.pdf; State of North Carolina, "Prior Authorization Criteria" (CVS Caremark Criteria), https://files.nc.gov/ncshp/documents/files/Xyrem-Policy-254-C-03-2020.pdf.

44. Motley Fool, Jazz Pharmaceuticals PLC (JAZZ) Q4 2020 Earnings Call Transcript, https://www.fool.com/earnings/call-transcripts/2021/02/23/jazz-pharmaceuticals-plc-jazz-q4-2020-earnings-cal/.

45. Robert Langreth and Ben Elgin, "'I Am Going to Die Without This': Regulators Target a Health-Care Lifesaver," *Bloomberg Businessweek*, November 21, 2017.

46. Jazz Pharmaceuticals Inc., "Citizen Petition," posted by FDA, May 28, 2012, https://www.regulations.gov/document/FDA-2012-P-0499-0001.

47. Petra Maresova et al., "Treament Cost of Narcolepsy with Cataplexy in Central Europe," *Therapeutics and Clinical Risk Management* 12 (2016): 1709–15.

48. Clare Dyer, "Judge Orders That Teenager with Narcolepsy Be Allowed to Try Sodium Oxybate," *BMJ* 353 (2016): i3413.

49. Giovanni Addolorato et al., "Post-marketing and Clinical Safety Experience with Sodium Oxybate for the Treatment of Alcohol Withdrawal Syndrome and Maintenance of Abstinence in Alcohol-Dependent Subjects," *Expert Opinion on Drug Safety* 19 (2020): 159–66.

50. Julien Guiraud et al., "Sodium Oxybate for the Maintenance of Abstinence in Alcohol-Dependent Patients: An International, Multicenter, Randomized, Double-Blind, Placebo-Controlled Trial," *Journal of Psychopharmacology* 269 (2022): 2698811221104063.

51. European Medicines Agency, Assessment report for Alcover, October 12, 2017, https://www.ema.europa.eu/en/documents/referral/alcover-article-294-referral-assessment-report_en.pdf.

52. Centers for Medicare and Medicaid Services, "Jazz Pharmaceuticals Inc.," Open Payments database, https://openpaymentsdata.cms.gov/company/100000005637.

53. Jazz Pharmaceuticals press release, "Jazz Pharmaceuticals Supports Narcolepsy Network Initiative to Launch the National Narcolepsy Awareness Campaign," March 5, 2007; Jazz Pharmaceuticals press release, "Wake Up Narcolepsy and Jazz Pharmaceuticals Announce Swinging for Sleep to Raise Funds for Narcolepsy Awareness," March 22, 2012. Jazz has supported Hypersomnia Foundation programs such as recording sessions at conferences, according to foundation board members.

54. From Jazz presentation on Xywav launch, October 13, 2021, p. 33.

55. Ed Silverman, "High Cost of Multiple Sclerosis Medicines Is Forcing Many Patients to Take 'Drastic Actions,'" *STAT+*, January 13, 2020.

56. Kathleen F. Villa et al., "Changes in Medical Services and Drug Utilization and Associated Costs After Narcolepsy Diagnosis in the United States," *American Health & Drug Benefits* 11 (2018): 137–45. Pharmacy costs can be calculated by subtracting medical services from total costs (medical plus pharmacy).

57. Y. Grace Wang et al., "Safety Overview of Postmarketing and Clinical Experience of Sodium (Xyrem): Abuse, Misuse, Dependence, and Diversion," *Journal of Clinical Sleep Medicine* 4 (2009): 365–71.

58. Zsófia Németh, Bernadette Kun, and Zsolt Demetrovics, "The Involvement of Gamma-Hydroxybutyrate in Reported Sexual Assaults: A Systematic Review," *Journal of Psychopharmacology* 24 (2010): 1281–87.

59. Deborah L. Zvosec, Stephen W. Smith, and B. J. Hall, "Three Deaths Associated with Use of Xyrem," *Sleep Medicine* 10: 490–93 (2009).

60. Deborah L. Zvosec et al., "Rigorous Study of Xyrem/Sodium Oxybate Use Among Patients with Obstructive Sleep Apnea and Other Conditions of Compromised Respiratory Function Is Critically Needed," *Sleep Medicine* 12 (2011): 103.

61. Food and Drug Administration, Warning Letter to Jazz Pharmaceuticals, October 11, 2011, http://wayback.archive-it.org/7993/20170112193124/http://www.fda.gov/ICECI/EnforcementActions/WarningLetters/2011/ucm275565.htm.

62. Y. Grace Wang et al., "Sodium Oxybate: Updates and Correction to Previously Published Safety Data," *Journal of Clinical Sleep Medicine* 7 (2011): 415–16.

63. Food and Drug Administration, "Drug Safety Communication: Warning Against Use of Xyrem (Sodium Oxybate) with Alcohol or Drugs Causing Respiratory Depression," December 17, 2012, https://www.fda.gov/drugs/drug-safety-and-availability/fda-drug-safety-communication-warning-against-use-xyrem-sodium-oxybate-alcohol-or-drugs-causing.

64. Institute for Safe Medication Practices, *QuarterWatch*, September 24, 2014, https://www.ismp.org/sites/default/files/attachments/2018–01/2013Q3.pdf.

65. Darren Scheer et al., "Prevalence and Incidence of Narcolepsy in a US Health Care Claims Database, 2008–2010," *Sleep* 42 (2019): zsz091.

66. Jed Black et al., "The Burden of Narcolepsy Disease (BOND) Study: Health-Care Utilization and Cost Findings," *Sleep Medicine* 15 (2014): 522–29.

67. Michael H. Silber et al., "The Epidemiology of Narcolepsy in Olmsted County, Minnesota: A Population-Based Study," *Sleep* 25 (2002): 197–202; William T. Longstreth et al., "Prevalence of Narcolepsy in King County, Washington, USA," *Sleep Medicine* 10 (2009): 422–26.

68. Heide Baumann-Vogel et al., "Narcolepsy Type 2: A Rare, yet Existing Entity," *Journal of Sleep Research* (September 2020): e13203; Ester Tio et al., "The prevalence of Narcolepsy in Catalunya (Spain)," *Journal of Sleep Research* 27 (2018): e12640. The Catalunya study used the stringent Brighton Collaboration case definition of narcolepsy.

69. Bernadette Villareal et al., "Diagnosing Narcolepsy in the Active Duty Military Population," *Sleep & Breathing* 25 (2021): 995–1002.

70. Xyrem International Study Group, "A Double-Blind, Placebo-Controlled Study Demonstrates Sodium Oxybate Is Effective for the Treatment of Excessive Daytime Sleepiness in Narcolepsy," *Journal of Clinical Sleep Medicine* 1 (October 2005): 391–97.

71. Jed Black and William C. Houghton, "Sodium Oxybate Improves Excessive Daytime Sleepiness in Narcolepsy," *Sleep* 29 (2006): 939–46.

72. Christian R. Baumann et al., "Challenges in Diagnosing Narcolepsy Without Cataplexy: A Consensus Statement," *Sleep* 37 (2014): 1035–42.

73. Rye consulted for Jazz every year between 2014 and 2018, receiving more than $8,000 in 2018, according to *Propublica*'s Dollars for Docs database.

74. Klemens Kaupmann et al., "Specific γ-Hydroxybutyrate-Binding Sites but Loss of Pharmacological Effects of γ-Hydroxybutyrate in GABA-B1-Deficient Mice," *European Journal of Neuroscience* 18 (2003): 2722–30.

75. S. P. Bessman and W. N. Fishbein, "Gamma-Hydroxybutyrate, a Normal Brain Metabolite," *Nature* 200 (December 1963): 1207–8.

76. Michel Maitre, Christian Klein, and Ayikoe G. Mensah-Nyagan, "Mechanisms for the Specific Properties of γ-Hydroxybutyrate in Brain," *Medicinal Research Reviews* 36 (2016): 363–88.

77. Ulrike Leurs et al., "GHB Analogs Confer Neuroprotection Through Specific Interaction with the CaMKIIα Hub Domain," *Proceedings of the National Academy of Sciences* 118 (2021): e2108079118.

78. Rafael Boscolo-Berto et al., "Narcolepsy and Effectiveness of Gamma-Hydroxybutyrate (GHB): A Systematic Review and Meta-analysis of Randomized Controlled Trials," *Sleep Medicine Reviews* 16 (2012): 431–43.

79. P. S. van Nieuwenhuijzen, I. S. McGregor, and G. E. Hunt, "The Distribution of Gamma-Hydroxybutyrate-Induced Fos Expression in Rat Brain: Comparison with Baclofen," *Neuroscience* 158 (2009): 441–55.

80. James K. Walsh et al., "Enhancing Slow Wave Sleep with Sodium Oxybate Reduces the Behavioral and Physiological Impact of Sleep Loss," *Sleep* 33 (2010): 1217–25.

81. Julie Vienne et al., "Differential Effects of Sodium Oxybate and Baclofen on EEG, Sleep, Neurobehavioral Performance, and Memory," *Sleep* 35 (2012): 1071–83.

82. Baclofen had undesirable effects on teenagers with narcolepsy in a Taiwanese study, increasing their sleepiness. Yu-Shu Huang and Christian Guilleminault, "Narcolepsy: Action of Two γ-Aminobutyric Acid Type B Agonists, Baclofen and Sodium Oxybate," *Pediatric Neurology* 41 (2009): 9–16.

83. Anne M. Morse, Kristin Kelly-Pieper, and Sanjeev V. Kothare, "Management of Excessive Daytime Sleepiness in Narcolepsy with Baclofen," *Pediatric Neurology* 93 (2019): 39–42.

84. Sarah Wurts Black et al., "GABA-B Agonism Promotes Sleep and Reduces Cataplexy in Murine Narcolepsy," *Journal of Neuroscience* 34 (2014): 6485–94.

16. BIOMARKERS OF SLEEPINESS—AND IH

1. A. Williamson and A. Feyer, "Moderate Sleep Deprivation Produces Impairments in Cognitive and Motor Performance Equivalent to Legally Prescribed Levels of Alcohol Intoxication," *Occupational and Environmental Medicine* 57 (October 2000): 649–55.

2. In the United States, people diagnosed with narcolepsy or IH do not necessarily have to relinquish their drivers' licenses; it varies state by state whether diagnoses must be reported to state agencies. Anand Bhat et al., "Drowsy Driving Considerations in Non-Commercial Drivers for the Sleep Physician," *Journal of Clinical Sleep Medicine* 15 (2019): 1069–71.

3. Fabio Pizza et al., "Car Crashes and Central Disorders of Hypersomnolence: A French Study," *PloS One* 10 (2015): e0129386; Pierre Philip et al., "Sleep Disorders and Accidental Risk in a Large Group of Regular Registered Highway Drivers," *Sleep Medicine* 11 (2010): 973–79.

4. Denise Bijlenga et al., "Comparing Objective Wakefulness and Vigilance Tests to On-the-Road Driving Performance in Narcolepsy and Idiopathic Hypersomnia," *Journal of Sleep Research* (2021): e13518.

5. Charles A. Czeisler, "Impact of Sleepiness and Sleep Deficiency on Public Health—Utility of Biomarkers," *Journal of Clinical Sleep Medicine* 7S (October 2011): S6–S8.

6. Janet M. Mullington et al., "Developing Biomarker Arrays Predicting Sleep and Circadian-Couples Risks to Health," *Sleep* 39 (April 2016): 727–36.

7. Jennifer M. Cori et al., "Eye Blink Parameters to Indicate Drowsiness During Naturalistic Driving in Participants with Obstructive Sleep Apnea: A Pilot Study," *Sleep Health* 7 (2021): 644–51.

8. Emma Laing et al., "Identifying and Validating Blood mRNA Biomarkers for Acute and Chronic Insufficient Sleep in Humans: A Machine Learning Approach," *Sleep* 42 (2019): zsy186.

9. David T. Plante, "Hypersomnia in Mood Disorders: A Rapidly Changing Landscape," *Current Sleep Medicine Reports* 1 (June 2015): 122–30.

10. Alexandros N. Vgontzas et al., "Differences in Nocturnal and Daytime Sleep Between Primary and Psychiatric Hypersomnia: Diagnostic and Treatment Implications," *Psychosomatic Medicine* 62 (2000): 220–26.

11. David T. Plante, Jesse D. Cook, and Michael L. Prairie, "Multimodal Assessment Increases Objective Identification of Hypersomnolence in Patients Referred for Multiple Sleep Latency Testing," *Journal of Clinical Sleep Medicine* 16 (2020): 1241–48.

12. David T. Plante et al., "Establishing the Objective Sleep Phenotype in Hypersomnolence Disorder with and Without Comorbid Major Depression," *Sleep* 42 (2019): zsz060.

13. David T. Plante, "Nocturnal Sleep Architecture in Idiopathic Hypersomnia: A Systematic Review and Meta-analysis," *Sleep Medicine* 45 (2018): 17–24.

14. Fabio Pizza et al., "Polysomnographic Study of Nocturnal Sleep in Idiopathic Hypersomnia Without Sleep Time," *Journal of Sleep Research* 22 (2013): 185–96.

15. Kate E. Sprecher et al., "Amyloid Burden Is Associated with Self-Reported Sleep in Nondemented Late Middle-Aged Adults," *Neurobiology of Aging* 36 (2015): 2568–76; Eun Y. Joo et al., "Analysis of Cortical Thickness in Narcolepsy Patients with Cataplexy," *Sleep* 34 (October 2011): 1357–64.

16. Stephanie G. Jones et al., "Regional Reductions in Sleep Electroencephalography Power in Obstructive Sleep Apnea: A High-Density EEG Study," *Sleep* 37 (2014): 399–407.

17. Another option could be experimental magnetic devices. Marcello Massimini et al., "Triggering Sleep Slow Waves by Transcranial Magnetic Stimulation," *Proceedings of the National Academy of Sciences* 104 (2007): 8496–501.

18. U.S. National Library of Medicine, "Neurophysiologic Correlates of Hypersomnia," https://www.clinicaltrials.gov/ct2/show/NCT01719315. Acoustic experiments were also proposed in Plante's National Institute for Neurological Disorders and Stroke grant application.

19. James K. Walsh et al., "Slow Wave Sleep Enhancement with Gaboxadol Reduces Daytime Sleepiness During Sleep Restriction," *Sleep* 31 (2008): 659–72.

20. Ransdell Pierson, "Merck, Lundbeck Scrap Insomnia Drug After Trials," Reuters, March 28, 2007.

21. Thien T. Dang-Vu et al. "Cerebral Correlates of Delta Waves During Non-REM Sleep Revisited," *Neuroimage* 28 (2005): 14–21.

22. Eun Y. Joo et al., "Cerebral Perfusion Abnormality in Narcolepsy with Cataplexy," *Neuroimage* 28 (November 2005): 410–16.

23. Thien T. Dang-Vu et al., "Sleep Deprivation Reveals Altered Brain Perfusion Patterns in Somnambulism," *PLoS One* 10 (2015): e0133474.

24. Robert A. Veselis et al., "Midazolam Changes Cerebral Blood Flow in Discrete Brain Regions: An H2(15)O Positron Emission Tomography Study," *Anesthesiology* 87 (1997): 1106–17.

25. Florence B. Pomares et al., "Beyond Sleepy: Structural and Functional Changes of the Default-Mode Network in Idiopathic Hypersomnia," *Sleep* 42 (2019): zsz156.

26. Lynn M. Trotti, "Waking Up Is the Hardest Thing I Do All Day: Sleep Inertia and Sleep Drunkenness," *Sleep Medicine Reviews* 35 (October 2017): 76–84.

27. Carlos H. Schenck, Erin C. Golden, and Richard P. Millman, "Treatment of Severe Morning Sleep Inertia with Bedtime Long-Acting Bupropion and/or Long-Acting Methylphenidate in a Series of 4 Patients," *Journal of Clinical Sleep Medicine* 17 (2021): 653–57.

28. Raphael Vallat et al., "Hard to Wake Up? The Cerebral Correlates of Sleep Inertia Assessed Using Combined Behavioral, EEG, and fMRI measures," *Neuroimage* 184 (2019): 266–78.

29. Patricia Tassi et al., "EEG Spectral Power and Cognitive Performance During Sleep Inertia: The Effect of Normal Sleep Duration and Partial Sleep Deprivation," *Physiology & Behavior* 87 (2006): 177–84.

30. Thomas J. Balkin et al., "The Process of Awakening: A PET Study of Regional Brain Activity Patterns Mediating the Re-establishment of Alertness and Consciousness," *Brain* 125 (2002): 2308–19.

31. Bedřich Roth et al., "Neurological, Psychological, and Polygraphic Findings in Sleep Drunkenness," *Schweizer Archiv für Neurologie, Neurochirurgie und Psychiatrie* 129 (1981): 209–22.

32. The Emory IH group fit ICSD-3 sleep latency criteria, but only a few reported long sleep periods. The experiments focusing on sleep drunkenness had not yet been analyzed.

33. Yves Dauvilliers et al., "18F-FDG-Positron Emission Tomography Evidence for Cerebral Hypermetabolism in the Awake State in Narcolepsy and Idiopathic Hypersomnia," *Frontiers in Neurology* 8 (2017): 350.

34. Rachel Danzig et al., "The Wrist Is Not the Brain: Estimation of Sleep by Clinical and Consumer Wearable Actigraphy Devices Is Impacted by Multiple Patient- and Device-Specific Factors," *Journal of Sleep Research* 29 (2020): e12926.

35. Jesse D. Cook, Michael L. Prairie, and David T. Plante, "Utility of the Fitbit Flex to Evaluate Sleep in Major Depressive Disorder: A Comparison Against Polysomnography and Wrist-Worn Actigraphy," *Journal of Affective Disorders* 217 (2017): 299–305. Jesse D. Cook, Michael L. Prairie, and David T. Plante, "Ability of the Multisensory Jawbone UP3 to Quantify and Classify Sleep in Patients with Suspected Central Disorders of Hypersomnolence: A Comparison Against Polysomnography and Actigraphy," *Journal of Clinical Sleep Medicine* 14 (2018): 841–48.

36. Anniina Alakujala et al., "Accuracy of Actigraphy Compared to Concomitant Ambulatory Polysomnography in Narcolepsy and Other Sleep Disorders," *Frontiers in Neurology* 12 (2021): 629709.

37. Michael T. Smith et al., "Use of Actigraphy for the Evaluation of Sleep Disorders and Circadian Rhythm Sleep-Wake Disorders: An American Academy of Sleep Medicine Clinical Practice Guideline," *Journal of Clinical Sleep Medicine* 14 (2018): 1231–37.

38. Jenny Gold, "The Sleep Apnea Business Is Booming, and Insurers Aren't Happy," *NPR Morning Edition*, January 16, 2012.

39. Massimiliano de Zambotti et al., "Sensors Capabilities, Performance, and Use of Consumer Sleep Technology," *Sleep Medicine Clinics* 15 (2020): 1–30.

40. Elisa Evangelista et al., "Alternative Diagnostic Criteria for Idiopathic Hypersomnia: A 32-Hour Protocol," *Annals of Neurology* 83 (2018): 235–47.

41. Elisa Evangelista et al., "Sleep Inertia Measurement with the Psychomotor Vigilance Task in Idiopathic Hypersomnia," *Sleep* (August 2021): zsab220.

17. THE FDA OPENS A DOOR

1.   Alyssa Cairns and Richard Bogan, "To Withhold or Not Withhold? Use of Psychiatric Medications on MSLT Outcomes in 1033 Patients," *Sleep* 42S (2019): A250–51.

2.   Lynn M. Trotti, Beth A. Staab, and David B. Rye, "Test-Retest Reliability of the Multiple Sleep Latency Test in Narcolepsy Without Cataplexy and Idiopathic Hypersomnia," *Journal of Clinical Sleep Medicine* 9 (2013): 789–95; Regis Lopez et al., "Test-Retest Reliability of the Multiple Sleep Latency Test in Central Disorders of Hypersomnolence," *Sleep* 40 (2017): zsx164; Chad Ruoff et al., "The MSLT Is Repeatable in Narcolepsy Type 1 but Not Narcolepsy Type 2: A Retrospective Patient Study," *Journal of Clinical Sleep Medicine* 14 (2018): 65–74.

3.   Rolf Fronczek et al., "To Split or to Lump? Classifying the Central Disorders of Hypersomnolence," *Sleep* 43 (2020): zsaa044.

4.   Lei Chen et al., "The Familial Risk and HLA Susceptibility Among Narcolepsy Patients in Hong Kong Chinese," *Sleep* 30 (2007): 851–58.

5.   Gert J. Lammers et al., "Diagnosis of Central Disorders of Hypersomnolence: A Reappraisal by European Experts," *Sleep Medicine Reviews* 50 (2020): 101306.

6.   Kiran Maski, Emmanuel Mignot, and Thomas E. Scammell, "Commentary on Lammers et al.," *Sleep Medicine Reviews* 52 (2020): 101327.

7.   Gert J. Lammers, "Diagnosing Hypersomnia Differently: A European Proposal," Hypersomnia Foundation, YouTube video posted May 19, 2021, https://www.youtube.com/watch?v=BHEKcbJQ7P4.

8.   Michel Billiard, "Diagnosis of Narcolepsy and Idiopathic Hypersomnia. An Update Based on the International Classification of Sleep Disorders, 2nd Edition," *Sleep Medicine Reviews* 11 (2007): 377–88.

9.   Emily Singer, "Diagnosis: Redefining Autism," *Nature* 491 (2012): S12–13; Karen Brown, " 'Schizophrenia' Still Carries a Stigma. Will Changing the Name Help?," *New York Times*, December 20, 2021.

10. Risperidone and aripiprazole were FDA approved for irritability associated with autism, but irritability is not considered a core symptom of autism. Sheena LeClerc and Deidra Easley, "Pharmacological Therapies for Autism Spectrum Disorder: A Review," *Pharmacy and Therapeutics* 40 (2015): 389–R397.

11. Bedřich Roth, *Narcolepsy and Hypersomnia* (Basel: Karger, 1980), 159.

12. Francesco P. Cappuccio et al., "Sleep Duration and All-Cause Mortality: A Systematic Review and Meta-analysis of Prospective Studies," *Sleep* 33 (2010): 585–92.

13. John Acquavella et al., "Prevalence of Narcolepsy and Other Sleep Disorders and Frequency of Diagnostic Tests from 2013–2016 in Insured Patients Actively Seeking Care," *Journal of Clinical Sleep Medicine* 16 (2020): 1255–63.

14. Yves Dauvilliers et al., "Safety and Efficacy of Lower-Sodium Oxybate in Adults with Idiopathic Hypersomnia: A Phase 3, Placebo-Controlled, Double-Blind, Randomised Withdrawal Study," *Lancet Neurology* 21 (2022): 53–65.

15. S. Nassir Ghaemi and Harry P. Selker, "Maintenance Efficacy Designs in Psychiatry: Randomized Discontinuation Trials—Enriched but Not Better," *Journal of Clinical and Translational Science*

1 (2017): 198–204; Lynn Marie Trotti, "Idiopathic Hypersomnia: Does First to Approval Mean First-Line Treatment," *Lancet Neurology* 21 (2022): 25–26.

16. Takeda Pharmaceutical Company press release, "Takeda Provides Update on TAK-994 Clinical Program," October 5, 2021.

17. Press releases from Harmony Biosciences, "Harmony Biosciences Initiates a Phase 2 Clinical Trial in Myotonic Dystrophy," June 29, 2021; "Harmony Biosciences Announces Plans to Initiate Phase 3 Clinical Trial for Treatment of Idiopathic Hypersomnia," December 1, 2021.

18. Jazz Pharmaceuticals, Sunosi investor update from July 2, 2019, downloaded from Jazz website, p. 33.

19. Rachel Sachs, "The FDA's Approval of Aduhelm: Potential Implications Across a Wide Range of Health Policy Issues and Stakeholders," *Health Affairs Forefront*, 2021, https://www.healthaffairs.org/do/10.1377/forefront.20210609.921363.

20. Through the Freedom of Information Act, I have requested FDA clinical review documents for Xywav and IH, which are available for other approved medications, such as solriamfetol and pitolisant.

21. Ragy Saad et al., "Clinical Presentation Prior to Idiopathic Hypersomnia Diagnosis Among US Adults: A Retrospective, Real-World Claims Analysis," *Sleep* 44S (2021): A196.

22. U.S. Food and Drug Administration, "FDA Drug Safety Communication: Warning Against Use of Xyrem (Sodium Oxybate) with Alcohol or Drugs Causing Respiratory Depression," December 17, 2012.

23. Ruzica Kovacević-Ristanović and Tomasz J. Kuźniar, "Use of Sodium Oxybate (Xyrem) in Patients with Dual Diagnosis of Narcolepsy and Sleep Apnea," *Sleep Medicine* 11 (2010): 5–6.

24. The *Times* article emphasized GHB's lurid past, prompting a backlash from the IH community online. Virginia Hughes, "F.D.A. Approves GHB, a 'Date Rape' Drug, for Rare Sleeping Disorder," *New York Times*, August 12, 2021.

25. Jazz Pharmaceuticals "Sleep Counts" website, viewed 2021, https://www.sleepcountshcp.com/patient-profiles-stories-videos.

26. Ragy Saad et al., "Utilization of Diagnostic Sleep Testing Prior to Idiopathic Hypersomnia Diagnosis Among US Adults: A Real-World Claims Analysis," *Sleep* 44S (2021): A197.

27. Thomas Roth et al., "Characteristics of Subjects Excluded from an Idiopathic Hypersomnia Randomized Clinical Trial (ARISE2)," *Sleep Medicine* 64S (2019): S327–28.

28. Sharon Begley, "'A Feature Not a Bug': George Church Ascribes His Visionary Ideas to Narcolepsy," *STAT*, June 8, 2017.

29. Judy Singer, *NeuroDiversity: The Birth of an Idea* (2017).

30. Celia Lacaux et al., "Increased Creative Thinking in Narcolepsy," *Brain* 142 (2019): 1988–99.

31. Michelle Emrich, "Broken by IH," https://www.youtube.com/watch?v=Ww_6LfS4dFc.

32. Jason C. Ong et al., "Developing a Cognitive Behavioral Therapy for Hypersomnia Using Telehealth: A Feasibility Study," *Journal of Clinical Sleep Medicine* 16 (2020): 2047–62.

33. Ronald D. Chervin, "Sleepiness, Fatigue, Tiredness, and Lack of Energy in Obstructive Sleep Apnea," *Chest* 118 (2000): 372–79.

34. Darren Scheer et al., "Prevalence and Incidence of Narcolepsy in a US Health Care Claims Database, 2008–2010," *Sleep* 42 (2019): zsz091.

35. Maunil K. Desai and Roberta Diaz Brinton, "Autoimmune Disease in Women: Endocrine Transition and Risk Across the Lifespan," *Frontiers in Endocrinology*, 2019, https://www.frontiersin.org/articles/10.3389/fendo.2019.00265/full.

36. Makoto Kawai et al., "Narcolepsy in African Americans," *Sleep* 38 (2015): 1673–81.

37. Katherine A. Dudley and Sanjay R. Patel, "Disparities and Genetic Risk Factors in Obstructive Sleep Apnea," *Sleep Medicine* 18 (2016): 96–102.

38. Hypersomnia Foundation, "A Call to Action," https://www.hypersomniafoundation.org/a-call-to-action-racial-disparities-in-hypersomnia-disorders/.

39. James Stevens, "A Disease of Dignity," *Ars Medica* 13 (2018).

40. Isabelle Arnulf, "Women and IH," Hypersomnia Foundation, YouTube video posted April 14, 2019, https://www.youtube.com/watch?v=L7TG5Z5rf3s.

41. British Generic Manufacturers Association, "Modafinil: Potential Risk of Congenital Malformations During Pregnancy," https://assets.publishing.service.gov.uk/media/5e43e03fe5274a6d34ddad60/Modafinil-Jan-2020.pdf.

42. Per Damkier and Anna Broe, "First-Trimester Pregnancy Exposure to Modafinil and Risk of Congenital Malformations," *JAMA* 323 (2020): 374–76.

43. Sigal Kaplan et al., "Pregnancy and Fetal Outcomes Following Exposure to Modafinil and Armodafinil During Pregnancy," *JAMA Internal Medicine* 181 (2021): 275–77.

44. K. F. Huybrechts et al., "Association Between Methylphenidate and Amphetamine Use in Pregnancy and Risk of Congenital Malformations," *JAMA Psychiatry* 75 (2018): 167–75.

45. Ilya Kolb et al., "Cleaning Patch-Clamp Pipettes for Immediate Reuse," *Scientific Reports* 6 (2016): 35001; Riley E. Perszyk et al., "Automated Intracellular Pharmacological Electrophysiology for Ligand-Gated Ionotropic Receptor and Pharmacology Screening," *Molecular Pharmacology* 100 (2021): 73–82.

46. Claire Wylds-Wright, "Wake Up Narcolepsy & Dr. Mignot," Wake Up Narcolepsy, YouTube video posted February 26, 2018, 1h18m–1h22m, https://www.youtube.com/watch?v=iovkibut2vI.

47. Peter J. H. Scott et al., "Investigation of Proposed Activity of Clarithromycin at $GABA_A$ Receptors Using [11C]Flumazenil PET," *ACS Medicinal Chemistry Letters* 7 (2016): 746–50.

48. Emmanuel Mignot was quoted as saying: "Marty Scharf is why we have this drug at all." In Gayle Greene, *Insomniac* (Piatkus, 2008), 209.

49. Low-Dose Naltrexone Trust, "LDN 2021 Conference," https://www.ldnrtevents.com/pages/ldn-2021.

50. Jarred Younger, Luke Parkitny, and David McLain, "The Use of Low-Dose Naltrexone (LDN) as a Novel Anti-inflammatory Treatment for Chronic Pain," *Clinical Rheumatology* 33 (2014): 451–59.

51. Pamela Weintraub, *Cure Unknown: Inside the Lyme Epidemic* (St. Martin's Griffin, 2013).

52. Fabian-Xose Fernandez, Julie Flygare, and Michael A. Grandner, "Narcolepsy and COVID-19: Sleeping on an Opportunity?," *Journal of Clinical Sleep Medicine* 16 (August 2020): 1415.

53. Destin Groff et al., "Short-Term and Long-Term Rates of Postacute Sequelae of SARS-CoV-2 Infection," *JAMA Network Open* 4 (2021): e2128568.

54. Lucie Barateau et al., "Comorbidity Between Central Disorders of Hypersomnolence and Immune-Based Disorders," *Neurology* 88 (2017): 93–100.

INDEX